Chromatography of Natural, Treated and Waste Waters

Chromatography of Natural, Treated and Waste Waters

T. R. Crompton

CRC Press
Taylor & Francis Group
Boca Raton London New York

CRC Press is an imprint of the
Taylor & Francis Group, an **informa** business
A TAYLOR & FRANCIS BOOK

CRC Press
Taylor & Francis Group
6000 Broken Sound Parkway NW, Suite 300
Boca Raton, FL 33487-2742

First issued in paperback 2019

© 2003 by Taylor & Francis Group, LLC
CRC Press is an imprint of Taylor & Francis Group, an Informa business

No claim to original U.S. Government works

ISBN-13: 978-0-415-28004-4 (hbk)
ISBN-13: 978-0-367-86348-7 (pbk)

Publisher's note:
This book was prepared from camera-ready-copy supplied by the author.

British Library Cataloguing in Publication Data
A catalogue record for this book is available from the British Library.

Library of Congress Cataloging in Publication Data
A catalog record for this book has been requested.

Visit the Taylor & Francis Web site at
http://www.taylorandfrancis.com

and the CRC Press Web site at
http://www.crcpress.com

Contents

Preface

Chromatographic techniques are becoming increasingly used for the determination of organic and organometallic compounds, cations and anions in all types of water ranging from non saline waters, ground and surface waters, potable water, sea and estuary waters and various industrial effluents. The techniques are then especially useful when mixtures of various substances are present and facilitate the determination in a single analysis. Extremely low concentrations can be determined.

Basically the techniques can be divided in three categories, those applicable only to relatively volatile substances such as gas chromatography, those applicable to non volatile substances such as thin layer chromatography and those applicable to both volatile and non volatile such as high performance liquid chromatography and other techniques and ion chromatography.

The purpose of this book is to draw together and systemise the body of information available in the world literature on the application of chromatographic procedures to the determination of all types of compounds and elements and mixtures thereof in non saline and seawater and treated waters. In this way reference to a very scattered literature can be avoided.

Methods are not presented in detail, space considerations alone would not permit this. Instead, the chemist is presented with details of methods available for a variety of types of water samples. Methods are described in broad outline, giving enough information for the chemist to decide whether he or she wishes to refer to the original paper. To this end information is provided on applicability of methods, advantages and disadvantages of one method compared to another, interferences, sensitivity, detection limits and data relevant to accuracy and precision.

The book commences with a chapter in which the principles and theory of various chromatographic techniques are discussed. Ion chromatography (Chapter 2) is a relatively recently introduced technique that has found extensive applications in the analysis of mixtures of anions and to a lesser extent of organic compounds and cations. Codetermination of anions and cations is possible. A variant of ion chromatography, namely electrostatic ion chromatography has to date found a very limited application to the determination of anions and is discussed in Chapter 3.

High performance liquid chromatography (Chapter 4) which is displacing conventional column chromatography (Chapter 6) has found numerous applications in the analysis of organic compounds, also metal organic compounds of mercury, tin, arsenic, manganese copper and lead. It has fewer applications in the determinations of anions and cations.

Various miscellaneous column chromatographic techniques including ion pair, micelle, ion exclusion, size exclusion and gel permeation are discussed in Chapters 7–11. All of these techniques have found limited selected applications in the determination of organic compounds, cations and anions.

Ion exchange chromatography (Chapter 12) is a well-established technique applicable to the determination of particular types of organic and organometallic compounds and anions and cations as has capillary column coupling isotachoelectrophoresis (Chapter 13).

Chapter 14 discusses the very useful technique of thin layer chromatography. It has extensive applications in the analysis of complex mixtures of organic compounds and also has found limited applications in the analysis of organometallic compounds. No applications to anions and cations have been reported to date. The technique can also be used to prepare extracts suitable for subsequent examination by infrared spectroscopy or mass spectrometry.

Gas chromatography has, of course, been used extensively in the analysis of many types of organic compounds with boiling points up to about 250°C, also to the analysis of organic compounds of lead, mercury, selenium, tin, manganese and silicon. Derivitisation of these compounds to produce compounds sufficiently volatile to be amenable to gas chromatography is frequently practised. Gas chromatography has also been applied to the determination of arsenic, antimony, selenium, tin, beryllium and aluminium and the common anions such as sulphate, nitrate, phosphate, sulphide, cyanide and thiocyanate.

Compounds that have been separated by way of gas chromatography and liquid chromatographic techniques cannot be identified by their retention time alone as many types of compounds have similar or identical retention times. Increasingly, in recent years, this problem has been overcome by connecting a mass spectrometer to the outlet of the separation column applications of mass spectrometry are discussed in Chapters 5 (high performance liquid chromatography) and 16 (gas chromatography).

The work has been written with the interests of the following groups of people in mind: management and scientists in all aspects of the water industry, river management, fishery industries, environmental management, sewage and trade effluent treatment and disposal, land drainage and water supply; also management and scientists in all branches of industry which produce aqueous effluents. It will also be of interest to agricultural chemists and to the medical profession, toxicologists, public health workers, public analysts, environmentalists and oceanographers.

Finally, it is hoped that the book will act as a spur to students and to industrial and academic staff concerned with the development of new analytical methods.

T.R. Crompton
October 2002

Chapter 1

Brief summary of methodologies

1.1 Ion chromatography

This technique developed by Small *et al.* in 1975 [1] is usually employed for the separation and determination of mixtures of inorganic anions in water. However, applications have also been found for the determination of many organic anions in water.

The technique uses specialised ion exchange columns and chemically suppressed conductivity detection. Advances in column and detection technologies have expanded this capability to include wider ranges of anions as well as organic ions. These recent developments, discussed below, provide the chemist with a means of solving many problems that are difficult, if not impossible, using other instrumental methods. Ion chromatography can analyse a wide variety of organic and inorganic anions more easily than either atomic absorption spectrometry or inductively coupled plasma techniques.

At the heart of the ion chromatography system is an analytical column containing an ion exchange resin on which various anions (and/or cations) are separated before being detected and quantified by various detection techniques, such as spectrophotometry, atomic absorption spectrometry (metals) or conductivity (anions).

Ion chromatography is not restricted to the separate analysis of only anions or cations, and, with the proper selection of the eluent and separator columns, the technique can be used for the simultaneous analysis of both anions and cations.

The original method for the analysis of mixtures of anions used two columns attached in series packed with ion exchange resins to separate the ions of interest and suppress the conductance of the eluent, leaving only the species of interest as the major conducting species in the solution. Once the ions were separated and the eluent suppressed, the solution entered a conductivity cell, where the species of interest were detected.

The analytical column is used in conjunction with two other columns, a guard column which protects the analytical column from troublesome contaminants, and a preconcentration column. The intended function of the preconcentration column is twofold. First, it concentrates the ions present in the sample, enabling very low levels of contaminants to be detected. Second, it retains non complexed ions on the resin, while allowing complexed species to pass through.

Dionex Corporation are a major producer of equipment for ion chromatography. Some of their equipment is discussed below.

Dionex Series 4000i Ion Chromatographs

Some of the features of this instrument are:

- chromatography module;
- up to six automated valves made of chemically inert, metal-free material eliminate corrosion and metal contamination;
- liquid flow path is completely compatible with all high performance liquid chromatography solvents;
- electronic valve switching, multi-dimensional, coupled chromatography or multi-mode operation;
- automated sample clean up or preconcentration;
- environmentally isolates up to four separator columns and two suppressors for optimal results;
- manual or remote control with Dionex Autoion 300 or Autoion 100 automation accessors;
- individual column temperature control from ambient to 100°C optional.

Dionex Ion-Pac Columns

Features are:

- polymer ion exchange columns are packed with pellicular resins for anion or cation exchange applications;
- 4μ polymer ion exchange columns have maximum efficiency and minimum operating pressure for high performance ion and liquid chromatography applications;
- ion exclusion columns with bifunctional cation exchange sites offer more selectivity for organic acid separations;
- neutral polymer resins have high surface area for reversed phase ion pair and ion suppression applications without pH restriction;
- 5 and 10μ silica columns are optimised for ion pair, ion suppression and reversed phase applications.

Micromembrane suppressor

The micromembrane suppressor (MMS) makes possible detection of non ultraviolet absorbing compounds such as inorganic anions and cations, surfactants, antibiotics, fatty acids and amines in ion exchange and ion pair chromatography. Two variants of this exist: the anionic (AMMS) and the cationic (CMMS) suppressor. The micromembrane suppressor consists of a low dead volume eluent flow path through altering layers of high-capacity ion exchange screens and ultra-thin ion exchange membranes. Ion exchange sites in each screen provide a site-to-site pathway for eluent ions to transfer to the membrane for maximum chemical suppression.

Dionex anion and cation micromembrane suppressors transform eluent ions into less conducting species without affecting sample ions under analysis. This improves conductivity detection, sensitivity, specificity and baseline stability. It also dramatically increases the dynamic range of the system for inorganic and organic ion chromatography. The high ion exchange capacity of the MMS permits changes in eluent composition by orders of magnitude, making gradient ion chromatography possible.

In addition, because of the increased detection specificity provided by the MMS sample, preparation is dramatically reduced, making it possible to analyse most samples after filtering and dilution.

Conductivity detector

Features include:

- high sensitivity detection of inorganic anions, amines, surfactants, organic acids, Group I and II metals, oxy-metal ions and metal cyanide complexes (used in conjunction with MMS);
- bipolar-pulsed excitation eliminates the non linear response with concentration found in analogue detectors;
- microcomputer-controlled temperature compensation minimises the baseline drift with changes in room temperature.

UVVis detector

Important factors are:

- high-sensitivity detection of amino acids, metals, silica, chelating agents, and other ultraviolet absorbing compounds using either post-column reagent addition or direct detection;
- non metallic cell design eliminates corrosion problems;
- filter-based detection with selectable filters from 214 to 800nm;
- proprietary dual wavelength detection for ninhydrin-detectable amino acids and PAR-detectable transition metals.

Optional detectors

In addition to the detectors discussed above, Dionex offer visible fluorescence and pulsed amperometric detectors for use with the series 4000i. Dionex also supply a wide range of alternative instruments, eg single channel (2010i) and dual channel (2020i). The latter can be upgraded to an automated system by adding the Autoion 100 or Autoion 300 controller to control two independent ion chromatograph systems. They also supply a 2000i series equipped with conductivity pulsed amperometric, ultraviolet-visible and refractometric detectors.

1.2 Electrostatic ion chromatography

Application of this technique has, so far, been limited to the determination of chloride bromide and iodide in seawater.

The desirability of an ion chromatography technique using only water as the mobile phase is summarised by Small [2]. There are a number of ion chromatographic techniques that, though not widely used, deserve mention because of their simplicity and their potential for further development [3,4]. All use water or other polar solvent as the sole component of the mobile phase. This not only eliminates the need for electrolytes and precise eluent make-up, but it avoids many of the detection problems that ionic eluents can impose. Detection can be expected to be very sensitive since the background is essentially deionised water.

Another approach for separating ions, also using water as the mobile phase but employing a zwitterionic stationary phase, has been developed by Hu *et al.* [5]. When a small amount of aqueous solution containing an analyte (cations and anions) is passed through a zwitterionic stationary phase, neither the analyte cations nor the analyte anions can get close to the opposite charge fixed on the stationary phase, because another charge on the same molecule, fixed on the stationary phase, repels the analyte ions simultaneously. The analyte cations and anions are forced into a new state of simultaneous electrostatic attraction and repulsion interaction in the column. This was termed an 'ion pairing-like form'. This method of separation was termed electrostatic ion chromatography [5]. The previous studies [5,6] demonstrated that the separation of inorganic ions (with the exception of cations having the same charge) using electrostatic ion chromatography is comparable to separations of the same ions obtained using conventional ion chromatographs. In previous studies [5,6], samples having high concentrations (mmol/L) of analyte ions were used in order to understand the mechanism. In those studies, samples with low concentrations of ions were not investigated.

Hu *et al.* [7] turned their attention to the determination of trace level inorganic ions using electrostatic ion chromatography. Initial results showing separate elution times for the same analyte gave new insights into the mechanism of electrostatic ion chromatography and led to the development of a new technique for simpler determination of trace level inorganic ions.

1.3 High performance liquid chromatography

There has been a growing interest in applying high performance liquid chromatography, to the determination of the not only volatile compounds, such as aliphatic and polyaromatic hydrocarbons, saturated and unsaturated aliphatic and polyaromatic hydrocarbons, saturated and unsaturated aliphatic halogen compounds, haloforms and some esters, phenols and others but also non volatile components of water.

High performance liquid chromatography has been developed to a very high level of performance by the introduction of selective stationary phases of small particle sizes, resulting in efficient columns with large plate numbers per litre.

There are several types of chromatographic columns used in high performance liquid chromatography.

Four basic types of elution system are used in high performance liquid chromatography. This is illustrated below by the systems offered by LKB, Sweden.

The isocratic system

This consists of a solvent delivery for isocratic reversed phase and gel filtration chromatography. The isocratic system (Fig. 1.1) provides an economic first step into high performance liquid chromatography techniques. The system is built around a high performance, dual-piston, pulse-free pump providing precision flow from 0.01 to 5mL min^{-1}.

Any of the following detectors can be used with this system:

- fixed wavelength ultraviolet detector (LKB Unicord 2510);
- variable ultraviolet-visible (190–600nm);
- wavelength monitor (LKB 2151);
- rapid diode array spectral detector (LKB 2140);
- refractive index detector (LKB 2142);
- electrochemical detector (LKB 2143);
- wavescan EG software (LKB 2146).

Basic gradient system

This is a simple upgrade of the isocratic system with the facility for gradient elution techniques and greater functionality (Fig. 1.1(b)). The basic system provides for manual operating gradient techniques such as reversed phase, ion exchange and hydrophobic interaction chromatography. Any of the detectors listed above under the isocratic system can be used.

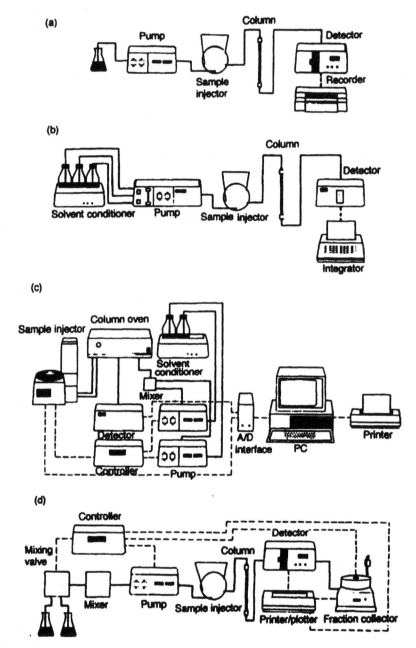

Fig. 1.1 Elution systems supplied by LKB, Sweden (a) isocratic bioseparation system; (b) basic system; (c) advanced chromatography system; (d) inert system
Source: Own files

Advanced gradient system

For optimum functionality in automated systems designed primarily for reversed phase chromatography and other gradient techniques, the LKB advanced-gradient system is recommended (Fig 1.1(c)). Key features include the following:

- a configuration that provides the highest possible reproducibility of results;
- a two-pump system for highly precise and accurate gradient formation for separation of complex samples;
- separation of complex samples;
- full system control and advanced method development provided from a liquid chromatography controller;
- precise and accurate flows ranging from 0.01 to 5ml min $^{-1}$.

This system is ideal for automatic method for development and gradient optimisation.

The inert system

By a combination of the use of inert materials (glass, titanium and inert polymers) this system offers totally inert fluidics. Primary features of the system include (Fig. 1.1(d)):

- the ability to perform isocratic or gradient elution by manual means;
- full system control from a liquid chromatography controller;
- precise and accurate flows from 0.01–5mL min $^{-1}$.

This is the method of choice when corrosive buffers, eg those containing chloride or aggressive solvents, are used.

Chromatographic detectors

The most commonly-used detectors are those based on spectrophotometry in the region 184-400nm, visible ultraviolet spectroscopy in the region 185–900nm, post-column derivativisation with fluorescence detection (see below), conductivity and those based on the relatively new technique of multiple wavelength ultraviolet detectors using a diode array system detector (described below). Other types of detectors available are those based on electrochemical principles, refractive index, differential viscosity and mass detection.

1.3.1 Reversed phase chromatography

The most commonly used chromatographic mode in high performance liquid chromatography is reversed phase chromatography. Reversed phase chromatography is used for the analysis of a wide range of neutral compounds such as carbohydrates and polar organic compounds. Most common reversed phase chromatography is performed using bonded silica-based columns, thus inherently limiting the operating pH range to 2.0–7.5. The wide pH range (0–14) of some columns (eg Dionex Ion Pac NSI and NS 1–5 μcolumns) removes this limitation, and consequently they are ideally suited for ion pairing and ion suppression reversed phase chromatography, the two techniques which have helped extend reversed phase chromatography to ionisable compounds.

High sensitivity detection of non chromophoric organic ions can be achieved by combining the power of suppressed conductivity detection with these columns. Suppressed conductivity is usually a superior approach to using refractive index or low-ultraviolet-wavelength detection.

Reversed phase ion pairing chromatography

Typically, reversed phase ion pairing chromatography is carried out using the same stationary phase as reversed phase chromatography. A hydrophobic ion of opposite charge to the solute of interest is added to the mobile phase. Samples which are determined by reversed phase ion pairing chromatography are ionic and thus capable of forming an ion pair with the added counter ion. This form of reversed phase chromatography can be used for anion and cation separations and for the separation of surfactants and other ionic types of organic molecules. An unfortunate drawback to using silica-based columns is that ion pairing reagents increase the solubility of silica in water, leading to loss of bead integrity and drastically reducing column life. Some manufacturers (eg Dionex) employ neutral macroporous resins, instead of silica, in an attempt to widen the usable pH range and eliminate the effect of ion pairing reagents. The technique has been applied to the analysis of organic compounds (thiols, cationic surfactants, sulphonated polyphenols and primary amines), also a wide range of cations (heavy metals, alkali metals, alkaline earths and lanthanides) and some anions (metalcyano complexes).

I.3.2 Size exclusion chromatography

Size exclusion techniques provide rapid gentle separations with a constant sample matrix and have generally been designed to separate large organic molecules eg copper(II) complexes of poly(amino carboxylic acids), polyarylamides and fulvic acids. Most separations have been restricted to relatively high molecular weight metal organic compounds and separations are usually carried out at high pressure. Thus Adamik and Bartok [8] used high pressure aqueous size exclusion chromatography with reverse pulse amperometric detection to separate copper II complexes of poly(aminocarboxylic acids), catechol and fulvic acids. The commercially available size exclusion chromatography columns were tested. Columns were eluted with copper(II) complexes of poly(amino-carboxylic acids), catechol and water derived fulvic acids. The eluent contained copper II to prevent dissociation of the labile metal complexes. Reverse pulse electrochemical measurements were made to minimise oxygen interferences at the detector. Resolution of a mixture of EDTA, NTA and DTP copper complexes was approximately the same on one size chromatography column as on Sephalex 9–25 columns but with a ten-fold increase in efficiency.

A linear detection response to amounts of water derived fulvic acids separated on the gas chromatography 100 column was obtained enabling direct measurement of amounts of fulvic acid in water samples. The detection limit was 40ng for copper EDTA and 12µg of fulvic acid.

Another approach for separation is to adjust chromatographic conditions to maximise chemical or apparent molecular size differences of dissolved components. Then, electrically different groups can be fractionated by size exclusion stationary phases even if molecular weights are similar. In the absence of salts or buffers (to reduce charge effects) sample components may be separated due to factors other than molecular weight. Distilled water may increase the apparent molecular size of ionic dissolved components due to hydration layer formation or other hydrogen bond bonding or ionic repulsion mechanisms. Sample column interactions, causing fractionation, may result from charge alterations or repulsions in the absence of salts or buffers in the mobile phase.

Gardner et al. [9] recognised that because component separations on size exclusion columns with distilled water are affected by chemical physical interactions as well as component molecular size, distilled water size exclusion chromatography will also fractionate dissolved metal forms. These workers interfaced distilled water size exclusion chromatography with inductively coupled argon plasma detection to fractionate and detect dissolved forms of calcium and magnesium in lake and river waters.

1.3.3 Ion exclusion chromatography

Unlike the pellicular packings used for ion exchange, the packings used in ion exclusion are derived from totally sulphonated polymeric materials. Separation is dependent upon three different mechanisms: Donnan exclusion, steric exclusion and adsorption/partitioning. Donnan exclusion causes strong acids to elute in the void volumes of the column. Weak acids which are partially ionised in the eluent are not subject to Donnan exclusion and can penetrate into the pores of the packing. Separation is accomplished by differences in acid strength, size and hydrophobicity. The major advantage of ion exclusion lies in the ability to handle samples that contain both weak and strong acids. A good example of the power of ion exclusion is the routine determination of organic acids in sea water. Without ion exclusion, the high chloride ion concentration would present a serious interference.

Ion exclusion chromatography provides a convenient way to separate molecular acids from highly ionised substances. The separation column is packed with a cation exchange resin in the H^+ form so that salts are converted to the corresponding acid. Ionised acids pass rapidly through the column while molecular acids are held up to varying degrees. A conductivity detector is commonly used. Tanaka *et al.* [10–12] have reported that the separations of orthophosphate from fluoride, sulphate and several condensed phosphates could be achieved by ion exclusion chromatography on a cation exchange resin in the H^+ form by elution with acetone–water and dioxan–water mixtures. As the separation mechanism of the ion exclusion chromatography by elution with water alone for numerous anions to their respective acids is based on the Donnon membrane equilibrium principle (ion exclusion effect), it is a highly useful technique for the separation of non-electrolytes such as carbonic acid from electrolytes such as hydrochloric acid and sulphuric acid. Ion exclusion chromatography has also been coupled with ion chromatography to determine simultaneously both weak and strong acids.

Ion exclusion chromatography has been applied to the determination of the following organic compounds and anions: ozonisation products, carboxylic acids; phosphate, nitrite, nitrate, silicate, bicarbonate, tartrate, malate, malonate, citrate, glycollate, formate and fumarate, arsenite, arsenate, chloride, bromide, iodide, thiocyanate and sulphate carbonate and also the cation arsenic.

1.3.4 Micelle exclusion chromatography

This technique has found limitations in the determination of bromide mixed halides, iodide, iodate, nitrite, nitrate, sulphide, sulphite, thio-cyanate, thiosulphate and isobutyrate in non saline waters.

The method is based on the partitions of the anions to a cationic micelle phase and shows different selectivity from ion exchange chromatography. In a kinetic method [13] for the determination of thiocyanate, sulphite and sulphide the anions are reacted with 5,5'dithiobis (2-nitrobenzoic acid) in aqueous cetyltrimethyl/ammonium bromide micelles.

1.3.5 Ion suppression chromatography

Ion suppression is a technique used to suppress the ionisation of compounds (such as carboxylic acids) so they will be retained exclusively by the reversed phase retention mechanism and chromatographed as the neutral species. Column packings with an extended pH range are needed for this application as strong acids or alkalis are used to suppress ionisation. In addition to carboxylic acids, the ionisation of amines can be suppressed by the addition of a base to the mobile phase, thus allowing chromatography of the neutral amine.

1.3.6 Gel permeation chromatography

This technique has been applied to the determination of organic compounds and cations including crude oils, fulvic acids, tannins, carbohydrates, trihalomethane precursors and lithium.

The technique separates components of a mixture in order of their molecular size, practionally their molecular weight. In a typical example [14], a sample of a crude oil was dissolved in toluene or tetrahydrofuran and pumped through a column of 60Å Styragel. The effluent was monitored by a differential refractometer. In the case of cations, separations based on molecular size have been achieved with alkali metals eluting in order potassium, sodium, lithium, magnesium and calcium.

1.3.7 Gel filtration chromatography

Gel filtration chromatography on a Sephadex G 100 column has been used to fractionate humic acid.

1.3.8 Supercritical fluid chromatography

Until recently the chromatographer has had to rely on either gas chromatography or high performance liquid chromatography for separations, enduring the limitations of both. Lee Scientific has created a new dimension in chromatography, one which utilises the unusual properties of supercritical fluids. With the new technology of capillary supercritical fluid chromatography (SFC) the chromatographer benefits from the best of both worlds – the solubility behaviour of liquids and the diffusion and viscosity properties of gases. Consequently, capillary SFC offers unprecedented versatility in obtaining high-resolution separations of difficult compounds. The technique, to date, has found limited but important applications in the determination of organic in waters. These include mixtures of chlorinated insecticides and polychlorobiphenyls, which are notoriously difficult separations to achieve by other forms of chromatography, and also separations of complex mixtures of different types of organic compounds.

Beyond its critical point, a substance can no longer be condensed to a liquid, no matter how great the pressure. As pressure increases, however,

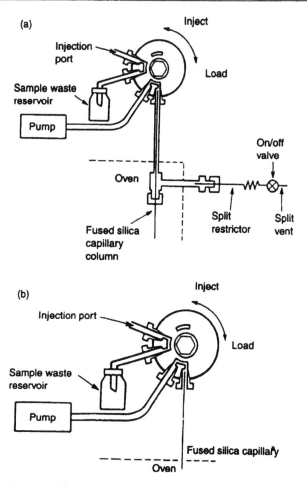

Fig. 1.2 Sample injectors: (a) split valve injector; (b) timed-split and direct valve injector
Source: Own files

the fluid density approaches that of a liquid. Because solubility is closely related to density, the solvating strength of the fluid assumes liquid-like characteristics. Its diffusivity and viscosity, however, remain. Supercritical fluid chromatography can use the widest range of detectors available to any chromatographic technique. As a result, capillary supercritical fluid chromatography has already demonstrated a great potential in application to water, environmental and other areas of analysis.

Supercritical fluid chromatography is now one of the fastest-growing analytical techniques. The first paper on the technique was by Klesper *et al.* [15], but supercritical fluid chromatography did not catch the analyst's attention until Novotny *et al.* [16] published the first paper on capillary supercritical fluid chromatography.

Supercritical fluid chromatography finds its applications in compounds that are either difficult or impossible to analyse by liquid chromatography or gas chromatography. Supercritical fluid chromatography is ideal for analysing either thermally labile or non volatile non chromatophoric compounds. The technique will be of interest to water chemists as a means of identifying and determining the non volatile components of water.

Most supercritical fluid chromatographs use carbon dioxide as the supercritical eluent, as it has a convenient critical point of 31.3°C and 72.5 atmospheres. Nitrous oxide, ammonia and *n*-pentane have also been used. This allows easy control of density between 0.2g mL^{-1} and 0.8g mL^{-1} and the utilisation of almost any detector from liquid chromatography or gas chromatography.

Wall [17] has discussed recent developments including timed-split injection, extraction and detection systems in supercritical fluid chromatography.

Timed-split injection

Capillary supercritical fluid chromatography utilises narrow 50μm or 100μm id columns of between 3 and 20m in length. The internal volume of a 3m × 50μm id column is only 5.8μL supercritical fluid chromatography operates at pressures from 15001b in^{-2} to 60001b in^{-2} (10335 kPa to 41340 kPa) and this means that gas chromatography injection systems cannot be used. High performance liquid chromatography injection systems are suitable for those pressure ranges, but even using small internal-loop injectors the volume introduced to the column is very large compared to the column's internal volume. To allow injections of about 10–50μL to be introduced to a capillary column, an internal-loop LC injector (Valco Inst Switzerland) has been used with a splitter (Fig. 1.2(a)) which was placed after the valve to ensure that a smaller volume was introduced into the column. This method works well for compounds which are easily soluble in carbon dioxide at low pressures.

Good reproducibility has been reported for capillary supercritical fluid chromatography using a direct injection method without a split restrictor. This method (Fig. 1.2(b)) utilises a rapidly rotating internal-loop injector (Valco Inst. Switzerland) which remains in-line with the column for only a short period of time. This then gives a reproducible method of injecting a small fraction of the loop into the column. For this method to be reproducible the valve must be able to switch very rapidly to put a small slug of sample into the column. To attain this a method called timed-split injection was developed (Lee Scientific). For timed split to operate it is essential that helium is used to switch the valve; air or nitrogen cannot provide sharp enough switching pulses. The injection valve itself must have its internal dead volumes minimised. Dead volumes prior to the valve allow some of the sample to collect prior to the loop, effectively allowing a double slug of sample to be injected which appears at the detector as a very wide solvent peak.

Detectors used in supercritical fluid chromatography include electron capture, photoionisation and sulphur chemiluminescence.

1.4 Ion exchange chromatography

Ion exchange chromatography is based upon the differential affinity of ions for the stationary phase. The rate of migration of the ion through the column is directly dependent upon the type and concentration of ions that constitute the eluent. Ions with low or moderate affinities for the packing generally prove to be the best eluents. Examples are hydroxide and carbonate eluents for anion separations.

The stationary phases commonly used in high performance liquid chromatography are typically derived from silica substrates. The instability of silica outside the pH range 2 to 7.5 represents one of the main reasons why ion exchange separations have not been extensively used in high performance liquid chromatography. To overcome this, some manufacturers (eg Dionex) supply columns under the trade name Ion-Pac which contain a packing which is derived from cross-linked polystyrene which is stable throughout the entire pH range. This pH stability allows eluents of extreme pH values to be used so that weak acids such as carbohydrates (and bases) can be ionised.

1.4.1 Anion exchange chromatography

Strong base anion exchange resins are manufactured by chloromethylation of sulphonated polystyrene followed by reaction with a tertiary amine:

They undergo the following reaction with anions:

$$\text{Resin} - \underset{\underset{R_1\ R_2\ R_3}{\diagup\ |\ \diagdown}}{CH_2}N^+Cl^- + M^+ + X^- \rightarrow \text{Resin} - \underset{\underset{R_1\ R_2\ R_3}{\diagup\ |\ \diagdown}}{CH_2} - N^+X^- + Cl^- + M^+$$

or

$$\text{Resin} - \underset{\underset{R_1\ R_2\ R_3}{\diagup\ |\ \diagdown}}{CH_2} - N^+\ Cl^- + MX^- \rightarrow \text{Resin} - \underset{\underset{R_1\ R_2\ R_3}{\diagup\ |\ \diagdown}}{CH_2} - N^+MX^- + Cl^-$$

where MX^- is a metal containing anion.

Weak base anion exchange resins are manufactured by chloromethylation of sulphonated polystyrene followed by reaction with primary or secondary amine:

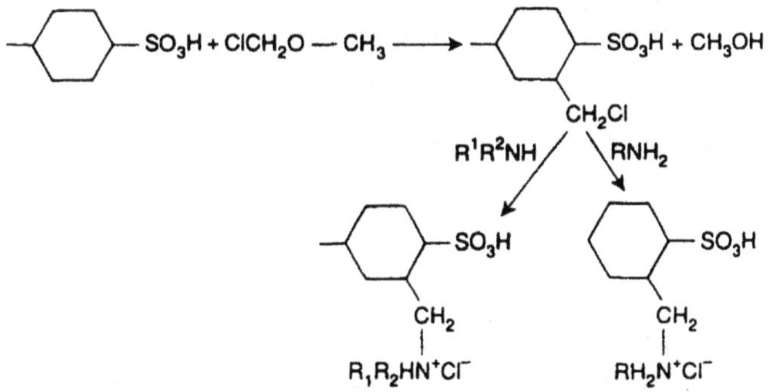

1.4.2 Cation exchange chromatography

Intermediate and highly polar types of resins are commonly referred to as ion exchange resins. Cationic exchange resins carry a negative charge and this reacts with positively charged metallic ions (cations) or cationic organic species. Anionic ion exchange resins carry a positive charge and this reacts with negatively charged anions or anionic organic species.

Strong acid cation exchange resins manufactured by the sulphonation of polystyrene or polydivinyl benzenes undergo the following reaction with cations:

$$\text{Resin } SO_3^-H^+ + M^+X^- \rightarrow \text{Resin } SO_3^-M^+ + H^+ + X^-$$

Weak acid cation exchange resins manufactured, eg by the polymerisation of methacrylic acid undergo the following reaction with cations:

$$\text{Resin } COO^-H^+ + M^+ + X^1 \rightarrow \text{Resin } COO^-M^+ + H^+ + X^-$$

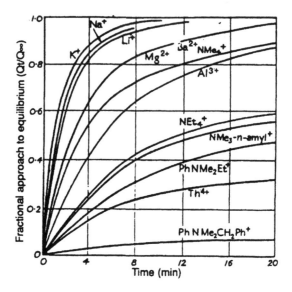

Fig. 1.3 Rate of exchange of phenolsulphonic acid resin (in ammonium form) with various cations
Source: Own files

Analysis is achieved by passing a large volume of water sample, suitably adjusted in pH and reagent composition down a small column of the resin. The adsorbed ions are then desorbed with a small volume of a suitable reagent in which the metals or metal complexes or anionic species dissolve. This extract can then be analysed by any suitable means.

1.4.3 Ion exchange chromatography theory

When an ion exchange reaction is carried out by a 'batch' method – that is, by putting a quantity of resin into a certain volume of solution – the reaction begins at once, but a certain time elapses before the equilibrium state is reached. It is generally a simple matter to determine the rate of exchange, for example, by sampling and analysing the solution at intervals or by making use of some physical property, such as electrical conductivity, which changes as the reaction proceeds. Such experiments show that rates of exchange vary very much from one system to another, the times of half-exchange ranging from fractions of a second to even months in certain extreme cases. Fig. 1.3 shows the process of exchange with a number of different cations exchanging with equal samples of the ammonium form of a phenolsulphonic acid resin under the same conditions.

Clearly, a knowledge of the rate of exchange is a prerequisite for the most effective use of resins. The factors which influence the rate of exchange and show how they account quantitatively for the form of kinetic curves such as those shown in Fig. 1.3 are listed below. By a series of controlled experiments in which one factor is varied at a time, it can readily be shown that, other things being equal, a high rate of exchange is generally favoured by the following choice of conditions:

(1) a resin of small particle size;
(2) efficient mixing of resin with the solution;
(3) high concentration of solution;
(4) a high temperature;
(5) ions of small size;
(6) a resin of low cross-linking.

The exchange reactions are generally much slower than reactions between electrolytes in solution, but there is no evidence to indicate the intervention of a slow 'chemical' mechanism such as is involved in most organic reactions, where covalent bonds have to be broken. The slowness of exchange reactions can be satisfactorily accounted for by the time required to transfer ions from the interior of the resin grains to the external solution, and vice versa. The rate of exchange is therefore seen to be determined by the rate at which the entering and leaving ions can change places. The process is said to be 'transport-controlled'; it is analogous, for instance, to the rate of solution of a salt in water. All the factors (1) to (6) listed above are such as to facilitate the transport of ions to or from or through the resin.

1.5 Conventional column chromatography

Although displaced to a large extent by high performance liquid chromatography many workers still use conventional chromatographic techniques in the analysis of organic and organometallic compounds and anions and cations. About half the papers dealing with organic and organic compounds were published after 1990, while all the papers concerned with conventional column chromatography of cations and anions were published prior to that date. This indicates the extent to which high performance liquid chromatography is displacing the conventional technique.

Classes of organic compounds analysed by this technique include non ionic surfactants, fatty acids, hydrocarbons (all types), organochlorine compounds, organosulphur and phosphorus compounds, substituted aromatic compounds, NTA, EDTA and insecticides and herbicides. Organometallic compounds studied include those of arsenic, lead, germanium, mercury and tin.

1.6 High performance liquid chromatography–mass spectrometry

The introductory paragraphs in section 1.3 mention a range of detectors that have been employed in the chromatography of various classes of compounds. While one or other of these types of detector will indicate when the separated substances are exiting the column, they do not provide any information on the identity of the separated compounds. In many instances where compounds of unknown identity in water samples are concerned it is vital to be able to identify the unknown separated organic compounds. For this reason, increasingly a mass spectrometer is being used to identify unknown organic compounds.

Applications of this technique are growing and include the analysis of mixtures of pesticides and herbicides, including organophosphorus, phenoxyacetic acid, carbamate, urea types, mixtures of various types of organic compounds, alkylbenzene sulphonates, polyethylene glycols, nonylphenyl ethoxylates, dioctadecylmethyl ammonium, ozonisation products and chlorination products.

Hewlett Packard supply the HP 5988A and HP 5987A mass-selective detectors for use with liquid chromatographs. This equipment has been used extensively for identifying and determining non volatile compounds. The technique has produced further improvements in liquid chromatography mass spectrometry. The particle–beam liquid chromatograph–mass spectrometer uses the same switchable electron impact chemical ionisation source and the same software and data system. Adding a gas chromatograph creates a versatile particle–beam liquid chromatograph–gas chromatography–mass spectrometry system that can be switched from liquid chromatography–mass spectrometry to gas chromatography–mass spectrometry in an instant.

Based on a new technology, particle-beam enhanced liquid chromatography–mass spectrometry expands a chemist's ability to analyse a vast variety of substances. Electron impact spectra from the system are reproducible and can be searched against standard or custom libraries for positive compound identification. A simple adjustment to the particle-beam interface is all it takes.

1.7 Capillary column coupling isotachoelectrophoresis

This technique has found limited applications to the determination of organic compounds, cations and anions in water. The technique offers many similar advantages to ion chromatography, namely multiple-ion analysis, little or no sample pretreatment, speed, sensitivity and automation.

Separation capillary columns are made in fluorinated ethylene–propylene copolymer. Detection is achieved by conductivity cells and an a.c. conductivity mode of detection is used for making the separations visible. The driving current is supplied by a unit enabling independent currents to be preselected for the preseparation and final analytical stages. The run of the analyser is controlled by a programmable timing and control unit. The zone lengths from the conductivity detector, evaluated electronically, can be printed on a line printer.

To satisfy the requirements for the properties of the leading electrolyte applied in the first stage and, consequently, to decide its composition, two facts had to be taken into account: the pH value of the leading electrolyte needs to be around four or less and at the same time the separations of the macroconstituents need to be optimised by means other than adjusting the pH of the leading electrolyte (anions of strong acids).

The choice of the leading electrolyte for the second stage, in which the microconstituents were finally separated and quantitatively evaluated, is straightforward, involving a low concentration of the leading constituent (low detection limit) and a low pH of the leading electrolyte (separation according to pK values).

At present, ion chromatography has a dominant position among the separation methods used for the analysis of inorganic anions [18–22]. However, when the separation efficiencies typically achieved by ion chromatography are compared to characterising capillary zone electrophoresis for analysis of a group of analytes, it is apparent that the latter technique has advantages. As the effective mobilities of inorganic anions are sufficiently due to differences in their ionic mobilities they can be influenced in a desired way via appropriately selecting (differences in pKa values) and/or via the use of suitable additives [22], the selectivity factors [23] are also favourable leading to rapid resolutions of the analytes by capillary zone electrophoresis.

Although in some instances it is possible to determine low concentrations of inorganic anions by capillary zone electrophoresis [24], problems arise when the concentrations of the sample constituents vary considerably. This is due to the fact that the determination of microconstituents may require the sample load to have an impact on both the migration velocities and the resolution of analytes [25]. The use of indirect detection, as preferred in the case of inorganic anions [24,26] is also less capable of achieving adequate load capacities.

Of the inorganic anions which are currently monitored chloride, sulphate, and in many instances also nitrate, are considered anionic macroconstituents. These are often accompanied by nitrite, fluoride and phosphate present in samples at 10^2–10^5-fold lower concentrations. Such a concentration span probably cannot be covered by capillary zone electrophoresis.

This technique offers very similar advantages to ion chromatography in the determination of anions in water, namely multipleion analysis, little or no sample pretreatment, speed, sensitivity and automation.

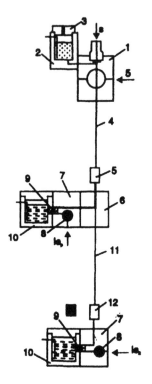

Fig. 1.4 Separation unit in a column coupling configuration as used for the analysis of anions in river water

1, sampling block with a 30µL sampling valve; 2, terminating electrolyte compartment with a cap (3); 4, 0.85mm id capillary tube (pre-separation column); 5 and 12, conductivity detectors; 6, bifurcation block; 7, refilling block with needle valve (8); 9, mechanically supported membranes; 10, leading electrolyte compartments; 11, 0.30mm id capillary tube; s, positions for the sample introduction (valve or microsyringe); lep and le, positions for refilling of the columns used for the first and second stages respectively

Source: Reproduced with permission from Elsevier Science, UK [30]

Very similar advantages are offered by capillary isotachophoresis [27,28]. However, with the exception of the work by Bocek *et al.* [29] in which the determinations of chloride and sulphate in mineral water were demonstrated, little attention has been paid to this subject.

Zelinski *et al.* [30] applied the technique to the determination of 0.02-0.1mg L^{-1} quantities of chloride, nitrate, sulphate, nitrite, fluoride and phosphate in river water. Approximately 25min was required for a full analysis. These workers used a column coupling technique which, by dividing the analysis into two stages, enables a high load capacity and a low detection limit to be achieved simultaneously without an appreciable increase in the analysis time.

The separation unit of the capillary isotachophoresis instrument used is shown in Fig. 1.4. A 0.85mm id capillary tube made of fluorinated ethylene propylene copolymer was used in the pre-separation (first) stage and a capillary tube of 0.30mm id made of the same material served for the separation in the second stage. Both tubes were provided with conductivity detection cells [31] and an a.c. conductivity mode of detection [27] was used for making the separations visible. The driving current was supplied by a unit enabling independent currents to be preselected for the preseparation and final analytical stages. The run of the analyser was controlled by a programmable timing and control unit. The zone lengths from the conductivity detector, evaluated electronically, were printed on a line printer.

The use of a column coupling configuration of the separation unit provides the possibility of applying a sequence of two leading electrolytes in one analytical run. Therefore, the choice of optimum separation conditions can be advantageously divided into two steps:

(1) the choice of a leading electrolyte suitable for the separation and quantitation of the macroconstituents in the first stage (preseparation column) simultaneously having a retarding effect on the effective mobilities of microconstituents (nitrite, fluoride, phosphate); and

(2) the choice of the leading electrolyte for the second stage in which only microconstituents are separated and quantified (macroconstituents were removed from the analytical system after their evaluation in the first stage).

To satisfy the requirements for the properties of the leading electrolyte applied in the first stage and, consequently, to decide its composition, two facts had to be taken into account, ie the pH value of the leading electrolyte needs to be around 4 or less (retardation of nitrite relative to the macroconstituents in this stage) and at the same time the separations of the macroconstituents need to be optimised by other means than adjusting the pH of the leading electrolyte (anions of strong acids). For the latter reason, complex equilibria and differentiation of anions through the charge number of the counter ions were tested at a low pH as a means of optimising the separation conditions for chloride, nitrate and sulphate in the presence of nitrite, fluoride and phosphate.

The retardation of chloride and sulphate through complex formation with cadmium enabling the separation of anions of interest [28] to be carried out, was found to be unsuitable, as the high concentration of cadmium ions necessary to achieve the desired effect led to the loss of fluoride and phosphate (probably owing to precipitation).

Similarly, the use of calcium and magnesium as complexing co-counter ions [32] to decrease the effective mobility of sulphate was found to be ineffective as a very strong retardation of fluoride occurred.

Table 1.1 Operational systems

| System No. | Parameter | Leading electrolytes | | Terminating electrolyte |
		1st stage	2nd stage	
1	Anion	Cl⁻	Cl⁻	CITR[b]
	Concentration (mmol L⁻¹)	8	1	2
	Counter ion	BALA[a]	BALA[a]	H⁺
	Co-counter ion	BTP[c]	–	–
	Concentration (mmol L⁻¹)	3	–	–
	Additive to the leading electrolyte	0.1% HEC[d]	0.1% HEC[d]	–
	pH	3.55	3.55	ca. 3
2	Anion	Cl⁻	Cl⁻	CITR[b]
	Concentration (mmol L⁻¹)	1	1	1
	Counter ion	H⁺	H⁺	H⁺
	Additive to the leading electrolyte	0.1% HEC[d]	0.1% HEC[d]	–
	pH	3.0	3.0	ca. 3

[a] β–alanine
[b] Citric acid
[c] Abbreviations: BALA = β-alanine; CITR = citric acid; BTP = aminopropane, 1,3–bis-tris(hydroxymethyl)methyl–aminopropane
[d] HEC = Hydroxyethylcellulose

Source: Reproduced with permission from Elsevier Science, Oxford [30]

Better results were achieved when a divalent organic cation was used as a co-counter ion in the leading electrolyte [33,34] employed in the first-separation stage when, simultaneously, the pH of the leading electrolyte was 4 or less, and the steady state configuration of the constituents to be separated was chloride, nitrate, sulphate, nitrite, fluoride and phosphate. The detailed composition of the operational system of this type used for quantitative analysis is given in Table 1.1 (system No. 1.)

The choice of the leading electrolyte for the second stage, in which the microconstituents were finally separated and quantitatively evaluated, was straightforward: a low concentration of the leading constituent (low detection limit) and a low pH of the leading electrolyte (separation according to pK values). The operational system used throughout this work in the second stage is given in Table 1.1.

An isotachopherogram from the analysis of a model mixture of anions obtained in the first separation stage is shown in Fig. 1.5. This isotachopherogram merely indicates the differences in the effective mobilities of the anions of interest, as the concentration of nitrite, fluoride and phosphate in river water are usually too low to be detected in the first stage.

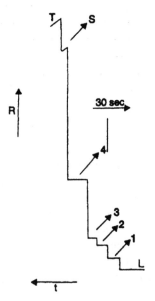

Fig. 1.5 Isotachopherogram for the separation of anions in the first stage using operational system No. I: I, nitrate (40mg L $^{-1}$); 2, sulphate (20mg L $^{-1}$); 3, nitrite (20mg L $^{-1}$); 4, fluoride (20mg L $^{-1}$); 5, phosphate (20mg L $^{-1}$). Driving current 250μA. L and T = leading and terminating anions, respectively; R = increasing resistance, t = increasing time. The sample was introduced with the aid of a 30μL valve

Source: Reproduced with permission from Elsevier Science, UK [30]

Both the cross sectional area of the capillary tube and the concentration of the leading constituent used in the second stage were optimised with respect to the determinations of microconstituents. In a search for optimal separating conditions, operational system No.2 (Table 1.1) was used for the determinations of microconstituents.

Kaniansky et al. [35] have also reported on a method for the determination of nitrate, sulphate, nitrite, fluoride and phosphate by capillary zone electrophoresis coupled with capillary isotacho-electrophoresis in the column coupling configuration. Such distributions of these anions are typical for many environmental matrices, and it is shown that capillary isotachophoresis–capillary zone electrophoresis tandem enables the capillary isotachoelectrophoresis determination of the macroconstituents, while capillary isotachophoresis–preconcentrated microconstituents cleaned up from the macroconstituents can be determined by capillary zone electrophoresis with a conductivity detector. This approach was effective when the concentration ratio of macro-microconstituents was less than $(2–3) \times 10^4$. The limits of detection achieved for the microconstituents (200, 300 and 600ppt for nitrite, fluoride and phosphate, respectively) enabled their determinations at low parts per billion concentrations in instances when the concentrations of sulphate or chloride in the samples were higher than 1000ppm. Although in such situations the load capacity of the capillary isotachophoresis column was not sufficient to provide the quantitations of the macro-constituents, the capillary isotachophoresis sample cleanup enabled the capillary zone electrophoresis determinations of the microconstituents present in the samples at 10^5–10^6-fold lower concentrations. Experiments with samples of such extreme compositions showed that anionic impurities from the capillary isotachophoresis electrolyte solutions trans-ferred together with the analytes into the capillary zone electrophoresis column are probably the main hindrances in extending this dynamic concentration range by decreasing the limits of the detection to a level associated with the noise characteristics of the conductivity detector.

The limited applications of this technique, to date include fluoride, chloride, bromide, nitrite, nitrate, sulphate, sulphide, phosphate, amino acetate, chlorodicarboxylic acids, volatile organic acids and chromium(VI).

1.8 Gas chromatography

Numerous organic compounds, organometallic compounds, cations and anions have been determined by gas chromatography.

The basic requirements of a high performance gas chromatograph are:

- the sample is introduced to the column in an ideal state, ie uncon-taminated by septum bleed or previous sample components, without modification due to distillation effects in the needle and quantit-atively, ie without hold-up or adsorption prior to the column;
- the instrument parameters that influence the chromatographic separation are precisely controlled;
- sample components do not escape detection; ie highly sensitive, reproducible detection and subsequent data processing are essential.

There are two types of separation column used in gas chromatography: capillary columns and packed columns.

Packed columns are still used extensively, especially in routine analysis. They are essential when sample components have high partition coefficients and/or high concentrations. Capillary columns provide a high number of theoretical plates, hence a very high resolution, but they cannot be used in all applications because there are not many types of chemically bonded capillary columns. Combined use of packed columns of different polarities often provides better separation than with a capillary column. It sometimes happens that a capillary column is used as a supplement in the packed-column gas chromatograph. It is best, therefore, to house the capillary and packed columns in the same column oven and use them routinely and the capillary column is used when more detailed information is required.

Conventionally, it is necessary to use a dual-column flow line in packed-column gas chromatography to provide sample and reference gas flows. The electronic baseline drift compensation system allows a simple column flow line to be used reliably.

Advances in capillary column technology presume stringent performance levels for the other components of a gas chromatograph as column performance is only as good as that of the rest of the system. One of the most important factors in capillary column gas chromatography is that a high repeatability of retention times be ensured even under adverse ambient conditions. These features combine to provide ±0.01min repeatability for peaks having retention times as long as 2h (other factors being equal).

Another important factor for reliable capillary column gas chromatography is the sample injection method. Various types of sample injection ports are available. The split/splitless sample injection port unit series is designed so that the glass insert is easily replaced and the septum is continuously purged during operation. This type of sample injection unit is quite effective for the analysis of samples having high-boiling-point compounds as the major components.

In capillary column gas chromatography, it is often required to raise and lower the column temperature very rapidly and to raise the sample injection port temperature. In one design of gas chromatography, the Shimadzu GC 14-A, the computer-controlled flap operates to bring in the external air to cool the column oven rapidly – only 6min from 500°C to 100°C. This computer-controlled flap also ensures highly stable column temperature when it is set to a near-ambient point. The lowest controllable column temperature is about 26°C when the ambient temperature is 20°C.

Perkin–Elmer supply a range of instruments including the basic models 8410 for packed and capillary work and the 8420 for dedicated capillary work, both supplied on purchase with one of the six different types of detection. The models 8400 and 8500 are more sophisticated capillary column instruments capable of dual detection operation with the additional features of keyboard operation, screen graphics method storage, host computer links, data handling and compatibility with laboratory automation systems. Perkin–Elmer supply a range of accessories for these instruments including an autosampler (AS–8300), an infrared spectrometer interface, an automatic headspace accessory (HS101 and H5–6), an autoinjector device (AI–I), also a catalytic reactor and a pyroprobe (CDS 190) and automatic thermal desorption system (ATD-50) (both useful for examination of sediments).

The Perkin–Elmer 8700, in addition to the features of the models 8400 and 8500, has the ability to perform multi-dimensional gas chromatography.

The optimum conditions for capillary chromatography of material heart-cut from a packed column demand a highly sophisticated programming system. The software provided with the model 8700 provides this, allowing methods to be linked so that pre-column and analytical column separations are performed under optimum conditions. Following the first run, in which components are transferred from the pre-column to the on-line cold trap, the system will reset to a second method and, on becoming ready, the cold trap is desorbed and the analytical run automatically started.

Other applications of the model 8700 system include fore-flushing and back-flushing of the pre-column, either separately or in combination with heart cutting, all carried out with complete automation by the standard instrument software.

1.9 Gas chromatography–mass spectrometry

The time has long since passed when one could rely on gas chromatographic or liquid chromatographic data alone to identify unknown compounds in water or other environmental samples. The sheer numbers of compounds present in such materials would invalidate the use of these techniques, and even in the case of simple mixtures the time required for identification would be too great to provide essential information in the case, for example, of accidental spillage of an organic substance into a water course or inlet to a water treatment plant where information is required very rapidly.

The practice nowadays is to link a mass spectrometer or ion trap to   the outlet of the gas chromatograph or high performance liquid chromatograph so that a mass spectrum is obtained for each chromatographic peak as it emerges from the separation column. If the peak contains a single substance then computerised library-searching facilities attached to the mass spectrometer will rapidly identify the substance. If the emerging peak contains several substances, then the mass spectrum will indicate this and in many cases will provide information on the substances present.

Finnigan MAT are one of the major suppliers of this equipment. They supply equipment for single stage quadruple mass spectrometry, mass spectrometry–mass spectrometry, high mass high resolution mass spectrometry, ion trap detection, and gas chromatography–mass spectrometry.

OWA–20130B organic in water gas chromatograph mass spectrometer (Finnigan MAT)

This system combines hardware and software features not found in any other low-cost gas chromatography–mass spectrometry system. The highly reliable 3000 series electron ionisation source and quadrupole analyser are used to provide superior mass spectrometer performance. The software is designed with the necessary automation to perform complete quantitative analysis of any target compounds. All routine system operating parameters are adjustable through the computer's graphics display terminal. The priority interrupt foreground/ background operation system allows all data-processing functions to be performed at any time with no limiting effects on data acquisition. Sophisticated data-processing programs are readily accessible through a simple commercial structure. The simplicity of the entire system allows complete analysis with minimal operator training. Standard features of this instrument include fully automated gas chromatography–mass spectrometry, automated compound analysis and quantification, software 4-800μ electron impact quadrupole mass spectrometer, high-capacity, turbidmolecular pump vacuum system, liquid-sample concentrators for volatile organic in water analysis, a sigma series programmable gas chromatograph, Grob-type split-splitless capillary column injector system, packed column injector with glass jet separator, Nova 4C/53k word, 16-bit minicomputer, graphics display terminal, 10Mbyte disk drive, a printer/plotter, an NBX 3100 spectra library, a full-scan or multiple-ion detector and a nine-track tape drive. Options include chemical ionisation ion source, direct inlet vacuum lock, programmable solids probe, direct exposure probe, various gas chromatographic detectors, autosampler, subambient gas chromatography operation and a 32Mbyte disk drive.

As an example of the application of gas chromatography–mass spectrometry, Fig. 1.6 shows a reconstructed ion chromatograph obtained for an industrial waste sample. The Finnigan MAT 1020 instrument was used in this work. Of the 27 compounds searched for, 15 were found. These data were automatically quantified. This portion of the report contains the date and time at which the run was made, the sample description, who submitted the sample to the analyst, followed by the names of the compounds. If no match for a library entry was found, the component was listed 'not found'. Also shown is the method of quantification and the area of the peak (height could also have been chosen).

The large peak at scan 502 (Fig. 1.6) does not interfere with the ability of the software to quantify the sample. Although the compound eluting at scan 502 was not one of the target compounds in the library being reverse-searched, it was possible to identify it by forward-searching the NBS library present on the system. The greatest similarity was in the comparison of the unknown with the spectrum of benzaldehyde.

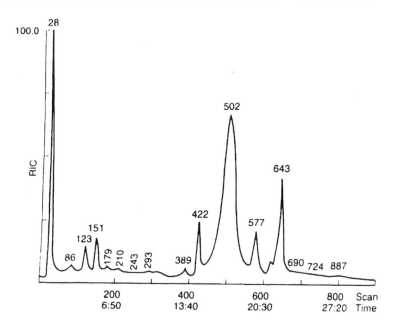

Fig. 1.6 Reconstructed chromatograph of industrial waste sample
Source: Own files

1.10 Thin layer chromatography

This technique has been used extensively for the separation and determination of mixtures of compounds in water by migration on thin layers,
usually, of silica or alumina. In the case of volatile compounds, such as
aliphatic hydrocarbons, care is needed as volatiles may be lost during the
separation process.

In general, the technique is limited to cases where the identity of the
substance to be determined is known, although, in some cases
identification of the separated compounds has been achieved by infrared
spectrometry or mass spectrometry of elutes of the individual separated
spots isolated from the thin layer plate. Types of compounds which have
been determined by thin layer chromatography are: chlorophylls and
degradation products, herbicides (carbamate, urea, triazine, phenoxy
acetic acid, organophosphorus and organochlorine types), hydrocarbons
(aliphatic, aromatic and polyaromatic), phenols, phenolic acids,
chlorinated hydrocarbons, amines, fluorescent agents, chlorinated
phenols and hydrocarbons, aldehydes, caprolacetone, quinones, polychlorobiphenyls, fungicides and amines, also organic compounds of lead,
tin and mercury.

1.11 Chromatographic detectors

Particularly in the case of ion chromatography and high performance liquid chromatography throughout the text, various types of detectors are discussed which when connected to the outlet of the separation detect effluents as they leave the column.

1.11.1 Visible spectrophotometric detector

This technique is only of value when the identity of the compound to be determined is known. There are also limitations on the sensitivity that can be achieved, usually milligrams per litre or occasionally, micrograms per litre.

Visible spectrophotometers are commonly used in the water industry for the estimation of colour in a sample or for the estimation of coloured products produced by reacting a colourless compound of the sample with a reagent, which produces a colour that can be evaluated spectrophotometrically.

Some commercially available instruments, in addition to visible spectrophotometers, can also perform measurements in the ultraviolet and near IR regions of the spectrum. These have not yet found extensive application in the field of water analysis.

1.11.2 Fluorescence detector

Generally speaking, concentrations down to the microgram per litre level can be determined by this technique with recovery efficiencies near 100%.

Potentially, fluorometry is valuable in every laboratory, including water laboratories, for the performance of chemical analysis where the prime requirements are selectivity and sensitivity. While only 5–10% of all molecules possess a native fluorescence, many can be induced to fluoresce by chemical modification or when tagged with a fluorescent molecule.

Luminescence is the generic name used to cover all forms of light emission other than that arising from elevated temperature (thermoluminescence). The emission of light through the absorption of ultraviolet or visible energy is called photoluminescence, and that caused by chemical reactions is called chemiluminescence. Light emission through the use of enzymes in living systems is called bioluminescence, the only known application of which to water analysis is the determination of adenosine triphosphate. Photoluminescence may be further subdivided into fluorescence, which is the immediate release (10^{-8}s) of absorbed light energy, as opposed to phosphorescence, which is delayed release (10^{-6}–10^{2}s) of absorbed light energy.

The excitation spectrum of a molecule is similar to its absorption spectrum, while the fluorescence and phosphorescence emissions occur at longer wavelengths than the absorbed light. The intensity of the emitted light allows quantitative measurement since, for dilute solutions, the emitted intensity is proportional to concentration. The excitation and emission spectra are characteristic of the molecule and allow qualitative measurements to be made. The inherent advantages of the technique, particularly fluorescence, are:

(1) Sensitivity, picogram quantities of luminescent materials are frequently studied.
(2) Selectivity, derived from the two characteristic wavelengths.
(3) The variety of sampling methods that are available, ie dilute and concentrated samples, suspensions, solids, surfaces and combination with chromatographic methods, such as that used in the high performance liquid chromatography separation of o-phthalyl dialdehyde derivatised amino acids in natural and sea water samples.

Fluorescence spectrometry forms the majority of luminescence analysis. However, the recent developments in instrumentation and room temperature phosphorescence techniques have given rise to practical and fundamental advances which should increase the use of phosphorescence spectrometry. The sensitivity of phosphorescence is comparable to that of fluorescence and complements the latter by offering a wider range of molecules of study.

The pulsed xenon lamp forms the basis for both fluorescence and phosphorescence measurement. The lamp has a pulse duration at half peak height of 10µs. Fluorescence is measured at the instant of the flash. Phosphorescence is measured by delaying the time of measurement until the pulse has decayed to zero.

Several methods are employed to allow the observation of phosphorescence. One of the most common techniques is to supercool solutions to a rigid glass state, usually at the temperature of liquid nitrogen (77K). At these temperatures molecular collisions are greatly reduced and strong phosphorescence signals are observed.

Under certain conditions phosphorescence can be observed at room temperature from organic molecules adsorbed on solid supports such as filter paper, silica and other chromatographic supports.

Phosphorescence can also be detected when the phosphor is incorporated into an ionic micelle. Deoxygenation is still required either by degassing with nitrogen or by the addition of sodium sulphite. Micelle-stabilised-room temperature phosphorescence (MS RTP) promises to be a useful analytical tool for determining a wide variety of compounds such as pesticides and polyaromic hydrocarbons.

Perkin–Elmer and Hamilton both supply luminescence instruments.

Perkin–Elmer LS–3B and LS–5B luminescence spectrometers

The LS–3B is a fluorescence spectrometer with separate scanning mono-chromators for excitation and emission, and digital displays of both monochromator wavelengths and signal intensity. The LS-5B is a ratioing luminescence spectrometer with the capability of measuring fluorescence, phosphorescence and bio- and chemiluminescence. Both instruments are equipped with a xenon discharge lamp source and have an excitation wavelength range of 230–720nm and an emission wavelength range of 250–800nm.

Perkin–Elmer LS–2B microfilter fluorometer

The model LS–2B is a low-cost easy-to-operate, filter fluorometer that scans emission spectra over the wavelength range 390–700nm (scanning) or 220–650nm (individual interferences filters). The essentials of a filter fluorometer are as follows:

- a source of ultraviolet/visible energy (pulsed xenon)
- a method of isolating the excitation wavelength
- a means of discriminating between fluorescence emission and excitation energy
- a sensitive detector and a display of the fluorescence intensity.

The model LS–2B has all these features arranged to optimise sensitivity for microsamples. It can also be connected to a highly sensitive 7µL liquid chromatographic detector for detecting the constituents in the column effluent. It has the capability of measuring fluorescence, time-resolved fluorescence and bio- and chemiluminescent signals. A 40-portion autosampler is provided. An excitation filter kit containing six filters – 310, 340, 375, 400, 450 and 480nm – is available.

1.11.3 Infrared and Raman spectrometric detectors

Fourier transform infrared spectrometry, a versatile and widely used analytical technique, relies on the creation of interference in a beam of light. A source light beam is split into two parts and a continually varying phase difference is introduced into one of the two resultant beams. The two beams are recombined and the interference signal is measured and recorded, as an interferogram. A Fourier transform of the interferogram provides the spectrum of the detected light. Fourier transform infrared spectroscopy, a seemingly indirect method of spectroscopy, has many practical advantages, as discussed below.

A Fourier transform infrared spectrometer consists of an infrared source, an interference modulator (usually a scanning Michelson interferometer), a sample chamber and an infrared detector. Interference signals measured at the detector are usually amplified and then digitised. A digital computer initially records and then processes the interferogram and also allows the spectral data that result to be manipulated. Permanent records of spectral data are created using a plotter or other peripheral device.

The principal reasons for choosing Fourier transform infrared spectroscopy are: first, that these instruments record all wavelengths simultaneously and thus operate with maximum efficiency; and, second, that they have a more convenient optical geometry than do dispersive infrared instruments. These two facts lead to the following advantages.

- Fourier transform infrared spectroscopy spectrometers achieve much higher signal-to-noise ratios in comparable scanning times.
- They can cover wide spectral ranges with a single scan in a short scan time, thereby permitting the possibility of kinetic time-resolved measurements.
- They provide higher-resolution capabilities without undue sacrifices in energy throughput or signal-to-noise ratios.
- They encounter none of the stray light problems usually associated with dispersive spectrometers.
- They provide a more convenient beam geometry – circular rather than slit shaped – at the sample focus.

Conventional Raman spectroscopy cannot be applied directly to aqueous extracts of sediments and soils, although it is occasionally used to provide information on organic solvent extracts of such samples. Fourier transform Raman spectroscopy, on the other hand, can be directly applied to water samples. The technique complements infrared spectroscopy in that some functional groups, eg unsaturation, give a much stronger response in the infrared. Several manufacturers (Perkin–Elmer, Digilab, Bruker) now supply Fourier transform infrared spectrometers.

1.11.4 Atomic absorption spectrometric detectors

Basically, the atomic absorption method was designed for the determination of cations. Since shortly after its inception in 1955, atomic absorption spectrometry has been the standard tool employed by analysts for the determination of trace levels of metals in water samples or column effluents. In this technique a fine spray of the analyte is passed into a suitable flame, frequently oxygen acetylene or nitrous oxide acetylene, which converts the elements to an atomic vapour. Through this vapour radiation is passed at the right wavelength to excite the ground state atoms to the first excited electronic level. The amount of radiation absorbed can then be measured and directly related to the atom concentration: a hollow cathode lamp is used to emit light with the characteristic narrow line spectrum of the analyte element. The detection system consists of a monochromator (to reject other lines produced by the lamp and background flame radiation) and a photomultiplier. Another key feature of the technique involves modulation of the source radiation so that it can be detected against the strong flame and sample emission radiation.

A limitation of this technique is its lack of sensitivity compared to that available by other techniques (eg inductively coupled plasma atomic emission spectrometry).

Suitable instrumentation is supplied by Thermoelectron, Perkin–Elmer, Varian Associates, GBC Scientific and Shimazu. All of these suppliers supply equipment with autosamplers and mercury and hydride attachments.

1.11.5 Inductively coupled plasma atomic emission spectrometric detectors

This technique has been found to be particularly useful for the determination in water of extremely low levels of a limited number of anions.

An inductively coupled plasma is formed by coupling the energy from a radiofrequency (1–3kW or 27–50MHz) magnetic field to free electrons in a suitable gas. The magnetic field is produced by a two- or three-turn water-cooled coil and the electrons are accelerated in circular paths around the magnetic field lines that run axially through the coil. The initial electron 'seeding' is produced by a spark discharge but, once the electrons reach the ionisation potential of the support gas, further ionisation occurs and a stable plasma is formed.

The neutral particles are heated indirectly by collisions with the charged particles upon which the field acts. Macroscopically the process is equivalent to heating a conductor by a radio-frequency field, the resistance to eddy-current flow producing joule heating. The field does not penetrate the conductor uniformly and therefore the largest current flow is at the periphery of the plasma. This is the so-called 'skin' effect and, coupled with a suitable gas-flow geometry, it produces an annular or doughnut-shaped plasma. Electrically, the coil and plasma form a transformer with the plasma acting as a one-turn coil of finite resistance.

The properties of an inductively coupled plasma closely approach those of an ideal source for the following reasons:

- The source accepts a reasonable input flux of the sample and is able to accommodate samples in the gas, liquid or solid phases.
- The introduction of the sample does not radically alter the internal energy generation process or affect the coupling of energy to the source from external supplies.

- The source is operable on commonly available gases and is available at a price that will give cost-effective analysis.
- The temperature and residence time of the sample within the source is such that all the sample material is converted to free atoms irrespective of its initial phase or chemical composition; such a source should be suitable for atomic absorption or atomic fluorescence spectrometry.
- If the source is to be used for emission spectrometry, then the temperature is sufficient to provide efficient excitation of a majority of elements in the periodic table.
- The continuum emission from the source is of a low intensity to enable the detection and measurement of weak spectral lines super-imposed upon it.
- The sample experiences a uniform temperature field and the optical density of the source should be low so that a linear relationship between the spectral line intensity and the analyte concentration can be obtained over a wide concentration range.

Greenfield *et al.* [36] were the first to recognise the analytical potential of the annular inductively coupled plasma. Wendt and Fassel [37] reported early experiments with a 'tear-drop' shaped inductively coupled plasma but later described the medium power (1–3kW), 18mm annular plasma now favoured in modern analytical instruments [38].

The current generation of inductively coupled plasma emission spectrometers provide limits of detection in the range of 0.1-500µg L^{-1} in solution, a substantial degree of freedom from interference and a capability for simultaneous multi-element determination facilitated by a directly proportional response between the signal and the concentration of the analyte over a range of about five orders of magnitude.

The most common method of introducing liquid samples into the inductively coupled plasma is by using pneumatic nebulisation in which the liquid is dispersed into a fine aerosol by the action of a high-velocity gas stream. The fine gas jets and liquid capillaries used in inductively coupled plasma nebulisers may cause inconsistent operation and even blockage when solutions containing high levels of dissolved solids, such as sea water or particulate matter, are used. Such problems have led to the development of a new type of nebuliser, the most successful being based on a principle originally described by Babington (US Patents). In these, the liquid is pumped from a wide-bore tube and thence conducted to the nebulising orifice by a V-shaped groove [39] or by the divergent wall of an over-expanded nozzle [40]. Such devices handle most liquids and even slurries without difficulty.

Nebulisation is inefficient and therefore not appropriate for very small liquid samples. Introducing samples into the plasma in liquid form reduces the potential sensitivity because the analyte flux is limited by the amount of solvent that the plasma will tolerate. To circumvent these problems a variety of thermal and electrothermal vaporisation devices have been investigated. Two basic approaches are in use. The first involves indirect vaporisation of the sample in an electrothermal vaporiser, eg a carbon rod or tube furnace or heated metal filament as commonly used in atomic absorption spectrometry [41–43]. The second involves inserting the sample into the base of the inductively coupled plasma on a carbon rod or metal filament support [44,45]. Instrumentation is available from Perkin–Elmer, Thermoelectron, Phillips, Baird and Spectroanalytical Ltd.

Mass spectrometry has also been linked to inductively coupled plasma detectors, see also Sections 1.6 and 1.9.

References

1 Small, H., Stevens, H.S. and Banman, W.C. *Analytical Chemistry*, **47**, 1801, (1975).
2 Small, H. *Ion Chromatography*. Plenum, New York, p. 132 (1989).
3 Blasius, E., Janzen, K.-P., Adrian, W., Klautke, G. *et al. Z. für Analytsch. Chemie*, **284**, 337 (1977).
4 Small, H., Soderquist, M.E. and Pischke, J.W. US Patent 4,732,686 (1988).
5 Hu, W., Takeuchi, T. and Haraguchi, H. *Analytical Chemistry*, **65**, 2204 (1993).
6 Hu, W., Tao H. and Haraguchi, H. *Analytical Chemistry*, **66**, 2514 (1994).
7 Hu, W., Mujasaki, A., Tao, H., Itoh, A., Unemura, T. and Haraguchi, H. *Analytical Chemistry*, **67**, 3713 (1995).
8 Adamik, M.L. and Bartak, D.E. *Analytical Chemistry*, **57**, 279 (1985).
9 Gardner, W.S., Lundrum, P.F. and Yates, D.A. *Analytical Chemistry*, **54**, 1198 (1982).
10 Tanaka, K. and Ishizaka, T. *Journal of Chromatography*, **19**, 7 (1980).
11 Tanaka, K. and Sunahara, H. *Bunseki Kagaku*, **27**, 95 (1978).
12 Tanaka, K., Nakajina, K. and Sunahara, H. *Bunseki Kagaku*, **26**, 102 (1977).
13 Gonzalez, V., Moreno, B., Silicia, D. and Rubio, S. *Perez-Bendito*, **65**, 1897 (1993).
14 Done, T.J. and Reid, W.K. *Separation Science*, **5**, 825 (1970).
15 Klesper, E., Corwin, A. and Turner, D. *Journal of Organic Chemistry*, **27**, 700 (1962).
16 Novotny, M., Springston, P.J. and Lee, M. *Analytical Chemistry*, **53**, 407A (1981).
17 Wall, R.J. In *Chromatography and Analysis*. John Wiley & Sons, Chichester (1988).
18 MacCarthy, P., Klusman, R.W. and Rice, J.A. *Analytical Chemistry*, **61**, 269R (1989).
19 Frankenberger, W.T. Jr., Mehra, H.C. and Gjerde, D.T. *Journal of Chromatography*, **504**, 211 (1990).
20 MacCarthy, P., Klusman, R.W., Cowling, S.W. and Rice, J.A. *Analytical Chemistry*, **63**, 310R (1991).
21 Tarter, J.D. ed. *Ion Chromatography*. M Dekker, New York and Basel (1987).
22 Fukushi, K. and Hiro, K. *Journal of Chromatography*, **518**, 189 (1990).
23 Foret, F. and Bocek, P. *Advances in Electrophoresis*, **3**, 271 (1989).
24 Jackson, P.E. and Haddad, P.R. *Journal of Chromatography*, **640**, 481 (1993).
25 Beckers, J.L. and Everaerts, F.M. *Journal of Chromatography*, **508**, 3 (1990).
26 Jones, W.R. and Jandik, P. *Journal of Chromatography*, **546**, 455 (1991).

27 Everaerts, D.M., Beckers, J.L. and Verheggen Th. P.E.M. In *Isotachophoresis – Theory Instrumentation and Applications*. Elsevier, Amsterdam, Oxford, New York (1976).
28 Hjalmarsson, S.G. and Baldesten, A. *Critical Reviews of Analytical Chemistry*, **11**, 261 (1981).
29 Bocek, P., Miedziak, I., Demi, M. and Janak, J. *Journal of Chromatography*, **137**, 83 (1987).
30 Zelinsky, J., Kaniansky, D., Havassi, P. and Lednarova, U. *Journal of Chromatography*, **294**, 317 (1984).
31 Kanianski, D., Koval, M. and Stankoviansky, S. *Journal of Chromatography*, **267**, 67 (1983).
32 Kaniansky, D. and Evereerts, F.M. *Journal of Chromatography*, **148**, 441 (1978).
33 Kaniansky, D., Madajova, V., Zelensky, I. and Stankoviansky, S. *Journal of Chromatography*, **194**, 11 (1980).
34 *The Testing of Water*. E. Merck, Darmstdt, Germany (1980).
35 Kaniansky, D., Zelinsky, I.M., Hyhenova, A. and Onuska, F.I. *Analytical Chemistry*, **66**, 4258 (1994).
36 Greenfield, S., Jones, L.L. and Berry, C.T. *Analyst (London)*, **89**, 713 (1964).
37 Wendt, R.H. and Fassel, U.A. *Analytical Chemistry*, **37**, 920 (1965).
38 Scott, R.H. *Analytical Chemistry*, **46**, 75 (1974).
39 Suddendorf, R.F. and Boyer, K.W. *Analytical Chemistry*, **50**, 1769 (1978).
40 Sharp, B.L. The Cone Spray Nebulizer. British Technology Group, Patent Assignment No 8432338 (1984).
41 Stepanova, M.I., Il'ina, R.H. and Shaposhnikov, Y.K. *Journal of Analytical Chemistry, USSR*, **27**, 1075 (1972).
42 World Health Organization. *International Standards for Drinking Water*, 3rd edn., Geneva, p. 37 (1971).
43 Cathrone, B. and Fielding, M. *Proc. Anal. Proceedings Chemical Society (London)*, **15**, 155 (1978).
44 Ogan, K., Katz, E. and Slavin, W. *Journal of Chromatographic Science*, **16**, 517 (1978).
45 Dunn, B.P. and Stich, M.E. *Journal of Fisheries Board, Canada*, **33**, 2040 (1976).

Chapter 2

Ion chromatography

This technique has been used predominantly for the determinations of anions and to a smaller extent cations (as their anionic complexes). Very few applications to organic compounds have been described.

2.1 Organic compounds

2.1.1 Non saline waters

2.1.1.1 Miscellaneous organic compounds

The limited applications of ion chromatography to the determination of organic compounds in non saline waters include sulphur-containing organic compounds [1,2], chlorophenols and bromophenols [2]. Brandt and Kettrup [1] showed it was possible to distinguish between individual halogens in the group parameters adsorbable organic halogens (AOX) and to determine the parameters of adsorbable organic sulphur compounds (AOS) by a technique employing pyrohydrolysis of organics followed by adsorption and ion chromatographic detection of the resultant anions (chloride, bromide and sulphate). Pyrohydrolysis apparatus, combustion conditions and adsorption of the inorganic anions were optimised for the complete conversion of several model compounds such as 4-chlorophenol, 4-bromophenol, 4-bromobenzoic acid, thiobenziamide and toluene sulphonic acid and maximal anion recoveries.

2.1.2 Waste waters

2.1.2.1 Carboxylic acids

Various workers have used ion chromatography to determine low molecular weight carboxylic acids in wastewater [3] and rainwater [4,5]. Detection limits in the three methods were less than $1\mu g \ mL^{-1}$.

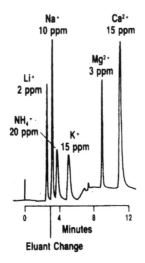

Fig. 2.1 Determination of alkali and alkaline earths
Source: Own files

2.2 Cations

2.2.1 Non saline waters

Ion chromatography as originally developed by Small and co-workers in 1975 provided a method for the separation and determination of inorganic anions and cations [6]. This original method used two columns attached in series packed with ion exchange resins to separate the ions of interest and suppress the conductance of the eluant, leaving only the species of interest as the major conducting species in the solution. Once the ions were separated and the eluent suppressed, the solution entered a conductivity cell where the species of interest were detected. Since its introduction, ion chromatography had advanced considerably and the technique is now routinely used for the analysis of organic and inorganic anions and cations and substances including organic acids and amines, carbohydrates, and alcohols.

The technique has since 1975 progressed rapidly and in 1978 a book was published on *Ion Chromatographic Analysis of Environmental Pollutants*. Other early papers on the application of ion chromatography include the determination of selected ions in geothermal well water [7,8], the determination of anions in potable water [9] and the separation of metal ions and anions [10] and anions and cations [11].

While atomic absorption spectroscopy and inductively coupled plasma techniques will continue to be the workhorse instrument in the metals analytical laboratory, an ever-growing need exists for the complementary

Table 2.1

Metal	Complexed form	Detector
Cr(VI), Mo, W	Naturally occurring oxides: CrO_4^{2-}, MoO_4^{2-}, WO_4^{2-}	Conductivity
Au(I), Au(III), Ag,	Cyano-complexes present in plating solutions:	Conductivity
	Co(III) $Au(CN)_2^-$, $Au(CN)_4^-$, $Ag(CN)_2^-$, $Co(CN)_6^{3-}$	
Pb, Cu, Zn, Ni	Prederivatised EDTA complexes: $Pb(EDTA)^{2-}$,	Conductivity
Cr(III)	$Cu(EDTA)^{2-}$, $Zn(EDTA)^{2-}$, $Ni(EDTA)^{2-}$, $Cr(EDTA)^-$	
Pd, Pt, Pb, Au	Chloro-complexes formed *in situ* in the column	UV or pulsed
	eluent: $PdCl_6^{2-}$, $PtCl_4^{2-}$, $PbCl_4^{2-}$, $AuCl_4^-$	amperometric

Source: Own files

capabilities of ion chromatography. The unique complements that ion chromatography brings to the metals analyst include the determination of inorganic anions, speciation of valence states, and the ability to work with difficult sample matrices. The arguments against the use of ion chromatography for metal determinations relate primarily to routine, high throughput analysis situations. If one is only interested in high through-put (element determinations per hour) for metals in ideal matrices, ion chromatography falls short of the speed offered by atomic spectroscopy. There can also be problems concerned with the specificity of ion chromatography when compared with atomic spectroscopic techniques.

However, as real-world analytical problems are most often neither ideal nor routine, analytical instrumentation supplemental to atomic spectroscopy would be a great advantage in developing methods for non routine applications. Thus, it is becoming apparent that the well-equipped analytical laboratory of the future will incorporate both atomic spectroscopy and ion chromatography.

Small and co-workers [6] at Dow Chemical discovered that alkali and alkaline earth metals could be separated and determined analogously to anions with a cation exchange separation and an anion 'stripper' now commonly known as a suppressor. With the latest developments in high efficiency columns and high capacity suppressors, it is possible
to determine the common alkali and alkaline earth metals along with ammonia in a single injection run of less than 15min (Fig. 2.1).

Since the cation suppression device relies on an addition of hydroxide in exchange for eluent anion (typically Cl⁻), this method precludes transition metal determinations because the metal hydroxides would precipitate out of solution before entering the conductivity detector. Thus, several years went by before the development and addition of transition metal capabilities to ion chromatography. Several methods have been reported for the determination of stable anionic metal complexes by

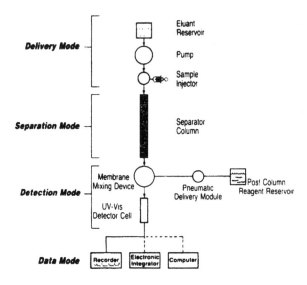

Fig. 2.2 Principles of post–column reaction ion chromatography for transition metals
Source: Own files

essentially the original anion exchange/chemical suppression/ conductivity configuration (Table 2.1).

2.2.1.1 Transition metals, alkaline and alkaline earth metals

Methods are now available to determine a broad range of transition metals with minimal sample pretreatment (avoiding prederivatisation). The principles involved in this method are described below and illustrated in Fig. 2.2.

Liquid samples are injected at the top of the ion exchange column through a fixed volume valve. Metal bands migrate through at differential rates determined largely by their affinity to complex with a ligand added to the eluent vs their electrostatic affinity for the stationary ion exchange sites. A strong metal completing colorimetric reagent is fed pneumatically and mixed with the column effluent. The metal bands are then typically determined at a visible wavelength in a flow-through cell absorbance detector.

Improvements in separations ability offered by the Dionex HPIC–CS5 column, which exhibits both anion and cation exchange capacity, have allowed the determination of ten metal ions in a single injection with a pyridine-2, 6-dicarboxylic acid (PCDA) eluent (Fig. 2.3). The post-column reagent used in the detection scheme was 4-(2-pyridylazo) resorcinol.

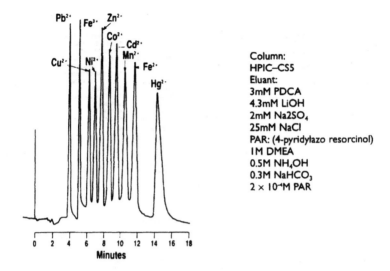

Column:
HPIC–CS5
Eluant:
3mM PDCA
4.3mM LiOH
2mM Na2SO₄
25mM NaCl
PAR: (4-pyridylazo resorcinol)
IM DMEA
0.5M NH₄OH
0.3M NaHCO₃
2 × 10⁻⁴M PAR

Fig. 2.3 Determination of ten metals in one injection
Source: Own files

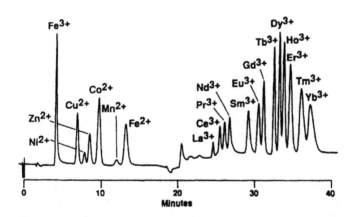

Fig. 2.4 Determination of heavier lanthanide series metals
Source: Reproduced by permission from International Science Communications, US [12]

Rubin and Heberling [12] have reviewed the applications of ion chromatography in the analyses of cations in non saline waters. Elements discussed include sodium, lithium, ammonium, potassium, magnesium, calcium, lead, copper, cadmium, cobalt, nickel, zinc, iron, manganese and the 14 lanthanide elements. A typical chromatogram is shown in Fig. 2.4.

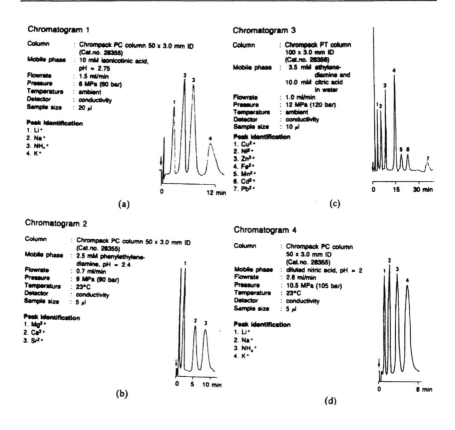

Fig. 2.5 (a) Ion chromatography of lithium, sodium, ammonium and potassium, (b) ion chromatography of magnesium, calcium and strontium, (c) ion chromatography of transition metals, (d) ion chromatography of lithium, sodium, ammonium and potassium
Source: Reproduced by permission from Elsevier Science, UK [13]

Basta and Tabakabi [13] used a Dionex Model 10 ion chromatograph for the simultaneous determination of potassium and sodium or of calcium and magnesium in different types of non saline waters, including soil extracts. The pH and specific conductance of the water samples are tabulated. Tabulated data are included comparing the results obtained by ion chromatography with those obtained by atomic absorption spectro-photometry and flame photometry, and showing the precision of ion chromatography for determination of the alkaline and alkaline earth metals. Ion chromatography gave results that were precise and accurate, and it could be used to determine concentrations as low as 0.1mg per litre. Only small (2 ml) samples were required, and analysis took only 6–7min.

Fig. 2.5 (a)–(d) shows some separations of metals that have been achieved using Chromopak PT and Chromopak PC columns. Lithium, sodium, potassium, ammonium, calcium and magnesium can all be determined by ion chromatographic techniques.

On the Chrompack PC columns the monovalent cations (lithium, sodium, ammonium and potassium) can be separated as illustrated in Fig. 2.5(a).

If ultraviolet absorbing acids are used in the mobile phase ultraviolet detection is possible. Operating conditions for this mode are:

10 mM isonicotinic acid
pH = 2.75
flowrate, 1.5 ml min $^{-1}$
2 mM picolinic acid
pH = 2.0
flowrate 2.6 ml min $^{-1}$

The common divalent cations can be analysed on the same column by using a phenylenediamine buffer (Fig. 2.5 (b)).

The cations of the transition metals which do not have sufficient retention on the Chrompack PC column can be separated on the Chrompack PT column (Fig. 2.5(c)).

2.2.1.2 Ammonium ion

Mizobuchi et al. [14] have described a method for the determination of ammonia in water based on reaction with fluorescamine to form a fluorophor which is then chromatographed on trichrosorb RP–18 using a mobile phase consisting of 0.05m phosphate buffer solution (pH2.0) and acetonitrite (65:35 v/v).

The analytical recovery was examined by spiking river water, effluent and rainwater with known amounts of ammonium. Average recoveries were from 95.8 to 100.7% on five replicate samples and their standard deviations were from 1.9 to 5.4. Reproducibility was measured with standard solution, river water, effluent and rainwater. The results showed that their relative standard deviations were 1.22 to 2.28%. The detection limit was 6 µg L $^{-1}$.

Typical chromatograms obtained for a rain water and a river water are shown in Fig. 2.6.

2.2.1.3 Total organic carbon

Fung et al. [15] have discussed a thermal combustion–ion chromatographic method for the determination of total organic carbon in industrial and potable waters. This method utilises a tube furnace and readily

44

Table 2.2 Ion chromatography or cations in non saline waters

Cation	Stationary phase	Eluent	Comments	Detection	LD	Ref.
Alkali earths Ca, Mg	Sulphonated poly (styrene–divinyl benzene)	0.12M HClO₄	Ions coupled with Arsenazo I @ pH10 in post column reaction	Spectrophoto- metric at 590nm	–	[16,17]
Alkali earth/alkali metal Na, K, Li,	2.1mm × 150mm Zorbox 51L	0.01M Li acetate	–	–	–	[18]
NH₄, Mg, Ca, Sr	Ion–Pac CS12	–	–	–	–	[19]
Li, Na, NH₄, K, Mg, Ca, Sr						
Ca, Mg, Mn, Zn	–	–	–	–	–	[20]
Na, K, Ca, Mg	–	–	–	–	–	[21]
Na, K, Mg, Ca	–	–	–	–	–	[13]
Na, K, Ca	–	–	–	–	0.1mg L⁻¹	[13]
Alkali and alkali earths	–	–	Comparison with AAS results	–	Alkali 0.05 –0.15µmol L⁻¹ Alkali earth	[22]
Al, Fe	Cation exchange Dionex CS2 resin	–	Post column derivitisation with Chrome Azurol S–acetyl methyl ammonium bromide (Trition X–100)	–	0.2–0.6µmol L⁻¹	[23]
Al	–	–	Fractionation of aluminium in fresh water	–	–	[24]
NH₄ and SO₄	–	–	Coupled to mass spectro- metric detector	–	–	[25]
Heavy metals Co, Cd, Mn	–	–	–	–	–	[26]

Table 2.2 continued

Cation	Stationary phase	Eluent	Comments	Detection	LD	Ref.
Cu, Ni, Zn, Mn	–	–	Post-column reaction studied interference by chelating agents	Spectrometric	–	[27]
Cu, Zn, Cd	Anion separation and suppression columns	0.03μM NaHCO$_3$ –0.3M Na$_2$CO$_3$	Metals as EDTA complexes	Conductio-metric	–	[28]
Fe(III)/Fe(II)	Cation exchange resin	–	As the bathophenothroline disulphonic acid complexes	Spectrophoto-metric at 530nm	–	[29]
Li	–	–	Linear calibration in range 2–3000μg L^{-1} RSD 4.4%	–	1μg L^{-1}	[30]

Source: Own files

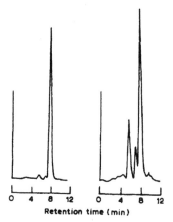

Fig. 2.6 Ion chromatograms, ammonia in river water (left) and rainwater (right). Measuring conditions were as follows: (river water) sample, 1 ml; range 2; injection volume, 5μL: (rainwater) sample, 1ml, range 1, injection volume, 20μL
Source: Reproduced by permission from American Chemical Society [14]

accessible high performance liquid chromatography equipment. To achieve complete oxidation, persulphate (0.25%) was added to oxidise non volatile organic compounds in solution and cupric oxide heated at 900°C to convert volatile organic compounds to carbon dioxide, which was scrubbed in a 20mL solution of 50mM potassium hydroxide with 10 drops of butanol added. The carbonate anion obtained was determined by non suppressed ion chromatography using 0.6mM potassium hydrogen phthalate as the eluent. Both surfactants and volatile and non volatile organic compounds commonly found in environmental waters give highly repeatable recoveries close to 100%. The detection limit (S/N = 2) and linear range for a 1L water sample are 2μg C L^{-1} and 10–2500μg C L^{-1}, respectively, and they can be adjusted using samples ranging from 100mL to 2L.

2.2.1.4 Other applications

Further applications of the determination of cations in non saline waters are reviewed in Table 2.2.

2.2.2 Aqueous precipitation

2.2.2.1 Alkali and alkaline earth metals and ammonium

Several groups of workers have applied ion chromatography to the analysis of rain [31–34].

Xiang and Chang [32] used ion chromatography to determine sodium, potassium and ammonium ions in rain water.

Small *et al.* [6] used this technique to determine ammonion ion in rain. Kadowaki [35] applied ion chromatography for the determination of sodium, potassium, ammonium, calcium and magnesium in rainwater samples. The method was compared to atomic absorption spectrometry and found to be less influenced by interferences.

2.2.3 Trade effluents and waste waters

2.2.3.1 Heavy metals

Ion chromatography has been applied to the determination of cobalt, nickel, copper, zinc and cadmium as their EDTA complexes using anion separation and suppressor columns and 0.03µm sodium bicarbonate–0.03µm sodium carbonate [28] eluant and a conductiometric detector.

2.2.3.2 Uranium

Uranium has been determined in process liquids by ion chromatography using an ammonium sulphate sulphuric acid element. Uranium was determined in the eluant. Uranium was determined in the eluate spectrophotometrically at 520 mm as the 4(-2-pyridylazo) resorcinol complex [36].

2.2.3.3 Ammonium

Merz and Oldeweme [37] described a selective ion chromatographic procedure coupled with fluorescence detection for the determination of ammonia in which ammonia is separated from amines and alkaline metals and reacted with *o*-phthalic acid dialdehyde on a separate column.

2.3 Anions

2.3.1 Ultraviolet versus conductivity detectors

Cochrane and Miller [38] state the sensitivity of an ultraviolet detector is about 10 times greater than that of an electrical conductivity detector. Ultraviolet detection of anions is not generally applicable except at very low wavelengths such as 215nm. An alternative method of detection of non ultraviolet absorbing species is to add a low level of ultraviolet absorbing substances to the eluant. The emergence of a component is then shown by a negative detector response.

The effect is easily achieved in the ion chromatography system where the preferred buffer solution (potassium hydrogen phthalate (5×10^{-3}M at pH 4.6) has a strong ultraviolet response with λmax at 280nm. At this

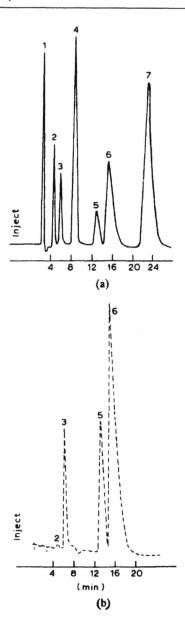

Fig. 2.7 Ion chromatogram of seven common anions (a) conductivity detectors (——)
peak 1, 3mg L^{-1} F$^-$; peak 2, 4mg L^{-1} Cl$^-$; peak 3, 10mg L^{-1} NO$_2^-$; peak 4, 50mg L^{-1} PO$_4^{3-}$;
peak 5, 10mg L^{-1} Br$^-$; peak 6, 30mg L^{-1} NO$_3^-$; peak 7, 50mg L^{-1} SO$_4^{2-}$; (b) ultraviolet
detector (- - -) at 192nm, position 2; peak 2, 4mg L^{-1} Cl$^-$; peak 3, 10mg L^{-1} NO$_2^-$; peak 5,
10mg L^{-1} Br$^-$; peak 6, 30mg L^{-1} NO$_3^-$ Dip at approx. 8min = PO$_4$
Source: Reproduced with permission from the American Chemical Society [39]

wavelength the excessively strong absorbance will not allow adequate zero suppression and the wavelength was therefore increased to a much less absorbing region (308nm) where the background response could be backed off sufficiently for the detector to be used at maximum sensitivity. A comparison of the conductivity and ultraviolet detectors showed that:

(1) The response is greater for ultraviolet detection by a factor of 5–30. A relatively stronger, negative ultraviolet response occurs for weakly acidic anions because there is no dependence on their ionisation.
(2) Linearity ranges using ultraviolet detection are greater than those for conductance.
(3) Conductance detection may give positive or negative peaks depending on eluant concentration and pH. Ultraviolet detection of non ultraviolet absorbing ions give a response in one direction only.

An ultraviolet detector used in series with a conductivity detector is the basis for a method [39] for the determination of anions in non saline waters. This combination of detectors greatly increases the amount of information that can be collected on a given sample. The application of ultraviolet detection has the following advantages:

(1) aid in the identification of unknown peaks;
(2) use in resolving overlapping peaks;
(3) help in eliminating problems associated with the carbonate dip;
(4) reduction of problems associated with ion exclusion in the suppressor column;
(5) ability to detect anions not normally detected by the conductivity detector, eg sulphide and arsenite.

For maximum sensitivity, the preferred position of the ultraviolet detector is after the suppressor column. The suppressor column also decreases the background absorbance of the other common ion chromatographic eluents such as $0.0002M$ $Na_2CO_3/0.002M$ NaOH, $0.006M$ Na_2CO_3 and $0.005M$ $Na_2B_4O_7 10H_2O$.

Fig. 2.7 shows a separation of seven common anions monitored using the conductivity detector. Fig. 2.7(b) was obtained with the ultraviolet detector after the suppressor column. As is illustrated in Fig. 2.7, nitrite, bromide, and nitrate absorb strongly in the ultraviolet, while fluoride, phosphate and sulphate do not show appreciable absorption above 190nm. Chloride absorbs weakly in the ultraviolet region below 200nm. Note that non absorbing anions are sometimes observed by the ultraviolet detector as negative peaks, as is shown in Fig. 2.7(b) for phosphate. Table 2.3 contains a more extensive list of ultraviolet active anions and also lists the detection limits for many of the anions.

Sulphide cannot be detected under normal ion chromatographic conditions. It is converted into hydrogen sulphide in the suppressor

Table 2.3 List of ultraviolet absorbing and non absorbing inorganic anions

Anion	Retention time (min)	UV active	UV detection limit* (mg L^{-1})	UV detector position	Wavelength (nm)
F$^-$	2.4	N*		2^c	195
S^{2-}	2.4	Y^b		1^d	205
IO$_3^-$	2.6	Y	0.08	2	195
ClO$_2^{1-}$	3.0	Y	0.2	2	195
H$_2$PO$_2^-$	3.0	N		2	195
NH$_2$SO$_3^-$	3.8	N		2	195
BrO$_3^-$	4.3	Y	0.16	2	195
Cl$^-$	4.6	Y	2	2	192
S^{2-}	5.4	Y		1	195
NO$_2^-$	5.9	Y	0.1	2	195
BF$_4^-$	6.6	N		2	195
HPO$_3^{2-}$	6.7	N		2	195
SeO$_3^{2-}$	6.9	Y	0.5	2	195
I$^-$	7.8	Y	0.15	2	195
PO$_4^{3-}$	8.4	N		2	195
SCN$^-$	10.8	Y	0.2	2	195
AsO$_4^{3-}$	10.9	Y	1.5	1	200
AsO$_4^{3-}$	12.2	Y	2	2	195
S$_2$O$_3^{2-}$	12.6	Y		2	195
Br$^-$	13.0	Y	0.1	2	195
AsO$_3^{3-}$	13.3	Y	1.2	1	200
SO$_3^{2-}$	13.4	Y		2	195
N$_3^-$	14.1	Y	0.3	2	195
NO$_3^-$	15.2	Y	0.1	2	195
ClO$_3^-$	15.2	Y	4	2	195
SeCN$^-$	18.8	Y	0.4	2	195
ClO$_4^-$	19.6	N		2	195
S^{2-}	21.2	Y	0.4	1	200
SO$_4^{2-}$	22.8	N		2	195
SeO$_4^{2-}$	27.2	Y	15	2	195

*N, not UV active. *Y, UV active. *2, after suppressor. *1, before suppressor. *100µL injection, peak height 2 × noise level

Source: Reproduced with permission from the American Chemical Society [39]

column. Hydrogen sulphide acid is a very weak acid that does not ionise sufficiently to be detected with the conductivity detector. The ultraviolet detector, however, is able to detect sulphide at low levels. Since sulphide is a weak acid anion, the ultraviolet detector was placed between the separator and suppressor columns.

Arsenous acid is a very weak acid and cannot be detected at low levels by the conductivity detector. However, like sulphide, it is easily detected with the ultraviolet detector. The simultaneous determination of arsenite and arsenate is possible.

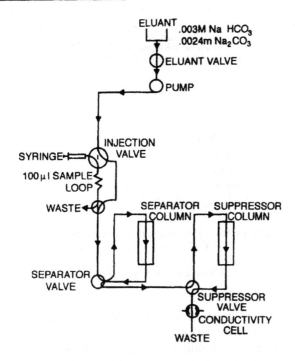

Fig. 2.8 Ion chromatographic flow diagram. Eluant conditions used for anion analysis
Source: Reproduced with permission from Elsevier Science, UK [40]

Arsenite also can be separated by ion chromatography using standard conditions. It elutes early, near the carbonate dip. For this application the ultraviolet detector should be placed between the separator and suppressor columns.

2.3.2 Non saline waters

2.3.2.1 Bromide, fluoride, chloride, nitrite, nitrate, sulphate, phosphate and ammonium

Smee *et al.* [40] used ion chromatography has been used for the measurement of background concentrations of fluoride, chloride, nitrate and sulphate in non saline waters.

Shown in Fig. 2.8 is a schematic diagram of the Dionex ion exchange liquid chromatograph used by Smee *et al.* [40]. The instrumentation consists essentially of a low capacity anion exchange column, the 'separator', a high capacity cation exchange column, the 'suppressor' and the detection system, a high sensitivity conductivity meter and recorder.

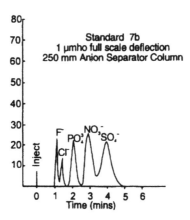

Fig. 2.9 Calibration chromatogram for a 250mm anion separator column using standard 7b containing 0.1µg mL $^{-1}$ F, 0.1µg mL $^{-1}$ Cl, 0.1µg mL $^{-1}$ PO $^{3-}$, 1.0µg mL $^{-1}$ NO $_3^-$ and 1.0µg mL $^{-1}$ SO $_4^{2-}$
Source: Reproduced with permission from Elsevier Science, UK [40]

Following the injection of a small volume (100µL) of sample, rapid exchange and separation of the anions is accomplished with a specially prepared low capacity resin (containing quaternary ammonium exchange sites existing as a thin film on the surface to facilitate fast equilibrium).

The otherwise highly conducting background of the eluant (0.003mol L $^{-1}$ sodium bicarbonate/0.0024mol L $^{-1}$ sodium carbonate) is virtually eliminated by passage through the high capacity suppressor column whereupon carbonic acid is formed together with the acid forms of the anions of interest. The conductivity meter used as a detector then only sees a small residual background due to the weakly dissociated carbonic acid and the conductivity of the acids of interest separated in time. Additional to this basic instrumentation, there are four reservoirs (2 eluant, 1 regenerant, 1 water), a sample injection valve with a 100µL loop, two Milton Roy fluid pumps with adjustable flow rates and an automatic timer for controlling the regeneration cycle of the cation exchange resin.

Approximately 1ml of untreated sample was injected into the entrance port by means of a hypodermic syringe; 100µL of sample was used to fill the sample loop, the remainder passing to waste. The determination began with a freshly regenerated separator column and proceeded sequentially through the sample series. Retention times of peaks from the resulting chromatograms were compared with known synthetic standards analysed under the same eluant strength and flow conditions. Measurements of anion concentrations in each sample were obtained by comparing peak heights with a standard curve generated from peak heights of mixed synthetic standard.

Table 2.4 Dionex replicate samples, measurements of precision non saline waters

Sample no.	F (250mm) (μg L^{-1})	F (500mm) (μg L^{-1})	Cl (250mm) (mg L^{-1})	Cl (500mm) (mg L^{-1})
7	852	860	20.0	14.5
18	852	860	20.0	14.7
22	870	880	19.7	14.5
26	852	920	20.0	13.5
30	852	960	20.0	13.5
Mean	855.6	896.0	19.9	14.1
S.D.	8.05	43.4	0.14	0.59
R.S.D.	0.9%	4.8%	0.7%	4.2%

NO$_3$ (250mm) (mg L^{-1})	NO$_3$ (500mm) (mg L^{-1})	SO$_4$ (250mm) (mg L^{-1})	SO$_4$ (500mm) (mg L^{-1})
<0.02	<0.05	27.3	25.0
<0.02	<0.05	27.5	25.0
<0.02	<0.05	27.3	24.2
<0.02	<0.05	27.3	24.6
<0.02	<0.05	27.5	24.2
0.02	0.05	27.4	24.6
–	–	0.11	0.4
–	–	0.5%	1.6%

Source: Reproduced with permission from Elsevier Science, UK [40]

Two separator columns were studied by Smee *et al.* [40] – one 250mm and one 500mm in length. Examples of chromatograms obtained from standards injected into a 250mm column are shown in Fig. 2.9.

Estimates of precision based on five analyses of the replicate non saline water samples are presented in Table 2.4. The 250mm column is more precise than the 500mm column for all anions analysed, mainly because the peak height of a response was measured rather than peak area, the shorter column producing the more narrow peaks. Fluoride determinations using the 500mm column exhibit a progressive upward increase in fluoride content caused by the aging of the column during the analytical run. This apparent increase in fluoride content may occur at the expense of the chloride ion, the next anion through the column, as there appears to be a gradual decrease in chloride content with time. No data on nitrate was obtained from the replicate sample. Precision for sulphate through both columns is excellent, although the 250mm column consistently produces the higher analytical values.

Table 2.5 Detection limits of chloride, nitrate and sulphate on the Zipax–SAX column with 1.4×10^{-3}mol L^{-1} Na$_2$–succinate as eluent

Ion	Detection limit (μg L^{-1})	
	200μL loop	Concentrator column
Chloride	10	2
Nitrate	30	4
Sulphate	30	10

Source: Reproduced by permission from American Chemical Society [41]

Van Os *et al.* [41] achieved complete separations in 6min of 1–30mg L^{-1} concentrations of bromide, chloride, nitrite, nitrate and sulphate using a Zipax SAX separation column, with eluent suppression and electrical conductivity detection [42]. The necessary high pressure packing techniques for packing the separation column have been described [43]. With sample preconcentration, detection limits were reduced to about 5μg L^{-1} but calibration graphs for chloride and nitrate were not linear. Sodium adipate and 1.4×10^{-3}M disodium succinate are used as eluants, both at pH7.

On Zipax SAX the ions are eluted in the order F$^-$ < Cl$^-$ < NO$_2^-$ ≈ H$_2$PO$_4^-$ < Br$^-$ < NO$_3^-$ < SO$_4^{2-}$. Fluoride is eluted together with the negative water peak. On Zipax SAX the optimal eluent temperature for obtaining good resolution in the minimal time is about 25°C.

Detection limits (three times the background noise) for chloride, nitrate and sulphate were determined for a 200μL loop and for a concentrator column. These are given in Table 2.5.

Tong and Shi [44] used ion chromatography with an electrical conductivity detector to determine fluoride, chloride, nitrate, sulphate and hydrogen phosphate (HPO$_4^{2-}$) in non saline water with detection limits of 0.8, 1.0, 8.0, 10.0 and 12.5μg L^{-1} respectively.

Marko Varga *et al.* [45] found that a cleanup column prior to the separation column, packed with a chemically bonded amine material (Nucleosil 5 HN$_2$) was found to be effective in removing interfering humic substances. No influence was found from humic substances in concentrations up to 45μg L^{-1} on ion chromatographic analysis of nitrate and sulphate (10–100mg L^{-1}) after passage through the cleanup column.

A short column packed with an amine bonded polymer (weak anion exchange material) connected prior to the anion separation column was found to be very effective for the selective removal of humic substances prior to anion separation by ion chromatography. About 500 sample injections could be performed before breakthrough of humic substances occurred as indicated by ultraviolet measurements made on the eluate at 225nm.

Bradfield and Cooke [46] described an ion chromatographic method using an ultraviolet detector for the determination of chloride, nitrate, sulphate and phosphate in water extracts of plants and soils. Plant materials are heated for 30min at 70°C with water to extract anions. Soils are leached with water and Dowex 50–X4 resin added to the aqueous extract which is then passed through a Sep–Pak C_{18} cartridge and the eluate then passed through the ion chromatographic column. The best separation of these ions was obtained using a 5×10^{-4} mol L^{-1} potassium hydrogen phthalate solution in 2% methanol at pH4.9. A reverse phase system was employed. Retention times were 5.5, 7.9, 12.6 and 18min for chloride, nitrate, phosphate and sulphate respectively. Recoveries ranged from 84 to 108% with a mean of 97%.

Figs. 2.10–2.12 show chromatograms obtained in the analysis of plant, soil and fertiliser extracts.

Conboy et al. [47] employed ion chromatography mass spectrometry to determine sulphate and ammonium compounds in non saline waters.

The ion spray liquid chromatography/mass spectrometry interface is coupled via a post-suppressor split with an ion chromatography system. The micromembrane suppressor selectively removes over 99.9% of the ion pair agents required for ion chromatography from the eluent. The resulting solution consists of analyte, organic modifier, and water, which is compatible with ion evaporation mass spectrometry. A flow rate of 0.8 or 1.0mL/min from the column was split after suppression such that approximately 10–20µL/min was directed to the ion spray liquid chromatography–mass spectrometry interface, which was coupled to an atmospheric pressure ionisation mass spectrometer. This system provided a convenient way to effect isocratic and gradient separations of organic ions under chromatographic conditions incompatible with most forms of mass spectrometric ionisation. This work describes the separation and positive ion detection of quaternary ammonium drugs and tetraalkylammonium compounds of industrial importance using both single and tandem mass spectrometric detection (eg ion chromatography–mass spectrometry and ion chromatography–mass spectrometry–mass spectrometry). The former easily provided the molecular weights of these compounds while the latter gave some structural information. The limit of detection of 40pg injected on-column for tetra–propylammonium cation is a factor of 10 better than that obtained with the conventional ion chromatography detector. The separation, detection, and identification of some alkyl sulphates and sulphonates are also shown with negative ion detection. Again, ion chromatography–mass spectrometry readily produced the molecular weights of these compounds while ion chromatography–mass spectrometry–mass spectrometry provided some structural information.

See also Table 2.7.

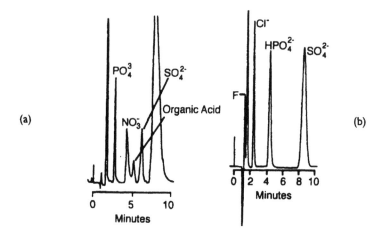

Fig. 2.10 Determination of anions in plant extracts. (a) fresh spinach sample; (b) fresh
anion sample. Aqueous extracts of macerate samples
Source: Reproduced with permission from the Royal Society of Chemistry [46]

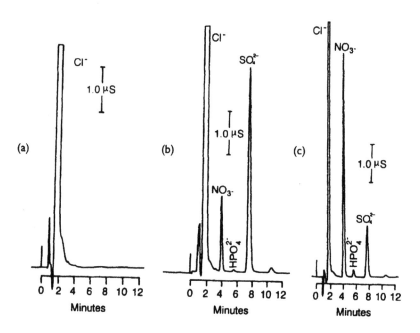

Fig. 2.11 Determination of anions in soil extracts: (a) blank 10mol L^{-1} KCl; (b) soil sample
A 10mol L^{-1} KCl extract; (c) soil sample B 10mmol L^{-1} KCl extract 1:5000dil.
AMPIC–NG1 should also be used in series to remove humic acids
Source: Reproduced with permission from the Royal Society of Chemistry [46]

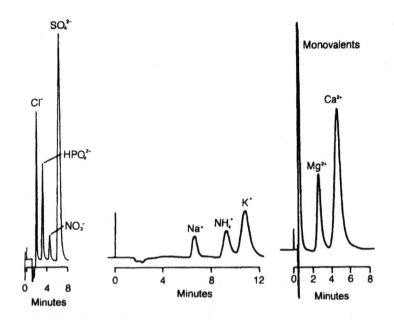

Fig. 2.12 Analysis of granulated fertiliser samples: (a) determination of anions; (b) determination of monovalent cations; (c) determination of divalent cations
Source: Reproduced with permission from the Royal Society of Chemistry [46]

2.3.2.2 Metal cyanide complexes

The cyanide ion in inorganic cyanides can be present as both complexed and free cyanide. In order to study the chromatography of metal cyanides, Rocklin and Johnson [48] prepared and assayed solutions of cadmium, zinc, copper, nickel, gold, iron and cobalt cyanides. Table 2.6 lists the percentage of total cyanide detected.

The results suggest that the complex cyanides can be grouped into three categories depending on the cumulative formation constant and stability of the complex.

Category 1 includes the weakly complexed and labile cyanides $Cd(CN)_4^{2-}$ (log B_4 = 18.78) and $Zn(CN)_4^{2-}$ (log B_4 = 16.7). These complexes completely dissociate under the chromatographic conditions used, the cyanide being indistinguishable from free cyanide.

Category 2 includes the moderately strong cyanide complexes $Ni(CN)_4^{2-}$ (log B_4 = 31.3) and $Cu(CN)_4^{3-}$ (log B_4 = 30.3). Although these complexes are labile, they are retained on the column and slowly dissociate during the chromatography. This slow dissociation produces tailing which lasts for several minutes as the free cyanide elutes and is

Table 2.6 Percentage of total cyanide in metal complexes determined as 'free' cyanide

Metal complex	log β_t	%
$Cd(CN)_4^{2-}$	18.8	102
$Zn(CN)_4^{2-}$	16.7	102
$Ni(CN)_4^{3-}$	31.3	81
$Cu(CN)_4^{3-}$	30.3	52
$Cu(CN)_3^{2-}$	28.6	42
$Cu(CN)_2^{-}$	24.0	38
$Au(CN)_2^{-}$	38.3	0
$Fe(CN)_6^{3-}$	42	0
$Co(CN)_4^{3-}$	64	0

Source: Reproduced with permission from the American Chemical Society [48]

detected. As the results presented in Table 2.6 demonstrate, the tailing and the non quantitative recovery of cyanide preclude the use of direct injection to determine total cyanide in samples containing all the anions listed except cadmium and zinc. These samples may be analysed after acid distillation and caustic trapping. The cyanide in the caustic solution can then be determined by ion chromatography with electrochemical detection.

Category 3 includes those cyanides which are inert and therefore totally undissociated, such as $Au(CN)_2^{-}$ (log $B_2 = 38.3$), $Fe(CN)_6^{3-}$ (log $B_6 = 42$) and $Co(CN)_6^{3-}$ (log $B_6 = 64$). No free cyanide was detected for these complexes. Although these complexes do not elute under the chromatographic conditions used, they can be eluted and determined by using different chromatographic conditions and conductivity detection.

Samples containing both free cyanide (or weakly complexed cyanide) and strongly complexed cyanide can be analysed for free cyanide by direct injection. The determination of total cyanide (both free and strongly complexed) requires distillation of the sample with caustic trapping. See also Table 2.7.

2.3.2.3 Cyanide, sulphide, iodide and bromide

Rocklin and Johnson [48] used an electrochemical detector in the ion chromatographic determination of cyanide and sulphide. They showed that by placing an ion exchange column in front of an electrochemical detector, using a silver working electrode, they were able to separate cyanide, sulphide, iodide and bromide and detect them in water samples at concentrations of 2, 30, 10 and 10µg L^{-1} respectively. Cyanide and sulphide could be determined simultaneously. The method has been applied to the analysis of complexed cyanides and it is shown that cadmium and zinc cyanides can be determined as total free cyanide while nickel and copper complexes can only partially be determined in this way. The strongly bound cyanide in gold, iron or cobalt complexes cannot be determined by this method.

This method is based on the work of Pihlar and Kosta [50,51] who showed that a silver working electrode has the ability to produce a current that is linearly proportional to the concentration of cyanide in an amperometric electrochemical flow through cell. The reaction for cyanide is:

$$Ag + 2CN^- \rightarrow Ag(CN)_{2-} + e^-$$

Under these conditions sulphides and halides produce insoluble precipitates rather than soluble complexes:

$$2Ag + S^2 \rightarrow Ag_2S + 2e$$
$$Ag + X^- \rightarrow AgX + e^-$$

Rocklin and Johnson [49] overcame the latter problem by placing an ion exchange column in front of the electrochemical detector [119]. Cyanide and sulphide are separated and thus are determined simultaneously. Although bromide and iodide can be determined by ion chromatography with conductivity detection, the use of electrochemical detection results in greater selectivity as well as increased sensitivity.

See also Table 2.7.

2.3.2.4 Arsenate, arsenite, selenate and selenite

Hoover and Jager [85] have described a procedure for the determination of traces of arsenite, selenite and selenate in the presence of major interferences (chloride, nitrate, and sulphate) in potable waters, surface waters and ground waters. By collecting a selected portion of the ion chromatogram, after suppression, on a concentrator column and re-injecting it under the original chromatographic conditions, it was possible to separate selenate, selenite and arsenite from chloride, nitrate and sulphate. Statistical detection limits varied from 0.02 to 1.2µg of trace element depending on the minor components to be separated and on the water matrix. The maximal reliably separated molar ratio was 2300 for sulphate/selenate.

Ion chromatography can readily determine arsenate, selenite, and selenate species in water in the absence of interferences. However, they will usually be completely obscured by the major anions. Arsenic and selenium are ordinarily at their highest oxidation state in surface waters but that is not necessarily the case in ground waters.

An eluent composition of 2.0m mol L $^{-1}$ sodium bicarbonate and 1.5m mol L $^{-1}$ sodium carbonate was found to elute arsenate about midway between nitrate and sulphate. Selenite eluted between chloride and nitrate in an eluent that was 3.0m mol L^{-1} in sodium bicarbonate and 2.0m mol L^{-1} in sodium carbonate.

Suppressed eluent retained in the concentrator columns produced a large carbonate peak on direct reinjection. This peak obscured or interfered with everything eluting before nitrate. In the determination of selenite, the concentrator was washed with deionised water for 5min after collection and before injection. The wash step was omitted in the determinations of arsenate and selenate where the carbonate did not interfere.

Figs. 2.13–2.15 show representative chromatograms for each of the trace anions at close to the statistical detection limit. The initial injection is shown on a logarithmic scale of conductance to include complete peaks of the major constituents. The recycle portion of the chromatogram was obtained at the linear scale, usually 1µs cm^{-1} full scale sensitivity. Fig. 2.14 shows a second collection and recycling of selenite. The process can be repeated indefinitely but there is evidence (below) of 10–15% loss of material each time.

See also Table 2.7.

2.3.2.5 Mixtures of organic and inorganic anions

2.3.2.5.1 Bromate, chloride, bromide, nitrite, nitrate, hypophosphite (HPO_2^{-}), selenite, selenate, sulphate, phosphate, pyrophosphate, arsenate, chromate, α-hydroxybutyrate, butyrate, formate, acetate, glycolate, gluconate, valerate, α-hydroxy valerate, pyruvate, monochloroacetate, dichloroacetate, trifluoroacetate, galactonurate, gluconurate, α-keto-glutarate, oxalate, fumarate, phthalate, oxalacetate, citrate, isocitrate, cis aconitate, trans aconitate, succinate, maleate, malonate, quinate, tartrate, hexane sulphonate, octane sulphonate, octane sulphate, decane sulphonate, dodecane sulphonate and dodecane sulphate

Dionex Corporation [86] have issued a technical note covering the application of anion exchange chromatography to the determination of the very wide range of anions, quoted above, in non saline waters.

Mobile phase ion chromatography is a form of reversed phase ion pair chromatography in which chemically suppressed conductivity detection is used. As in reversed phase high performance liquid chromatography, gradient elution is accomplished by varying the percentage of organic solvent in the eluant during the run.

Although the ionic strength of the eluant may remain the same throughout the run, the background conductivity can decrease due to the changing dielectric constant. These baseline changes can be compensated either by chemical means or by computer baseline subtraction. Often, mobile phase ion chromatography is used to elute ions which are very

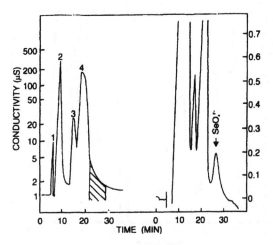

Fig. 2.13 Selenate in drinking water: sample size 5ml, 10ng of Se(V) added – major peaks: (1) fluoride; (2) chloride; (3) nitrate; (4) sulphate. Shaded area was collected for recycle
Source: Reproduced with permission from the American Chemical Society [85]

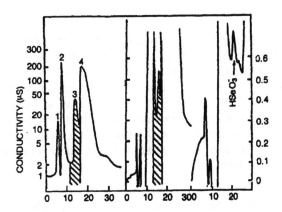

Fig. 2.14 Selenite in potable water, recycled twice; sample size 10ml, 100ng of Se(IV) added – major peaks: (1) fluoride; (2) chloride; (3) nitrate; (4) sulphate. Shaded areas were collected for recycle
Source: Reproduced with permission from the American Chemical Society [85]

strongly retained. When this is the case, the baseline conductivity change can be minimised without affecting the separation by adding very weakly retained ions directly to the stronger eluant (higher percentage of organic solvent).

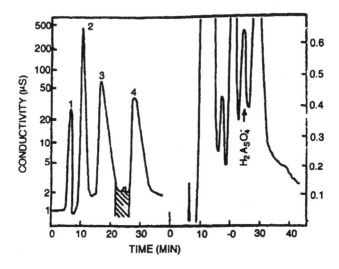

Fig. 2.15 Arsenate in river water: sample size 5ml, 50ng of As(V) added – major peaks: (1) fluoride; (2) chloride; (3) nitrate; (4) sulphate. Shaded area was collected for recycle
Source: Reproduced with permission from the American Chemical Society [85]

The analytical conditions recommended by Dionex for analysing mixtures of anions listed under the heading to this section are given below.

Although many columns and eluants can be used for gradient elution, the conditions used in Figs. 2.16 and 2.17 are recommended as a good starting point. The HPIC–ASSA (5μ) separator with sodium hydroxide eluant provides the optimum combination of efficiency, selectivity, and speed without an unacceptable baseline slope. If fewer ions than the 36 shown in Fig. 2.16 need to be separated, the gradient steepness can be increased to reduce the run time.

2.3.2.5.2 Lactate, glycolate, succinate, chloride, malate, sulphate, tartrate, maleate, fluoride, α-hydroxybutyrate, hydroxy valerate, formate, valerate, pyruvate, monochloroacetate, bromate, galaconurate, nitrite, gluconurate, dichloroacetate, trifluoroacetate, hypophosphite, selenite, bromide, nitrate, oxalate, selenate, α-ketoglutarate, fumarate, phthalate, oxalacetate, phosphate, arsenate, chromate, citrate, isocitrate, cis aconitate and transaconitrate

Some typical separations of these anions achieved are shown in Fig. 2.18. Fig. 2.18(a),(b) illustrates ion chromatographic separations of mono and diprotic organic acids by ion exchange using anodic AMMS and CMMS micromembrane suppressors.

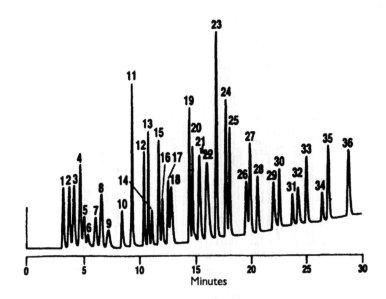

Fig. 2.16 Ion chromatography of inorganic and organic anions

All anions 10ppm unless noted

1 F⁻ (1.5ppm)	19 HPO₃²⁻
2 α-Hydroxybutyrate	20 SeO₃²⁻
3 Acetate	21 Br⁻
4 Glycolate	22 NO₃⁻
5 Butyrate	23 SO₄²⁻
6 Gluconate	24 Oxalate
7 α-Hydroxyvalerate	25 SeO₄²⁻
8 Formate (5ppm)	26 α-Ketoglutarate
9 Valerate	27 Fumarate
10 Pyruvate	28 Phthalate
11 Monochloroacetate	29 Oxalacetate
12 BrO₃⁻	30 PO₄³⁻
13 Cl⁻ (3ppm)	31 AsO₄³⁻
14 Galacturonate	32 CrO₄²⁻
15 NO₂⁻ (5ppm)	33 Citrate
16 Glucuronate	34 Isocitrate
17 Dichloroacetate	35 cis-Aconitate
18 Trifluoroacetate	36 trans-Aconitate

Source: Reproduced with permission from Dionex Corporation [86]

A further development is the Dionex HPIC AS5A–5µ analytical anion separator column. This offers separation efficiency previously unattainable in ion chromatography. When combined with a gradient pump and an anion micromembrane suppressor the AS5A–5µ provides

Gradient elution using AS6

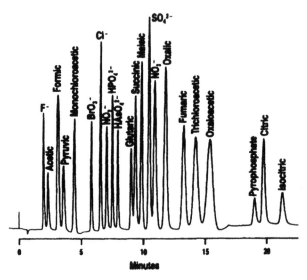

Fig. 2.17 Ion chromatography of inorganic and organic anions

Conditions:

Eluant 1:	50mM Mannitol, 2% CH$_3$CN
Eluant 2:	35mM p–cyanophenol, 50mM NH3, 2% CH$_3$CN
Eluant 3:	% CH$_3$CN
Flow rate:	20mL min^{-1}
Columns:	ATC, AS6, AMMS
Regenerant:	10mM H$_2$SO$_4$, 200mM H$_3$BO$_3$
Range 30μS	

Gradient program

Time	0	3	3.1	7	13	15
%1:	45	40	40	30	25	0
%2:	7	15	30	53	53	100
%3:	48	45	30	17	22	0

Source: Reproduced with permission from Dionex Corporation [86]

an impressive profile of inorganic ions and organic acid anions from a single injection of sample (Fig 2.19). Note that phosphate and citrate, strongly retained trivalent ions, are efficiently eluted in the same run that also gives baseline resolution of the weakly retained monovalents fluoride, acetate, formate and pyruvate. Quantitation of all the analytes shown in Fig. 2.19 using conventional columns would require at least three injections under different eluent conditions.

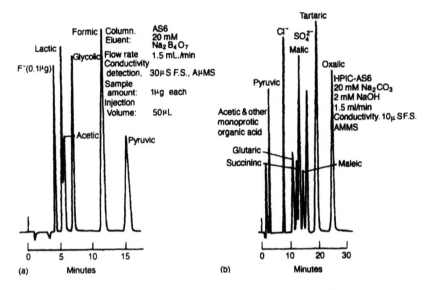

Fig. 2.18 Ion chromatograms obtained with Dionex instrument using (anodic) AMMS and CMMS micromembrane suppression; (a) monoprotic organic acids by anion exchange; (b) diprotic organic acids by anion exchange
Source: Own files

Another benefit of using the AS5A–5μ gradient pump combination is the ability to change easily the order of elution of ions with different valencies simply by changing the gradient profile. For example, if nitrate were present in high enough concentration to interfere with a malate peak, the malate peak could be moved ahead of the nitrate peak by using a slightly different gradient (Fig 2.19c).

See also Table 2.7.

2.3.2.6 Other applications

Further applications of ion chromatography to the determination of anions in non saline waters are summarised in Table 2.7.

2.3.3 Aqueous precipitation

2.3.3.1 Bromide

Ion chromatography using an amperometric detector has been used to determine down to 10μg L^{-1} bromide in rainwater and ground waters [87]. The equipment features an automated ion chromatograph, including

Table 2.7 Ion chromatography of anions in non saline waters

Anion	Stationary phase	Eluent	Comments	Detection	LD	Ref.
AsO_3^{3-}/SO_3^{2-}	—	—	—	—	—	[52]
$SO_3^{2-}/S_2O_3^{2-}$	—	4.8mmol L^{-1} NaHCO$_3$, then 6.9mmol L^{-1} Na$_2$CO$_3$	Gradient column	—	—	[53]
Cl^-, NO_3^{-1}, SO_4^{2-}	—	—	—	—	—	[54]
CO_3^{-2}/HCO_3^{-2}	Zipax, Sax and Westcan	—	—	Conductivity and UV	—	[55]
SiO_3^{-2}	—	—	—	—	—	[56]
SO_4^{-2}	Nucleosit GB ion exchange resin	Sulphosalicylic acid	—	Radio ion chromatography	—	[57]
Misc. anion	—	—	—	—	—	[58]
NO_2^-/NO_3^{-1}	Polystyrene–divinyl benzene copolymer	—	—	Conductiometric	—	[59]
NO_2^{-1}, Br^{-1}, NO_3^{-1}, SO_4^{2-}, I^{-1}, SCN^{-1}	Sodium naphthalene trisulphonate	—	—	Spectrophotometric	—	[60]
Fl^-, ClO^{1-}, Cl^{1-}, NO_2^{-1}, HPO_2^{2-}, AsO_4^{3-}, NO_3^{-1}, ClO_3^{1-}, SO_4^{2-}, CrO_4^{2-}, BF_4^{1-}, Glycolate acetate, lactate, propionate, aceto acetate, α hydroxy butyrate, chloroacetate, isobutyrate, butyrate	—	—	—	Spectrofluorometric	—	[61]

Table 2.7 continued

Anion	Stationary phase	Eluent	Comments	Detection	LD	Ref.
(as Ruthenium II, I, 10 phenanthroline or ruthenium II, 2,2' bipyridy ion interaction complexes) FI⁻, Cl⁻, AsO₄³⁻, NO₃⁻						
Cl^-, NO_2^-, Br^-, NO_3^-, PO_4^{-3}, SeO_3^{-2}, SO_4^{-2}, lactate, acetate, propionate, butyrate, isobutyrate	Mixed mode stationary phase	—	Careful control of eluent pH and ionic strength	—	—	[62]
NO_2^{-1}, NO_3^{-2}	—	—	—	Conductiometric	—	[63]
F^{-1}, BO_3^{-3}, glycolate, malonate, Cl^{-1}, AsO_4^{-4}, NO_3^{-1}, SO_3^{-2}	—	—	—	—	—	[140]
PO_4^{-3}, SO_4^{-2}, Cl^{-1}	—	—	—	—	—	[64]
SO_4^{-2}, Cl^{-1}	—	—	—	—	—	[65]
Benzoate, chloro-benzoate, arsenate, dimethyl benzoate, salicylate, hydro-xybenzoate, resorcylate, propionate, gallate	—	—	—	—	—	[66]
Various anions and cations	—	Benzene carboxylic acids	Comparison of various benzene carboxylic acids eluants	—	—	[67]

Table 2.7 continued

Anion	Stationary phase	Eluent	Comments	Detection	LD	Ref.
S^{-2}, CN^{-1}	—	Basic medium	—	Electrochemical amperometric	1mg L^{-1}	[68]
Misc. anions	—	—	Suppressed ion chromatography	—	—	[69]
SO_3^{-2}, SeO_3^{-3}	—	—	—	—	—	[70]
I^{-1}, PO_4^{-3}	—	—	—	—	—	[71]
SO_4	—	—	Suppressed ion chromatography	—	—	[72]
Formate	—	—	—	—	0.5µmol L^{-1}	[73]
HCO_3^{-1}	—	—	—	—	—	[74,75]
CO_3^{-2}, HCO_3^{-1}	—	—	—	—	—	[76]
PO_4	—	—	—	—	—	[77–79]
AsO_4^{-3}, SO_3^{-2}	—	—	—	—	—	[52]
AsO_3^{-3}, AsO_4^{-3}	—	—	—	Direct current plasma	—	[80]
S^{-2}	—	—	Conversion to methylene blue	—	<1µg L^{-1}	[81]
S^{-2}	—	—	Gas dialysis ion chromatography, precision 1.9µg L^{-1}	—	1.9µg L^{-1}	[82]
BO_3^{-3}	—	—	—	—	—	[83]
BO_3^{-3}	—	—	—	Conductivity	—	[84]

Source: Own files

1. F⁻ (1.5 ppm)
2. α-Hydroxybutyrate
3. Acetate
4. Glycolate
5. Butyrate
6. Gluconate
7. α-Hydroxyvalerate
8. Formate (5 ppm)
9. Valerate
10. Pyruvate
11. Monochloroacetate
12. BrO₃⁻

13. Cl⁻ (3 ppm)
14. Galacturonate
15. NO₂⁻ (5 ppm)
16. Glucuronate
17. Dichloroacetate
18. Trifluoroacetate
19. HPO₃²⁻
20. SeO₃²⁻
21. Br⁻
22. NO₃⁻
23. SO₄²⁻
24. Oxalate

25. SeO₄²⁻
26. α-Ketoglutarate
27. Fumarate
28. Phthalate
29. Oxalacetate
30. PO₄³⁻
31. AsO₄³⁻
32. CrO₄²⁻
33. Citrate
34. Isocitrate
35. cis-Aconitate
36. trans-Aconitate

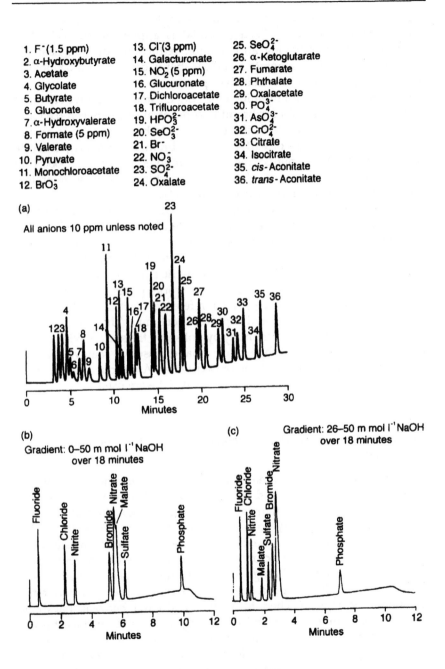

Fig. 2.19 Multi-component analysis by ion chromatography
Source: Own files

a programme controller, an automatic sampler, an integrator and an amperometric detector. A fixed potential is applied to the cell. Any electroactive species having an oxidation–reduction potential near the applied potential will generate a current that is directly proportional to the concentration of the electroactive species.

The detection limits for bromide is 0.01mg L^{-1} and the relative standard deviation is less than 5% for bromide concentrations between 0.05 and 0.5mg L^{-1}. Chloride interferes if the chloride to bromide ratio is greater than 1000:1 for a range of 0.01–0.1mg L^{-1} bromide. Similarly, chloride interferes in the 0.1–1.0mg L^{-1} range if the ratio is greater than 5000:1. In the latter case, a maximum of 2000mg L^{-1} chloride can be tolerated. Recoveries of bromide ranged from 97 to 110%.

This analysis is performed using a Dionex Model 12 ion chromatograph with 200µL sample loop, an eluent flow rate of 2.3ml min^{-1} (30% of full capacity) and a sample pump flow rate of 3.8ml min^{-1} (50% of full capacity). Pulse damper is installed just before the injection valve to reduce flow pulsation. A Dionex amperometric detector consisting of cell and potentiostat. The cell contains a silver working electrode, a platinum counter electrode and an Ag/AgCl reference electrode. This cell is installed between separator and suppressor columns. The working silver electrode should be cleaned monthly by polishing the surface with a small amount of an abrasive toothpaste on a paper tissue to remove the grey tarnish and then rinsing with deionised water. An applied potential of 0.10V and a detector output range of 100nA V^{-1} were used.

See also Table 2.12.

2.3.3.2 Chloride, bromide, fluoride, nitrate, nitrite, sulphate, sulphite and phosphate

The application of ion chromatography to the determination of nitrite, nitrate, halogens, sulphate and phosphate (also calcium and magnesium) sodium and potassium in rainwater is discussed below. Wetzel et al. [88] used ion chromatography to obtain determinations down to µg L^{-1} concentrations of chloride, phosphate, nitrate and sulphate in rain water samples. Conductivity detector response is linear with concentrations over the mg L^{-1} to µg L^{-1} range for the determinands examined. It was shown that a 3 × 50mm pellicular anion exchange resin with tetraethyl ammonium groups attached to a styrene divinyl benzene backbone concentrator column can be loaded remotely and stored for at least seven days before analysis without significantly affecting ionic determinations.

The response for phosphate, nitrate and sulphate was linear over four orders of magnitude (2–10^{4}µg L^{-1}). Response for chloride ion was linear over three orders of magnitude (2–10^{3}µg L^{-1}). Correlation coefficients for a linear least squares fit were between 0.9877 and 0.9999.

Table 2.8 Anion determination conditions by ion chromatography

Separator column	3 × 500mm, filled with low-capacity anion exchange resin
Suppressor column	6 × 250nm, filled with high-capacity cation exchange resin
Eluent	Mixture of 2m mol L^{-1} Na$_2$ CO$_3$ + 5m mol L^{-1} NaOH
Pumping speed	156ml h^{-1}
Sample loop	100μL

Source: Reproduced with permission from Elsevier Science, UK [89]

Oikawa and Saitoh [89] reported studies of the application of ion chromatography to the determination of fluoride, chloride, bromide, nitrite, nitrate, sulphate, sulphite and phosphate ions in 3ml samples of rainwater. The results show that the most suitable eluent for this purpose is 2m mol L^{-1} sodium carbonate/5m mol L^{-1} sodium hydroxide. The reproducibility of the determination was satisfactory for standard solutions of all the ions except nitrite. This problem was solved by preparing standard and sample solutions with the same composition as the eluent.

Filter the rainwater sample with a membrane filter having a pore size of 0.45μm. Use the filtrate as the sample for the test. Inject approximately 2ml of the sample into a microloop with a syringe. Table 2.8 lists the determination conditions. Obtain the peak areas of the chromatogram by the half value width method (width mm) in half value of peak height × peak height (mm) and prepare a calibration curve.

Using this procedure, except for nitrite, the coefficient of variation for each ion was less than 5%, and reproducibility was relatively good (Table 2.9). The peak height of nitrite was elevated as the number of sample injections were increased, and after nine tests the increase in peak height was as high as 20%.

Using the following conditions, Schwabe *et al.* [90] determined chloride, fluoride, nitrate, phosphate and sulphate in 20min in rainwater and potable water in amounts down to 0.5mg L^{-1}.

Type	Dionex D12
Sensitivity	1ks
Analytical column	Fast run anion separator
Suppressor column	Anion suppressor
Sample loop	100μL
Pump flow	115ml h^{-1}
Pressure	22 bar
Run time	11min per sample
Eluent	0.0048mol L^{-1} Na HCO$_3$
	0.0023mol L^{-1} Na$_2$CO$_3$

Table 2.9 Reproducibility of determination of anions in standard mixed solution

Ion No	F^- 3mg L^{-1}	Cl^- 5mg L^{-1}	NO_2^- 10mg L^{-1}	NO_3^- 30mg L^{-1}	SO_3^{2-} 20mg L^{-1}	SO_4^{2-}
1	16.4	11.2	10.0	4.3	3.1	9.4
2	7.0	11.5	10.5	4.4	3.2	9.9
3	16.7	11.3	10.9	4.6	3.2	10.5
4	16.9	11.5	11.0	4.4	3.2	9.7
5	16.1	11.8	11.3	4.5	3.3	10.6
6	17.1	11.9	11.6	4.6	3.3	10.7
7	17.4	12.3	11.8	4.6	3.3	10.3
8	16.8	12.1	11.9	4.5	3.2	10.6
9	16.9	11.9	11.9	4.6	3.2	10.3
X	16.8	11.7	11.2	4.5	3.2	10.2
CV%	2.3	3.2	6.0	2.5	2.0	4.5

Source: Reproduced with permission from Elsevier Science, UK [89]

Bucholz et al. [91] have described a procedure for the determination of less than 1μg L^{-1} of chloride, nitrite and sulphate in 2ml of rainwater sample using non-suppressed ion chromatography. Detection limits are less than 0.1mg L^{-1} for chloride and nitrate and 0.25mg L^{-1} for sulphate. The method can accomplish the simultaneous analysis of chloride, nitrate, nitrite and sulphate in less than 25min.

The injection of a 2ml sample greatly improved detection sensitivity. However, the injection of a sample of this magnitude onto a column with a bed volume of only 4ml did present a problem. The reduced conductivity of the sample, compared to the eluent buffer, resulted in a large drop in the baseline as the sample front entered the conductivity cell. At a recommended eluent buffer concentration of 4m mol L^{-1} potassium hydrogen phthalate (pH 4.5) as suggested by the manufacturer, the chloride and nitrate peaks were on an upward sloping baseline, which prevented accurate determination of the peak area with the recording integrator. The sulphate peak, whose retention time was much longer than that of the other two anions, always exhibited a baseline suitable for accurate quantitation. The chloride and nitrate peaks were shifted from the sloping portion of the baseline by a reduction in the concentration of the eluent buffer. Reduction in buffer concentration increased the retention time of all the anion peaks. All peaks also experienced broadening and, in the case of sulphate, this was significant enough to limit the reduction in buffer concentration to 2m mol L^{-1} potassium hydrogen phthalate (pH 4.5). At this concentration the chloride and nitrate peaks were shifted from the sloping baseline for more accurate quantitation, while the sulphate

Table 2.10 Comparison of anion concentrations: chromatographic values vs EPA values

	Concentrations (mg L^{-1})		
	Cl$^-$	NO$_3^-$	SO$_4^-$
I. EPA value	4.6	0.34	1.8
Chromatography value			
Run I	5.0	0.32	1.83
Run II	4.84	0.34	1.72
Mean	4.93	0.33	1.78
2. EPA value	2.3	0.17	0.90
Chromatography value			
Run I	2.27	0.16	0.88
Run II	2.39	0.14	0.89
Mean	2.33	0.15	0.89

Source: Reproduced by permission from Kluwer Academic Plenum, Amsterdam [91]

peak had not broadened to the point where detection sensitivity was too greatly diminished.

The accuracy of the method was considered by determination of the concentration of chloride, nitrate and sulphate in an EPA nutrient/ mineral standard. The results are shown in Table 2.10. The proximity of the chromatographic values to the EPA values indicates that corrections for matrix effects, due to additional constituents present in this standard, are unnecessary.

The use of concentration columns in conjunction with ion chromato- graphy is one means of improving the sensitivity of this procedure. Wetzel *et al.* [88] used this procedure to obtain determinations down to µg L^{-1} concentrations of chloride, phosphate, nitrate and sulphate in rainwater samples. Conductivity detector response is linear with concentrations over the mg L^{-1} to µg L^{-1} range for the determinands examined. It was shown that a 3 × 50mm peculiar anion exchange resin with tetraethyl ammonium groups attached to a styrene divinyl benzene backbone concentrator column can be loaded remotely and stored for at least 7 days before analysis without significantly affecting ionic determinations.

Yamamoto *et al.* [7] and Matsuchita [92] have described a method for the determination of chloride, phosphate, nitrate, nitrite, sulphate, calcium and magnesium in rainwater. They used 1m mol L^{-1} EDTA as an eluent and a silica based anion exchange column.

Apparatus

Ion chromatographic system consisting of a Toyo Soda Model HLC–6DL ion chromatograph [92]. A Toyo Soda Model CM–8 conductivity detector

Table 2.11 Adjusted retention times

Anions	Rt/min $^{-1}$	Cations	Rt/min $^{-1}$
Cl$^-$	3.4	Ca^{2+}	7.4
H$_2$PO$_4^-$	3.6	Mg^{2+}	9.0
NO$_2^-$	3.8		
NO$_3^-$	5.0		
SO$_4^{2-}$	14.2		

Source: Reproduced with permission from the American Chemical Society [7]

with a range of 5μs cm $^{-1}$. The sample loop volume was 0.1ml and a flow rate of 1.0ml min $^{-1}$ was employed. The separation column, 4.6mm id 50mm long, made of Teflon, was packed with porous silica based anion exchanger of TSK gel IC–Anion SW (Toyo Soda, particle size 5 ± 1 μm capacity 0.1mequiv g $^{-1}$. The column and the conductivity detector was maintained at 27 ± 1°C.

EDTA eluent (1m mol L $^{-1}$) was prepared by dissolving ethylene diaminetetraacetic acid disodium salt in deionised water and adjusting to pH 6.0 with 0.1mol L $^{-1}$ sodium hydroxide.

Table 2.11 shows the retention times of chloride, phosphate (H$_2$PO$^-$), nitrite, nitrate, sulphate, calcium and magnesium. All of these ions were eluted within 15min. The calibration curves of these ions by peak height measurements were linear up to the concentration of 20mg L $^{-1}$. The detection limits defined as the concentration corresponding to twice the value of noise of the base line were 0.04mg L $^{-1}$ for chloride, 0.05mg L $^{-1}$ for nitrite, nitrate, magnesium and calcium and 0.1mg L $^{-1}$ for phosphate and sulphate.

Wagner *et al.* [93] and Rowland [94] have also determined chloride, nitrate and sulphates in amounts down to 0.5mg L $^{-1}$.

Rowland [94] pointed out that before the installation of the chromatograph, some of the solutions might contain substances that could reduce the ion exchange efficiency of the column. In particular, in the studies on the effects of acid rain on conifer species, rain collected beneath the dense tree canopy (through-fall) and solutions running down the tree stems (stemflow), was found to contain colourless soluble organic compounds that gradually fouled the column. Such a sample may influence the chromatogram obtained for the subsequent sample. The 'pseudo carry-over' was eliminated by passing these sample types through a clean-up cartridge (Sep–pak C18, Waters Associates). It was found that the guard column protects the separator column from long-term contamination and an initial clean-up is not required. Leachates from the lower mineral soil horizons may contain iron and aluminium colloidal complexes, but column efficiency does not appear to be adversely affected by these solutions. See also Table 2.12.

2.3.3.3 Other applications

Other ion chromatographic methods for the determination of anions in rain are listed in Table 2.12.

2.3.4 Seawater

2.3.4.1 Iodide

Ito *et al.* [95] have described an ion chromatographic method for determining trace iodide in concentrated salt solutions. The method had a detection limit of 5μg of iodide L^{-1}, and a relative standard deviation (n = 5) of 3% at 0.1mg of iodide L^{-1}.

In a more recent work Ito [101] has described a simple and highly sensitive ion chromatographic method with ultraviolet detection for determining iodide in seawater. A high-capacity anion exchange resin with polystyrene–divinylbenzene matrix was used for both preconcentration and separation of iodide. Iodide in artificial seawater (salinity, 35‰) was trapped quantitatively (98.8 ± 0.6%) without peak broadening on a preconcentrator column and was separated with 0.35M sodium perchlorate + 0.01M phosphate buffer (pH 6.1). On the other hand, the major anions in seawater, chloride and sulphate ions, were partially trapped (5–20%) and did not interfere in the determination of iodide. The detection limit for iodide was 0.2μg L^{-1} for 6mL of artificial seawater. This method was applied to determination of iodide (ND – 18.3μg L^{-1}) and total inorganic iodine (I$^-$ + IO$_3^-$ – I, 50.0–52.7μg L^{-1}) in seawater samples taken near Japan.

2.3.4.2 Nitrate and phosphate

Tyree and Bynum [55] describe an ion chromatographic method for the determination of nitrate and phosphate in seawater. The pretreatment comprised vigorous mixing of the sample with a silver-based cation exchange resin, followed by filtration to remove the precipitated silver salt.

Peschet and Tinet [102] have also discussed the determination of phosphate in seawater.

2.3.4.3 Sulphate

Singh *et al.* [72] have determined sulphate in deep sub-surface waters by suppressed ion chromatography.

Table 2.12 Ion chromatography of anions in rain and ice

Cation	*Stationary phase*	*Eluent*	*Comments*	*Detection*	*LD*	*Ref.*
Hydroxymethane sulphonate form of sulphur(IV)	–	–	Analysis of ice	–	0.1µg L^{-1}	[96]
SO_3^{-2}	–	–	–	–	0.125mg L^{-1}	[97]
Cl^{-1}	–	–	–	–	–	[7,78,87,88,89 90,91,93,94,98 99]
F^{-1}, Cl^{-1}, NO_3^{-1}, SO_4^{-2}	–	–	–	–	F 0.03 Cl 0.04 NO_2 0.10 SO_2 0.25mg L^{-1}	[100]
Br^{-1}, Cl^{-1}, F^{-1}	–	–	–	–	–	[98,99]
SO_4^{2-1}, Cl^{-1}, NO_3^{-1}	–	–	–	–	–	[78]
PO_4^{-3}, Cl^{-1}, F^{-1}, Br^{-1}, NO_2^{-1}, BrO_2^{-1}	–	–	–	–	–	[90]
PO_4^{-3}, Cl^{-1}, NO_2^{-1}, NO_3^{-1}, SO_4^{-2}	–	–	–	–	–	[98]
SO_4^{-2}, Cl^{-1}, PO_4^{-3}, SO_4^{-2}	–	–	–	–	–	[99]
SO_4^{-2}, Cl^{-1}, NO_2^{-1}, PO_4^{-3}, SO_4^{-2}	–	–	–	–	–	[7]
SO_4^{-2}, Cl^{-1}, SO_4	–	–	–	–	–	[93,94]
SO_4^{-2}, Cl^{-1}, NO_2^{-1}, SO_4^{-2}	–	–	–	–	–	[91]
SO_4^{-2}, Cl^{-1}, F^{-1}, Br^{-1}, NO_2^{-1}, SO_4^{-2}, SO_3^{-2}, PO_4^{-3}	–	–	–	–	–	[89]
SO_4^{-2}, Cl^{-1}, Br^{-1}, F^{-1}, NO_2^{-1}, PO_4^{-3}, BRO_3^{-1}, SO_4^{-1}	–	–	–	–	–	[87]

Source: Own files

Xiao-Hua Yang *et al.* [103] determined nanomolar concentrations of individual low molecular weight–carboxylic acids (and amines) in seawater. Diffusion of the acids across a hydrophobic membrane was used to concentrate and separate carboxylic acids from inorganic salts and most other organic compounds prior to the application of ion chromatography. Acetic acid, propionic acid, butyric acid–1, butyric acid–2, valeric and pyruvic acid, acrylic acid and benzoic acid were all found in reasonable concentrations in seawater.

2.3.5 Potable water

2.3.5.1 Bromide, chloride, nitrite, nitrate, sulphate and phosphate

Morrow and Minear [104] have developed an ion chromatographic procedure using a concentrator column to improve sensitivity. This analysis, which was carried out on a Dionex model 125 ion chromatograph equipped with concentrator column in place of an injection loop, coupled with an autosampler, allowed detection at the µg L^{-1} level. A concentrator column is a short separator column used to strip ions from a measured volume of matrix leading to a lowering of detection limits by several orders of magnitude. The concentrated sample was flushed into the system using 0.003M sodium bicarbonate/0.0024M sodium carbonate eluant.

When using conventional ion chromatographic separation techniques, it is possible that other matrix anions also common to non saline waters may coelute with bromide. For example, bromide and nitrate elute simultaneously using a standard anion separator column (Dionex No. 30065), standard anion suppressor (Dionex No. 30366) and standard eluant (0.003M sodium bicarbonate/0.0024M sodium carbonate).

A representative chromatogram is shown in Fig. 2.20.

Nitrate in excess of 17mg L^{-1} interfered with the elution of bromide and sulphate in excess of 25mg L^{-1} interfered with the elution of bromide.

See also Table 2.13.

2.3.5.2 Chloride, bromide and iodide

Bachmann and Matusca [105] have described a gas chromatographic method for the determination of µg L^{-1} quantities of chloride, bromide and iodide in potable water. This method involves reaction of the halide with an acetone solution of 7-oxabicyclo–(4,1,0)heptane in the presence of nitric acid to form halogenated derivatives of cyclohexanol and cyclohexanol nitrate.

The column effluent passes through a pyrolysis chamber at 800°C and then through a conductivity detector. The solution is injected on to a gas chromatographic column (OV–10–Chromosorb W–HPl, 80–100 mesh,

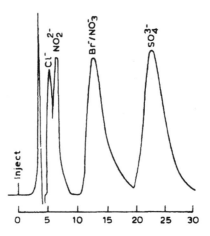

Fig. 2.20 Ion chromatography on anions in potable water. Simultaneous elution of bromide and nitrate using standard eluant
Source: Reproduced with permission from Elsevier Science, UK [104]

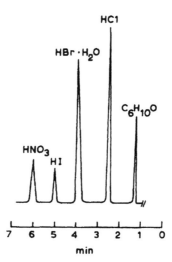

Fig. 2.21 Gas chromatogram of halogen mixture in potable water
Source: Reproduced with permission from Springer Verlag Heidelberg [105]

150cm, 0.2cm dia.) operated at 50°C. Hydrogen is used as carrier gas. Chloride contents determined by this method were $5.0 \pm 1.8\mu g$ L^{-1} chloride in potable water, $20.0 \pm 1.5\mu g$ L^{-1} in drinking water. A typical gas chromatogram is shown in Fig. 2.21 showing peaks for the three halogens. See also Table 2.13.

2.3.5.3 Chlorite and chlorate

Chlorine dioxide (ClO_2) is a widely used disinfectant and bleaching agent that is currently being used by many drinking water treatment utilities in the United States, Canada and Europe for oxidation and disinfection. It is frequently applied as an initial oxidant during treatment, followed by chlorination (gaseous Cl_2 or HOCl) as a final disinfectant. When used as an oxidant, chlorine dioxide reacts to form chlorite (ClO_2^-) and chlorate (ClO_3^-) which have been shown to cause haemolytic anaemia in laboratory animals. Both ClO_2^- and ClO_3^- concentrations are under consideration for regulation by EPA and because of possible adverse health effects will likely be regulated with a maximum contaminant limit of <1.0mg L^{-1} for combined chlorine dioxide, chlorite and chlorate.

In response to these needs Dietrich et al. [108] used flow injection analysis with iodometric detection and ion chromatography with conductiometric detection in two methods for the determination of chlorite and chlorate in chlorinated and chloroaminated potable water.

The two methods were accurate and effective for reagent water. The ion chromatographic method was accurate for measurement of chlorite and chlorate concentrations in drinking water even in the presence of other oxidants including chloramines. However, flow injection analysis was affected by chloramines and other oxidants in drinking water, resulting in inaccurate determinations. While chlorite concentrations were unstable in chlorinated drinking water, addition of sodium oxalate increased the stability to up to three days and addition of ethylenediamine increased stability up to 18 days. Chlorate concentrations were stable in drinking water for up to 18 days with or without a preservative.

2.3.5.4 Sulphide and cyanide

Bond et al. [68] have described a method for the simultaneous determination of down to 1mg L^{-1} free sulphide and cyanide by ion chromatography, with electrochemical detection. These workers carried out considerable exploratory work on the development of ion chromatographic conditions for separating sulphide and cyanide in a basic medium (to avoid losses of toxic hydrogen cyanide and hydrogen sulphide) and on the development of a suitable amperometric detector.

Table 2.13 Ion chromatography of anions in potable waters

Anion	Stationary phase	Eluent	Comments	Detection	LD	Ref.
Co^{-1}, Br^{-1}, NO_3^{-1}, PO_4^{-3}, SO_4^{-2}	—	—	—	—	—	[77,90]
Cl^{-1}, F^{-1}, NO_3^{-1}, PO_4^{-3}, SO_4^{-2}	—	—	—	—	—	[106,141]
AsO_3^{-3}, SeO_3^{-2}, SeO_4^{-2}	—	—	—	—	—	[85]
SO_3^{-2}	—	—	Gas dialysis–ion chromatography	—	$2\mu g\ L^{-1}$	[82]

Source: Own files

2.3.5.5 Other applications

Further applications of ion chromatography to the analysis of potable waters are listed in Table 2.13.

2.3.6 Waste waters

2.3.6.1 Cyanide, free and total

Nonomura [107] determined free cyanide and metal cyanide complexes in wastewaters by ion chromatography with conductive detection.

The cyanide is not detected by the conductivity detector of the ion chromatograph due to its low dissociation constant ($pK = 9.2$). Nonomura [107] described an ion chromatographic method for the determination of free cyanide and metal cyanide complexes that uses a conductivity detector. It is based on the oxidation of cyanide ion by sodium hypochlorite to cyanate ion ($pK = 3.66$). Therefore, cyanide ion can now be measured indirectly by the conductivity detector. In this procedure, optimum operating conditions were examined. In addition, the interferences from anions and reducing agents were investigated. The method was applied to the determination of metal cyanide complexes. The coefficients of variation (%) for CN^- (1.05mg L^{-1}), $Zn(CN)_4^{2-}$ (CN^-, 0.80mg L^{-1}) and $Ni(CN)_4^{2-}$ (CN^-, 0.96mg L^{-1}) were 1.1%, 1.5% and 0.5%, respectively.

Lui et al. [109] have described an automated system for determination of total and labile cyanide in water samples. The stable metal–cyanide complexes such as $Fe(CN)_6^{3-}$ are photo-dissociated in an acidic medium with an on-line Pyrex glass reaction coil irradiated by an intense mercury lamp. The released cyanide is separated from most interferences in the sample matrix and is collected in a dilute sodium hydroxide solution by gas diffusion using a hydrophobic porous membrane separator. The cyanide ion is then separated from remaining interferences such as sulphide by ion exchange chromatography and is detected by an amperometric detector. The characteristics of the automated system were studied with solutions of free cyanide and metal–cyanide complexes. The results of cyanide determination for a number of wastewater samples obtained with this method were compared with those obtained with the standard method. The sample throughput of the system is eight samples per hour and the detection limit for total cyanide is 0.1µg L^{-1}.

See also Table 2.13.

2.3.6.2 Borate

Hill and Lash [83] have described an ion chromatographic method for the determination of down to 0.05mg L^{-1} of borate in environmental waters, nuclear fuel dissolvent solutions and effluents.

Table 2.14 Ion chromatographic conditions

Eluent	0.003mol L^{-1} NaHCO$_3$/0.004mol L^{-1} Na$_2$CO$_3$
Flow rate	138ml/h, 30%
Analytical column	3 × 500mm brine anion separator
Suppressor column	6 × 250mm anion suppressor
Detector sensitivity	3μmho full scale
Injection volume	100μL

Source: Reproduced by permission from the American Chemical Society [83]

Borate is selectively concentrated on Amberlite XE–243 ion exchange resin and converted to tetrafluoroborate using 10% hydrofluoric acid. Tetrafluoroborate is strongly retained by the resin, thus allowing excess fluoride to be eluted without loss of boron. The tetrafluoroborate is eluted with 1mol L^{-1} sodium hydroxide and is determined in the eluent by ion chromatography. Boron is quantified to a lower limit of 0.05mg L^{-1}.

A Model 14 ion chromatograph (Dionex Corporation) was used in this work. Clear polymethylpentane plastic ware (Nagel Company) was used in the preparation of all reagents and standards. Separating columns; standard 3 × 150 and 3 × 500mm anion separator columns, as well as a 3 × 500mm brine anion separator were used. Suppressor column-standard; 6 × 250mm anion suppressor column, from Dionex Corporation.

Samples and calibration standards containing between 1 and 20mg L^{-1} tetrafluoroborate are injected into the ion chromatograph using the conditions shown in Table 2.14. Results are calculated on a peak height ratio basis.

2.3.6.3 Sulphide

Goodwin *et al.* [82] determined trace sulphides in turbid waters by a gas dialysis–ion chromatographic method. The sulphide is converted to hydrogen sulphide which is then isolated from the sample matrix by diffusion through a gas dialysis membrane and trapped in a dilute sodium hydroxide solution. A 200μL portion of this solution is injected into an ion chromatograph for determination with an electrochemical detector. Detection limits were ≥1.9ng/mL^{-1}.

2.3.6.4 Other applications

Other applications of ion chromatography to the analysis of waste waters are listed in Table 2.15.

Table 2.15 Ion chromatography of anions in wastewaters

Polyonic	Stationary phase	Eluent	Comments	Detection	LD	Ref.
SO_3^{2-}, Cl^{-1}	—	—	—	—	—	[110]
Cl^{-1}, F^{-1}, NO_3^{-1}, NO_2^{-1}, PO_4^{-3}, SO_4^{-2}	—	—	—	—	—	[106]
SO_4^{-2}, F^{-1}, Cl^{-1}, Br^{-1}, NO_2^{-1}, PO_4^{-3}	—	—	SD <3%	—	≥40µg L^{-1}	[111]
Cl^{-1}, F^{-1}, SO_4^{-2}, PO_4^{-3}	—	—	—	—	—	[112]
F^{-1}, Cl^{-1}, Br^{-1}, NO_3^{-1}, NO_2^{-1}, PO_4^{-3}, SO_4^{-2}	—	—	Suppressed ion chromatography RSD 0.3% (Br^{-1} and NO_2^{-1}, to 1.6% (NO_3^{-1})	—	—	[113]
Cl^{-1}, F^{-1}, NO_2^{-1}, NO_3^{-1}, PO_4^{-3}, SO_4^{-2}	—	—	—	—	—	[106]
Cl^{-1}, SO_3^{-2}	—	—	—	—	—	[20]
F^{-1}, Cl^{-1}, Br^{-1}, SO_4^{-2}, SO_3^{-2}, CNS^{-1}, $S_2O_3^{-2}$	—	—	—	—	—	[114]
Free CN and metal cyanide complexes	—	—	—	—	—	[107,109,115]
Anionic CN complexes of Zn, Cu, Ni, Au, Fe, Co	—	—	—	—	—	[49]
Dithionite	—	—	—	—	—	[118]

Source: Own files

2.3.7 Trade effluents

2.3.7.1 Chloride, bromide, fluoride, nitrite, nitrate, sulphate and phosphate

The work of Mosko [116] is important in that he is one of the few workers who have given serious consideration to the determination of nitrite in water. His paper is concerned with the determination of chloride, sulphate, nitrate, nitrite, orthophosphate, fluoride and bromide in industrial effluents, waste water and cooling water. Two types of analytical columns were evaluated (standard anion and fast run series). Chromatographic conditions, sample pretreatment and the results of interference, sensitivity, linearity, precision, comparative and recovery studies are described. The standard column provided separation capabilities which permitted the determination of all seven anions. The fast run column could not be used for samples containing nitrite or bromide owing to resolution problems.

Mosko [116] used a Dionex Auto Ion System 12 Analyzer (Dionex Corp., Sunnyvale, California) consisting of a Model 12 ion chromatograph (Dionex Corp.) coupled with a Gilson autosampler, a Columbia Scientific Industries Supergrator 3–A integrator and a Gilford Instrument Laboratories dual pen strip chart recorder. A concentrator/ precolumn (4 × 50mm, Dionex 30825) and separator column (4 × 250mm, Dionex 30827) were used for the standard seven anion determination. Two anion concentrator precolumns (4 × 50mm Dionex 30825) connected in series (fast run series) were employed for the five anion analyses. A high capacity suppressor column (9 × 100mm Dionex 30828) containing a strong acid cation exchange resin in the hydrogen form was used for all chromatography. The ion chromatograph was equipped with a 250µL sample loop. A conductivity full scale sensitivity setting of 30µmho was routinely used.

Eluent pump flow rate of 35–40% (161–184ml h^{-1}) maintained system pressure between 500 and 700psi. Both the sample and regeneration pumps were operated at 50% flow rate (230ml h^{-1}).

See also Table 2.18.

2.3.7.2 Borate

Hill and Lash [83] have described an ion chromatographic method for the determination of down to 0.05mg L^{-1} of borate in environmental waters, nuclear fuel dissolvent solutions and effluents.

Borate is selectively concentrated on Amberlite XE–243 ion exchange resin and converted to tetrafluoroborate using 10% hydrofluoric acid. Tetrafluoroborate is strongly retained by the resin, thus allowing excess fluoride to be eluted without loss of boron. The tetrafluoroborate is eluted

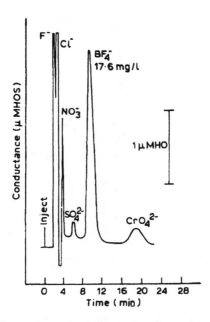

Fig. 2.22 Effects of diverse ions on IC determination of tetrafluoroborate ion
Source: Reproduced by permission from the American Chemical Society [83]

with 1M sodium hydroxide and is determined in the eluant by ion chromatography. Boron is quantified to a lower limit of 0.05mg L^{-1}.

None of the common ions interferes at a 100 to 1 mole ratio to boron. In addition, at least a 280 to 1 mole ratio of fluoride to boron, as well as a 100 to 1 mole ratio of chloride to boron, can be tolerated. Fig. 2.22 shows an ion chromatogram of a sample which initially contained sulphate, chloride, nitrate, fluoride and chromate at a 100 to 1 mole ratio and which was prepared according to the standard procedure. As shown, tetrafluoroborate peaks are well resolved from all the residual amounts of these anions not completely removed by the XE–243 column treatment.

2.3.7.3 Cyanide, sulphide, iodide and bromide

The Rocklin and Johnson [48] procedure described in Section 2.3.2.3 (non saline waters) has been applied to the determination of cyanide and sulphide in trade effluents.

Electrochemical detection was performed with an Ion Chrom/Amperometric Detector (P/N 35221). The cell consists of a silver rod working electrode 1.3cm long × 0.178cm in diameter, an Ag/AgCl reference

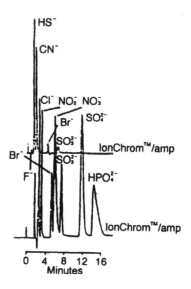

Fig. 2.23 Simultaneous analysis by using electrochemical and conductivity detection. Concentrations are 300μg L^{-1} sulphide, 500μg L^{-1} cyanide, 1mg L^{-1} fluoride, 4mg L^{-1} chloride, 10mg L^{-1} nitrites, 10mg L^{-1} bromide, 25mg L^{-1} nitrate, 30mg L^{-1} sulphite, 25mg L^{-1} sulphate and 50mg L^{-1} phosphate
Source: Reproduced by permission from the American Chemical Society [49]

electrode separated from the flowing stream by a Nafion cation exchange membrane and platinum counter electrode. (Nafion is a registered trade mark of E.I. du Pont de Nemours & Co.) The cell geometry is based on one previously reported by Lown *et al.* [119]. The working electrode was occasionally cleaned by mechanical polishing. The applied potential was 0.V for cyanide and sulphide, 0.20V for iodide, and 0.30V for bromide.

Fig. 2.23 shows the separation achieved on a 12 anion standard by this procedure. Sulphide, cyanide, bromide, and sulphite are detected at the silver electrode while nitrite, nitrate, phosphate and sulphate produce no response. Due to the low dissociation of hydrogen sulphide and hydrogen cyanide following protonation by the suppressor column, they are not detected by the conductivity detector.

2.3.7.4 Metal cyanide complexes

The Rocklin and Johnson procedure described in section 2.3.2.2 has been applied to trade effluents.

Table 2.16 Chromatographic conditions and method detection limits in reagent water

Analyte	Retention[1] time (min)	Relative retention time	Method[2] Detection limit mg L^{-1}
Fluoride	1.2	1.0	0.005
Chloride	3.4	2.8	0.015
Nitrite–N	4.5	3.8	0.004
O–phosphate–P	9.0	7.5	0.061
Nitrate–N	11.3	9.4	0.013
Sulphate	21.4	17.8	0.206

Standard conditions:
Sample loop 100μL. Pump volume 2.30mL min^{-1}
[1]Concentrations of mixed standard (mg L^{-1})

Fluoride	3.0	O–phosphate–P	9.0
Chloride	4.0	Nitrate–N	30.0
Nitrite–N	10.0	Sulphate	50.0

[2]MDL calculated from data obtained using an attenuator setting of 1μMHO full scale. Other settings would produce an MDL proportional to their value

Source: Reproduced with permission from the Environmental Protection Agency, Cinncinati [106]

2.3.7.5 Other applications

Further applications of ion chromatography to the analysis of trade effluents are reviewed in Table 2.18.

2.3.8 Ground and surface waters

The detection limit for bromide is 0.01mg L^{-1} and the relative standard deviation is less than 5% for bromide concentrations between 0.05 and 0.5mg L^{-1}. Chloride interferes if the chloride to bromide ratio is greater than 1000:1 for a range of 0.01–0.1mg L^{-1} bromide; similarly, chloride interferes in the 0.1–1.0mg L^{-1} range if the ratio is greater than 5000:1. In the latter case, a maximum of 2000mg L^{-1} of chloride can be tolerated. Recoveries of bromide ranged from 97 to 110%.

2.3.8.1 Chloride, fluoride, nitrate, nitrite, phosphate and sulphate

The Environmental Protection Agency US [106] have published a standard method for the determination of anions in surface waters.

The recommended operating conditions are listed in Table 2.16. Single operator accuracy and precision are listed in Table 2.17.

Table 2.17 Single-operator accuracy and precision, determinations in surface water

Analyte	Spike (mg L^{-1})	No. of replicates	Mean recovery (%)	Standard deviation (mg L^{-1})
Chloride	1.0	7	105.0	0.139
Fluoride	0.50	7	74.0	0.0038
Nitrate–N	0.50	7	100.0	0.0058
Nitrite–N	0.51	7	88.2	0.0053
O–phosphate–P	0.51	7	94.1	0.020
Sulphate	10.0	7	111.6	0.709

Source: Reproduced with permission from the Environmental Protection Agency, Cinncinati [106]

2.3.8.2 Bromide

Ion chromatography using an amperometric detector has been used to determine down to 10µg L^{-1} bromide in rain water and groundwaters [120]. The equipment features an automated ion chromatograph, including a programme controller, an automatic sampler, an integrator and an amperometric detector. A fixed potential is applied to the cell. Any electroactive species having an oxidation–reduction potential near the applied potential will generate a current that is directly proportional to the concentration of the electroactive species.

2.3.8.3 Selenate and selenite

Reddy *et al.* [121] studied the speciation of selenium in groundwaters by adsorption of selenite and selenate onto copper oxide particles followed by hydride generation atomic absorption spectrometry and ion chromatography.

Hoover and Jager [85] have discussed the determination of selenate together with other anions (selenite and arsenite) in potable and ground waters.

Other anions which have been determined by ion chromatography in groundwaters include sulphide, hypochlorite and cyanide [49,142].

2.3.9 Boiler waters

2.3.9.1 Chloride, sulphate and nitrite

Roberts *et al.* [122] have described a single column ion chromatographic method for the determination of chloride and sulphate in steam condensate and boiler feed water. This was shown to be a valuable

Table 2.18 Ion chromatography of anions in trade effluents

Anion	Stationary phase	Eluent	Comments	Detection	LD	Ref.
SO_4^{-2}, NO_2^{-1}, NO_3^{-1},	–	–	–	–	–	[116]
PO_4^{-3}, F^{-1}, Br^{-1}						
SO_4^{-2}, Cl^{-1}, F^{-1}	–	–	–	–	–	[40]
SO_3^{-2}	–	–	–	–	–	[117]
SO_3^{-2}, dithionite	–	–	–	–	–	[118]

Source: Own files

technique for analysing µg L $^{-1}$ levels of chloride and sulphate in very pure waters. The anions are concentrated on a short precolumn, separated on a low capacity ion exchange column and detected by an electrical conductivity detector. The apparatus is simple and no 'suppressor' column is needed. Adaptation to on-line analysis would be inexpensive and automation would require control of only the load/ inject valve.

The preliminary work by Roberts et al. [122] was carried out using a resin of very low exchange capacity (0.003mequiv g $^{-1}$) and a 7.5×10^{-4}M solution of benzoic acid as the eluant. Unusually good sensitivity can be obtained in single column anion chromatography using this eluant. To increase the sensitivity further, the size of the sample loop was increased to 500µL from the usual 100µL. This method adequately separated chloride and sulphate with detection limits of 10mg L $^{-1}$ chloride and 100mg L $^{-1}$ sulphate. However, a prolonged dip in the base line occurred sometime after the sulphate peak and made this procedure inefficient for repetitive analyses. Chloride must be separated from the water dip before it can be analysed reliably in the same run that sulphate is being determined. This was accomplished by carefully constructing a concentrator column and by careful choice of the column dimensions, resin capacity and eluant strength. The effect of column and eluant parameters on anion retention times has been discussed previously by Gjerde et al. [123,124].

2.3.9.2 Carbonate

Carbonate has been determined in high priority water in amounts down to 0.02µg L $^{-1}$ by ion chromatography on Zipax SAX phthalic acid or mucomic acid at pH 6.5 and 7.2 as eluting agent and an indirect ultraviolet detector set at 321µm [125].

2.3.9.3 Other applications

Other anions which have been determined in boiler feed or high purity waters include chloride, nitrite and sulphate in amounts down to 1µg L $^{-1}$ [126], sulphate and phosphate [127], phosphate and sulphate in boiler blow down waters in amounts down to 1mg L $^{-1}$ [127].

2.4 Mixtures of anions and cations

2.4.1 Introduction

Ion chromatography is not restricted to the separate analysis of only anions or cations. With the proper selection of eluant and separation columns, the technique can be used for the simultaneous analysis of both anions and cations.

Table 2.19 Instrumental conditions

Instruments	Dionex model 2010i ion chromatograph
	Dionex electrochemical detector
	Houston Instrument Omniscribe chart recorder
Electrochemical applied potential	0.4V
Chart recorder speed	0.25cm min $^{-1}$
Flow rate	1.5ml min $^{-1}$
Suppressor column	Dionex AMS, preproduction prototype,
	Membrane suppressor
Injection volume	0.10ml
Eluant	0.0016mol L $^{-1}$, Li$_2$CO$_3$ + 0.0024mol L $^{-1}$ LiCOOH.H$_2$O
	pH = 10.4 Monovalent cations and anions
Separator columns	
Anion separator	150 × 4mm id Dionex HPIC–AS3 anion
Cation separator	200 × 4mm id Dionex HPIC–CS1 cation
	0.0033mol L $^{-1}$ Cu phthalate divalent cations and anions
Eluant	
Separator columns	
Anion separator	250 × 4.6mm id Vydac 3021C4.6 anion
Cation separator	200 × 4mm id Dionex HPIC–CS1 cation

Source: Reproduced with permission from International Scientific Communications Inc, Shelton, USA [138]

The simultaneous determinations of anions and cations using single injection ion chromatography was introduced by Yamamoto and co-workers in 1984 [7]. This technique determined cations and anions simultaneously using a complexing agent, ethylenedinitrilotetraacetic acid, to complex the divalent metals. These divalent metals were later separated and detected as anions along with the uncomplexed inorganic anions.

Since then, various workers have discussed the codetermination of anions and cations by ion chromatography [11,64,128–137] using various types of detectors [137].

2.4.2 Aqueous precipitation

2.4.2.1 Alkali metals, alkali earths and anions

Ion chromatographic methods have been described for the co-determination of anions and cations in rainwater. Thus Jones and Tarter [138], using the conditions given in Table 2.19 reported determinations down to 1mg L $^{-1}$ of anions (chloride, bromide and sulphate) and cations (sodium, potassium, magnesium and calcium) in rainwater without converting the cations to anion complexes prior to detection [139]. The technique uses a cation separator column, a conductivity detector, an anion suppressor column, and either a second conductivity detector or an electrochemical detector in sequence. The use of different eluants provides a means for the detection of monovalent cations and anions and divalent cations and anions in each of the samples. Using an eluant with a basic pH, it is possible to separate simultaneously and detect the monovalent cations (with the exception of the ammonium ion) and anions, while an eluant with an acidic pH allows for the separation and detection of divalent cations and anions.

The equipment and operating conditions used by Jones and Tarter [138] are described below. The instrumental set-up consists primarily of two separator columns, two detectors and one suppressor column. The analyte flows through the injection valve into the cation separator, where the cations are separated. The separated cations are then detected using a conductivity detector (this is, in effect, a single column ion chromato-graphic analysis at this stage). The anions, which are essentially unretained on the cation column, are separated on the anion separator column. The ions then travel through the anion suppressor column, where the previously separated and detected cations are removed. Finally, the separated anions pass through the second detector, which may be a conductivity detector or an electrochemical detector. The electrochemical detector responds to pH changes in the eluant as the dissociated acids pass through the detector.

Two different eluants were used, lithium carbonate–lithium acetate dihydrate, and copper phthalate. A stock solution of the lithium carbonate–lithium acetate dihydrate eluant was prepared from ACS Certified salts using distilled deionised water with the appropriate dilution to obtain the working eluant.

The copper phthalate eluant was prepared by mixing a solution of cupric acetate with an excess of potassium hydroxide and filtering the resulting cupric hydroxide precipitate. The cupric hydroxide precipitate was then mixed with an equimolar amount of phthalic acid and heated gently overnight to produce copper phthalate.

Apparatus
Dionex 2010i ion chromatograph, Dionex electrochemical detector, Houston Instrument Omniscribe chart recorder.

Conditions
Electrochemical applied	
potential	0.4V
Chart recorder speed	0.25cm min^{-1}
Flow rate	1.5ml min^{-1}

Suppressor column	Dionex 4M S₁ membrane suppressor
Injection volume	0.1mL

Reagents
For monovalent ions:

Eluant	$0.0016mol\ L^{-1}\ Li_2CO_3 + 0.0024mol\ L^{-1}\ LiCooCH$, $H_2O\ pH = 10.4$ monovalent cations and anions

Separator columns

Anion separator	150 × 4mm id Dionex HPLC −A 53 anion
Cation separator	200 × 4mm id Dionex HPLC −CS 2 cation
For divalent ions:	
Eluant	$0.0033mol\ L^{-1}$ copper phthalate divalent cations and anions

Separator columns

Anion separator	250 × 4.6mm id Vydac 3021 C 4.6 anion
Cation separator	200 × 4mm id Dionex HPLC–CS 1 cation

2.4.3 Potable waters

2.4.3.1 Alkali metals, alkaline earths and anions

Jones and Tarter [11] have applied this technique to the simultaneous determination of metals (sodium, potassium, calcium, magnesium) and anions (chloride, sulphate, nitrate, bromide) in potable waters. The technique uses a cation separator column, a conductivity detector, an anion separator column and an anion suppressor column. Two different eluants were used: lithium carbonate–lithium acetate dihydrate, and copper phthalate.

All water samples were injected and analysed as received without prior preparative steps.

The normal procedure for the use of the ion chromatograph was followed, with the exception of the use of the electrochemical detector after the anion suppressor. The solutions were injected and the detector and chart recorder were adjusted to provide peaks of appropriate height.

The operating parameters for the two eluants are listed in Table 2.20. For the detection of monovalent actions and anions, a basic eluant, lithium carbonate–lithium acetate, is used. Analysis of divalent cations and anions involves the use of an acidic copper phthalate eluant.

Simultaneous analysis of both anions and cations indicates that water samples from various localities contain many of the same ions but in differing amounts. Fig. 2.24 illustrates typical chromatograms of tap water and rain water.

Table 2.20 Instrumental conditions

Instruments	Dionex model 2010i ion chromatograph
	Dionex electrochemical detector
	Houston Instrument Omniscribe chart recorder
Electrochemical applied potential	0.4 V
Chart recorded speed	0.25 cm/min
Flow rate	1.5 mL/min
Suppressor column	Dionex AMS, preproduction prototype, Membrane Suppressor
Injection volume	0.10 mL
Eluant	0.0016 M Li_2CO_3 + 0.0024 M $LiCH_3CO_2$ $2H_2O$
	pH = 10.4 monovalent cations and anions
Separator columns	
Anion separator	150 × 4 mm i.d. Dionex HPIC–AS3 anion
Cation separator	200 × 4 mm i.d. Dionex HPIC–CS1 cation
Eluant	0.0033 M Cu phthalate divalent cations and anions
Separator columns	
Anion separator	250 × 4.6 mm i.d. Vydac 3021C4.6 anion
Cation separator	200 × 4 mm i.d. Dionex HPIC–CS1 cation

Source: Reproduced by permission from International Science Communications Inc, Shelton, USA [11]

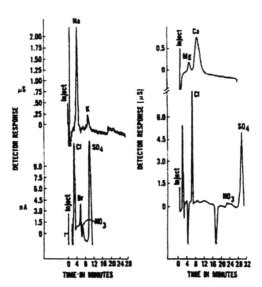

Fig. 2.24 Chromatograms of tap water (Denton,Tex.) Monovalent cations and anions: Na^+ = 30 ppm; K^+ = 5 ppm; Cl^- = 23 ppm; Br^- = 0.7 ppm; NO_3^- = 15 ppm; SO_4^{2-} = 44 ppm. For the divalent cations and anions: Mg^{2+} = 4 ppm; Ca^{2+} = 38 ppm
Source: Reproduced by permission from International Science Communications Inc., Shelton, USA [11]

References

1 Brandt, G. and Kettrup, A. *International Journal of Environmental Analytical Chemistry*, **31**, 129 (1987).
2 Brandt, G. and Kettrup, A. *Fresenius Zeitschrift für Analytische Chemie*, **327**, 213 (1987).
3 Ivanov, A.A., Shpigun, O.A. and Zolotov, Z.A. *Zhur Anal Khim*, **42**, 694 (1987).
4 Backman, S.R. and Penden, M.E. *Water, Air and Soil Pollution*, **33**, 191 (1987).
5 Ye, R., Zhang, C. and Wu, Z. *Huanijing Huaxue*, **6**, 36 (1987).
6 Small, H., Stevens, T.S. and Bauman, W.C. *Analytical Chemistry*, **47**, 1801 (1975).
7 Yamamoto, M., Yamamoto, H. and Yamamoto, Y. *Analytical Chemistry*, **56**, 822 (1984).
8 Lash, R.P. and Hill, C.J. *Analytica Chimica Acta*, **114**, 405 (1980).
9 Pohlandt, C. *South Africa Journal of Chemistry*, **33**, 87 (1980).
10 Cassidy, M. and Elchuk, S. *Analytical Chemistry*, **54**, 1558 (1982).
11 Jones, V.K. and Tarter, J.G. *American Laboratory*, **17**, 48 (1985).
12 Rubin, R.B. and Heberling, S.S. *International Laboratory*, September, 54 (1987).
13 Basta, N.T. and Tabakabi, N.A. *Journal of Environmental Quality*, **14**, 450 (1985).
14 Mizobuchi, M., Tamase, K., Kitada, Y., Sasaki, M. and Tanigawa, K. *Analytical Chemistry*, **56**, 603 (1984).
15 Fung, Y.S. and Zucheng Wu Duo, K.L. *Analytical Chemistry*, **68**, 2186 (1996).
16 Smith, D.L. Ames Lab. Report, IS–T–1318; Order No. DE870 13601, 73 pp. Avail. NTIS from Energy Res. Abstr. 1987, 12(21), Abstr. No. 43657 (1987).
17 Smith, D.L. and Fritz, J.S. *Analytica Chimica Acta*, **204**, 87 (1988).
18 Iwashido, T., Ishimaru, K. and Motomizu, S. *Analytical Science*, **4**, 81 (1988).
19 Gros, N., Gorene, B. *Chromatographia*, **39**, 448 (1994).
20 Holcombe, L.J. and Meserole, F.B. *Water Quality Bulletin*, **6**, 37 (1981).
21 Yan, D. and Schwedt, G. *Fresenius Zeitschrift für Analytische Chemie*, **320**, 121 (1985).
22 Hill, R. and Lieser, K.H. *Fresenius Zeitschrift für Analytische Chemie*, **327**, 165 (1987).
23 Yan, D.R. and Schwedt, G. *Fresenius Zeitschrift für Analytische Chemie*, **320**, 252 (1985).
24 Rowland, A.P. Institute of Terrestrial Ecology, Abbots Ripton, Cumbria, UK. *Chemical Analysis in Environmental Research*. ITE Symposium No. 18. Merlewood Research Station, 19–20 September (1985).
25 Conboy, J.J., Henion, J.A., Martin, M.W. and Zweigenbaum, J.A. *Analytical Chemistry*, **62**, 800 (1990).
26 Cheam, V. and Li, E.X. *Journal of Chromatography*, **450**, 361 (1988).
27 Vasconcelos, M.T. and Gomes, G.A.R. *Journal of Chromatography*, **696**, 227 (1995).
28 Tanaka, T. *Fresenius Zeitschrift für Analytische Chemie*, **320**, 125 (1985).
29 Saitoh, H. and Oikawa, K.T. *Chromatography*, **329**, 247 (1985).
30 Hoshika, Y., Murayama, N. and Muto, G. *Bunseki Kagaku*, **36**, 174 (1987).
31 Cox, J.A., Dabek, E., Zlatorzynski, R.S. and Tanaka, N. *Analyst (London)*, **113**, 1401 (1988).
32 Xiang, D. and Chang, Y. *Fenxi. Ceshi. Tongbao*, **6**, 15 (1987).
33 Seinfeld, J.H. *Atmospheric Chemistry and Physics of Air Pollution*, John Wiley & Sons, New York (1986).
34 Tanabe, K. *Shokubai No Hataraki (Action of Catalyzer)*, Kagakudojin, Kyoto, Japan, Chapter 13, pp. 148–50 (1988).

35 Kadowaki, S. and Kogaal, I. *Taisaku*, **23**, 1167 (1987).
36 Byerley, J.J., Scharer, J.M. and Atkinson, G.F. *Analyst (London)*, **112**, 41 (1987).
37 Merz, W. and Oldeweme, J. *Vom Wasser*, **69**, 95 (1987).
38 Cochrane, R.A. and Miller, D.E. *Journal of Chromatography*, **21**, 392 (1982).
39 William, R.G. *Analytical Chemistry*, **57**, 851 (1985).
40 Smee, B.W., Hall, G.E.M. and Koop, D.T. *Journal of Geochemical Exploration*, **10**, 245 (1978).
41 Van Os, M.J., Slanina, J., De Ligny, C.L., Hammers, W.E. and Agenderbos, J. *Analytical Chemistry*, **54**, 73 (1982).
42 Slanina, J., Lingerak, W.A., Ondelmanm, J.E. *et al.* (eds.) *Ion Chromatographic Analysis and Environmental Pollutants*. Volume 2. Ann Arbor Science Publishers, Michigan (1979).
43 Asshauer, J. and Halazz, I. *Journal of Chromatographic Science*, **12**, 139 (1974).
44 Tong, B. and Shi, S. *Yankuang Ceshi*, **6**, 202 (1987).
45 Marko Varga, G., Csiky, I. and Jonsson, J.A. *Analytical Chemistry*, **56**, 2066 (1984).
46 Bradfield, G. and Cooke, D.T. *Analyst (London)*, **110**, 1409 (1985).
47 Conboy, J.J., Henion, J.D., Martin, M.W. and Zweigenbaum, J.A. *Analytical Chemistry*, **22**, 800 (1990).
48 Rocklin, R.D. and Johnson, E.L. *Analytical Chemistry*, **55**, 4 (1983).
49 Sunden, T., Lingron, M. and Cedergren, A. *Analytical Chemistry*, **53**, 1691 (1981).
50 Pihlar, B. and Kosta, L. *Analytica Chimica Acta*, **114**, 275 (1980).
51 Pihlar, B., Kosta, L. and Hrvstoviski, B. *Talanta*, **26**, 805 (1979).
52 Hansen, L.D., Richter, B.E., Rollins, D.K., Lamb, J.D. and Eatough, J. *Analytical Chemistry*, **51**, 633 (1979).
53 Marko Varga, G., Csiby, I. and Jonsson, J.A. *Analytical Chemistry*, **56**, 2066 (1984).
54 Wilken, R.D. and Kock, H.H. *Fresenius Zeitschrift für Analytische Chemie*, **320**, 477 (1984).
55 Tyree, S.Y. and Bynum, M.A.G. *Limnology and Oceanography*, **29**, 1337 (1984).
56 Okada, T. and Kuwamoto, T. *Analytical Chemistry*, **57**, 258 (1985).
57 Hordijk, C.A. and Cappenberg, T.E. *Journal of Microbiological Methods*, **3**, 205 (1985).
58 Sawicki, E., Mullik, J.D. and Wittgenstein, E. In *Ion Chromatographic Analysis of Environmental Pollutants*. Ann Arbor Science, Ann Arbor, Michigan (1978).
59 Iskandarani, Z. and Pictrzyk, J.D. *Analytical Chemistry*, **54**, 2061 (1982).
60 Maki, S.A. and Danielson, N.P. *Analytical Chemistry*, **63**, 699 (1991).
61 Rigas, P.G. and Piebrzyk, D.J. *Analytical Chemistry*, **60**, 1650 (1988).
62 Saari-Nordhaus, R. and Anderson, J.M. *Analytical Chemistry*, **64**, 2283 (1992).
63 Masengo, E., Gennaro, M.C. and Abrigo, C. *Analytical Chemistry*, **64**, 1885 (1992).
64 Simon, N.S. *Analytical Letters*, **21**, 319 (9188).
65 Cheam, V. and Chan, A.S.Y. *Analyst (London)*, **112**, 993 (1987).
66 Hirijama, N. and Fuwamoto, T. *Analytical Chemistry*, **65**, 141 (1993).
67 Motomizu, S., Sawatani, I., Hironaka, T., Oshima, M. and Toal, K. *Bunseki Kagaku*, **36**, 77 (1987).
68 Bond, A.M., Heritage, I.D., Wallace, A.C. and McCormick, M.J. *Analytical Chemistry*, **54**, 582 (1982).
69 Singh, F.P., Abbas, N.M. and Smesko, S.A. *Journal of Chromatography, A*, **733**, 73 (1996).
70 Sunden, T., Lingrmo, M. and Cedergren, A. *Analytical Chemistry*, **55**, 2 (1983).
71 Dils, J.S. and Smeenk, G.M.M. H_2O, **18**, 7 (1985).

72 Singh, R.P., Pambid, E.R. and Abbas, N.M. *Analytical Chemistry*, **63**, 189 (1991).
73 Kieber, D.J., Vaughan, G.M. and Mopper, K. *Analytical Chemistry*, **60**, 1654 (1988).
74 Montiel, A. and Dupont, J.J. *Tribune de Cebedeau*, **27**, 27 (1974).
75 Tanaka, K. and Fritz, J.S. *Analytical Chemistry*, **59**, 708 (1987).
76 Brandt, G. and Kittrup, A. *Fresenius Zeitschrift für Analytische Chemie*, **320**, 485 (1985).
77 Smee, B.W., Hall, G.E.M. and Koop, D.J. *Journal of Geochemical Exploration*, **10**, 245 (1978).
78 Dits, J.S. and Smeenk, G.N.M. *H_2O*, **18**, 7 (1985).
79 Karison, M. and Frankenberger, W.T. *Soil Science Society of American Journal*, **51**, 72 (1987).
80 Urasa, T. and Ferede, F. *Analytical Chemistry*, **59**, 1563 (1987).
81 Haddad, P.R. and Heckenberg, L. *Journal of Chromatography*, **447**, 415 (1988).
82 Goodwin, L.R., Franscom, D., Urso, A. and Dieken, F.P. *Analytical Chemistry*, **60**, 216 (1988).
83 Hill, C.J. and Lash, R.P. *Analytical Chemistry*, **52**, 24 (1980).
84 Erkelens, C., Billiet, H.A.H., De Galan, L. and De Leer, E.W.B. *Journal of Chromatography*, **404**, 67 (1987).
85 Hoover, T.B. and Jager, G.D. *Analytical Chemistry*, **56**, 221 (1984).
86 Dionex Corporation Technical Note. TN19. January 1987. *Gradient Elution in Ion Chromatography. Anion Exchange with Conductivity Detection* (1987).
87 Pyen, G.S. and Erdmann, D.E. *Analytica Chimica Acta*, **149**, 355 (1983).
88 Wetzel, R.A., Anderson, C.L., Schleicher, H. and Crook, D.G. *Analytical Chemistry*, **51**, 1532 (1979).
89 Oikawa, K. and Saitoh, H. *Chemosphere*, **11**, 933 (1982).
90 Schwabe, R., Darimont, T., Mohlman, T., Pabel, E. and Sonneborn, M. *International Journal of Environmental Analytical Chemistry*, **14**, 169 (1983).
91 Bucholz, A.E., Verplough, C.J. and Smith, J.L. *Journal of Oceanographic Science*, **20**, 499 (1982).
92 Matzuchita, S., Tada, Y., Baba, D. and Hosako, J. *Journal of Chromatography*, **259**, 459 (1983).
93 Wagner, F., Valenta, P. and Nurnberg, W. *Fresenius Zeitschrift für Analytische Chemie*, **320**, 470 (1985).
94 Rowland, A.P. *Analytical Proceedings (London)*, **23**, 308 (1986).
95 Ito, K. and Sunahara, H. *Bunseki Kagaku*, **37**, 292 (1988).
96 Davies, D.M. and Ivey, J.P. *Analytica Chimica Acta*, **194**, 275 (1987).
97 Tao, D. and Zhang, X. *Shanghai Huanjing Kexue*, **6**, 29 (1987).
98 Matzuchita, S., Baba, N. and Ireshiga, T. *Analytical Chemistry*, **56**, 822 (1984).
99 Yamamoto, S., Toda, Y., Baba,D. and Hosak, J.J. *Journal of Chromatography*, **259**, 459 (1983).
100 Xiang, L., Luo, Z., Wang, S. and Yang, C. *Shanghai Huaning Kexue*, **6**, 34 (1987).
101 Ito, K. *Analytical Chemistry*, **59**, 3028 (1997).
102 Peschet, J.L. and Tinet, C. *Techniques Sciences Methods*, **81**, 351 (1986).
103 Xiao-Hua Yang, Lee C. and Schantor, M.I. *Analytical Chemistry*, **65**, 857 (1993).
104 Morrow, C.M. and Minear, R.A. *Water Research*, **18**, 1165 (1984).
105 Bachmann, K. and Mutsusca, P. *Fresenius Zeitschrift für Analytische Chemie*, **315**, 243 (1983).
106 O'Dell, J., Pfaff, J.P., Gates, M.E. and McKee, G.D. Environmental Protection Agency EPA 600/4–84–017 March 1984. *The Determination of Inorganic Anions in Water by Ion Chromatography Method 300.0* (1984).
107 Nonomura, M. *Analytical Chemistry*, **59**, 2073 (1987).

108 Dietrich, A.M., Ledder, J.D., Gallagher, D.L., Grobeel, M.N. and Hocha, R.C. *Analytical Chemistry*, **64**, 496 (1992).
109 Lui, Y., Rocklin, R.D. and Doyle, M.J. *Analytical Chemistry*, **62**, 766 (1990).
110 Halcombe, D.J. and Meserole, V.T. *Water Quality Bulletin*, **6**, 37 (1981).
111 Cao, D. and Luo, Y. *Sepu*, **5**, 189 (1987).
112 Green, L.W. and Woods, J.R. *Analytical Chemistry*, **53**, 2187 (1981).
113 Cameron, A., Pohlandt, C. Report MINTEK–M22D. Order No PB88–139324, 11pp. Available NTIS from Gov. Dep. Announce, Index (US) 1988, 88 Abstract No. 819,860 (1987).
114 Nadkarni, R.A. and Brewer, M. *American Laboratory*, **19**, 50 (1987).
115 Pohlandt-Watson, C. Report MINTEK M283, 6 pp. (1986).
116 Mosko, J. *Analytical Chemistry*, **56**, 629 (1984).
117 McCormick, M.J. and Dixon, L.M. *Journal of Chromatography*, **322**, 478 (1985).
118 Petrie, L.M., Jakely, M.E., Brandwig, R.L. and Kroenings, J.G. *Analytical Chemistry*, **65**, 952 (1993).
119 Lown, J.A., Koile, R. a nd Johnson, D.C. *Analytica Chimica Acta*, **116**, 33 (1980).
120 Fishman, M.J. and Skoustad, M.W. *Analytical Chemistry*, **35**, 146 (1963).
121 Reddy, K.J., Zhang, Z., Blalock, M. and Vance, G.F. *Environmental Science and Technology*, **29**, 1754 (1995).
122 Roberts, K.M., Gjerde, D.T. and Fritz, J.S. *Analytical Chemistry*, **53**, 1691 (1981).
123 Gjerde, D.T., Schmuckler, G. and Fritz, S.S. *Journal of Chromatography*, **35**, 187 (1980).
124 Gjerde, D.T. Ph.D. Dissertation, Iowa State University (1980).
125 Brandt, G., Matuschek, G. and Keltrup, A. *Fresenius Zeitschrift für Analytische Chemie*, **321**, 653 (1985).
126 Tretter, H., Paul, G., Blum, T. and Schrank, H. *Fresenius Zeitschrift für Analytische Chemie*, **361**, 650 (1985).
127 Stevens, T.S. and Turkleson, V.T. *Analytical Chemistry*, **49**, 1176 (1977).
128 Umile, C. and Huber, J.F.K. *Journal of Chromatography*, **640**, 27 (1993).
129 Shatyk, W. *Journal of Chromatography*, **640**, 309 (1993).
130 Shatyk, W. *Journal of Chromatography*, **640**, 316 (1993).
131 Kumac, T., Sugawara, K. and Machida, K. *Nippon Eiseigaku Zasshi*, **42**, 1037 (1988).
132 Schmitz, F. *GIT–Suppl.*, **3**, 37 (1988).
133 Kanal, Y. *Chishitsu Chosasho Geppo*, **38**, 587 (1987).
134 Barak, P. and Chen, Y. *Journal of Soil Science of America*, **51**, 257 (1987).
135 Mullins, F.G.P. *Analyst (London)*, **112**, 665 (1987).
136 Iegas, C.A.A. *Analyst (London)*, **118**, 1039 (1993).
137 Frenzel, W., Schepers, D., Schulz, G. *Analytica Chimica Acta*, **277**, 103 (1993).
138 Jones, U.K. and Tarter, J.G. *International Laboratory*, **36**, September (1985).
139 Jones, U.K. and Tarter, J.G. *Journal of Chromatography*, **312**, 456 (1984).
140 Berglund, I., Dasgupta, P.T., Lopez, J.L. and Nara, O. *Analytical Chemistry*, **65**, 1192 (1983).
141 Environmental Protection Agency Cincinnati, U.S. Report EDA–600/4–84–017 March 1984. The determination of inorganic ions in water by ion chromatography method 300.0 (1984).
142 Peschet, J.L. and Tinet, C. *Techniques Sciences Methods*, **81**, 351 (1986).

Chapter 3

Electrostatic ion chromatography

3.1 Introduction

The desirability of an ion chromatography technique using only water as the mobile phase is summarised by Small [1]. There are a number of ion chromatographic techniques that, though not widely used, deserve mention because of their simplicity and their potential for further development. All use water or another polar solvent as the sole component of the mobile phase. This not only eliminates the need for electrolytes and precise eluent make-up, but it avoids many of the detection problems that ionic eluents can impose. Detection can be expected to be very sensitive since the background is essentially deionised water.

In an earlier paper dealing with the results of ion separation achieved using water as the mobile phase, published in 1977 [2], the stationary phase used was a crown ether bonded one. Later, Small et al. [3] reported that inorganic ions could also be separated using water as the mobile phase when a very weak ion exchange resin stationary phase was used. The separations achieved using both of these methods, however, were poor compared with the results of the separation of the same ions using the conventional ion chromatographs (using an ion exchange stationary phase with a mobile phase containing the replacing ions).

Hu et al. [4] developed another approach for separating ions, also using water as the mobile phase but employing a zwitterionic stationary phase. When a small amount of aqueous solution containing an analyte (cations and anions) is passed through a zwitterionic stationary phase, neither the analyte cations nor the analyte anions can get close to the opposite charge fixed on the stationary phase, because another charge on the same molecule, fixed on the stationary phase, repels the analyte ions simultaneously. The analyte cations and anions are forced into a new state of simultaneous electrostatic attraction and repulsion interaction in the column. This was termed an 'ion-pairing-like form'. This method of separation was termed electrostatic ion chromatography [4,5]. The previous studies [4,5] demonstrated that the separation of inorganic ions (with the exception of cations having the same charge) using electrostatic ion chromatography is comparable to separations of the same ions obtained using conventional ion chromatographs. In previous studies [4,5], samples having high concentrations (mmol/L) of analyte ions were used in order to understand the mechanism. In those studies, samples with low concentrations of ions were not investigated.

In a series [4,6–13] employing stationary phases coated with zwitterionic surfactants (ie those containing both positively and negatively charged functional groups but carrying no formal net charge), it has been demonstrated that inorganic anions can be separated using pure water as eluent, with unique separation selectivity. This method has been termed electrostatic ion chromatography. The separation has been attributed to a simultaneous electrostatic attraction and repulsion mechanism occurring at both the positive and the negative charges on the stationary phase. A drawback of electrostatic ion chromatography is that when the sample contains multiple anions and cations, each analyte anion may be eluted as more than one peak, with each peak being a specific combination of the anion with one of the cations of the sample. Recently, Hu *et al.* [12] showed that addition of a small quantity of a suitable electrolyte to the eluent causes analyte anions to be eluted only as a single peak, irrespective of the number and type of cations in the sample.

3.2 Determination of anions

Hu *et al.* [13] investigated the separation mechanism in more detail and have applied the method to the determination of nitrate, bromide, and iodide in seawater.

The determination of nitrate, bromide, and iodide in seawater is of importance to oceanographic research [14–16]. Ion exchange chromatography is generally inapplicable to this analysis for several reasons. First, the large amount of matrix ions (chloride, sulphate) saturate the active sites of the stationary phase and thereby impede the separation of the target analytes. Second, the high ionic strength of the sample causes self-elution of the sample band during injection, leading to peak broadening and loss of separation efficiency. Third, the levels of the target analytes are often very low so that detection becomes a major problem, especially when the eluted peaks are poorly defined in shape.

To overcome these difficulties, Ito and Sunahara [17] suggested the use of the matrix ions as the eluent ions, for example, the use of relatively high concentrations of sodium chloride solution.

In the most recent method described by Hu *et al.* [13] for the direct determination of ultraviolet-absorbing inorganic anions in saline matrixes an octadecylsilica column modified with a zwitterionic surfactant (3-(*NN*–dimethylmyristylammonio)propanesulfonate) is used as the stationary phase, and an electrolytic solution is used as the eluent. Under these conditions, the matrix species (such as chloride and sulphate) are only retained weakly and show little or no interference. It is proposed that a binary electrical double layer is established by retention of the eluent cations on the negatively charged (sulphonate) functional groups of the zwitterionic surfactant (forming a cation–binary electrical double layer) and by retention of eluent anions on the positively charged (quaternary ammonium) functional groups of the zwitterionic surfactant (forming an anion–binary electrical double layer). Sample anions are able to distribute into the cation–binary electrical double layer and to form pairs with the binary electrical double layer cations, while at the same time experiencing repulsion from the anion–binary electrical double layer. Anions are therefore eluted in order of increased propensity to form ion pairs. The method has been applied to the determination of bromide, nitrate, and iodide in artificial seawater, giving detection limits of 0.75µg L $^{-1}$ for bromide, 0.52µg L $^{-1}$ for nitrate, and 0.8µg L $^{-1}$ for iodide using ultraviolet absorbance detection at 210nm and relative standard deviations of <1.2%. Real seawater samples have also been analysed successfully.

References

1 Small, H. *Ion Chromatography*. Plenum Press, New York p. 132 (1989).
2 Blasius, E., Janzen, K.P., Adrian, W. *et al. Zeitschrift für Analytische Chemie*, **284**, 337 (1977).
3 Small, H., Soderquist, M.E. and Pischke, J.W. U.S. Patent No. 4,732,686 (1988).
4 Hu, W., Takeuchi, T. and Haraguchi, H. *Analytical Chemistry*, **65**, 2204 (1993).
5 Hu, W., Tao, H. and Haraguchi, H. *Analytical Chemistry*, **66**, 2514 (1994).
6 Hu, W., Tao, H. and Haraguchi, H. *Analytical Chemistry*, **66**, 765 (1994).
7 Hu, W., Miyazaki, A., Tao, H. *et al. Analytical Chemistry*, **67**, 3713 (1995).
8 Hu, W. and Haraguchi, H. *Journal of Chromatography*, **723**, 251 (1996).
9 Hu, W., Hasebe, K., Reynolds, D.M. *et al. Journal of Liquid Chromatography and Related Technology*, **20**, 1903 (1997).
10 Hu, W. and Haddad, P.R. *Trends in Analytical Chemistry*, **17**, 69 (1998).
11 Hu, W., Hasebe, K., Reynolds, D.M. and Haraguchi, H. *Analytica Chimica Acta*, **358**, 143 (1997).
12 Hu, W. and Haddad, P.R. *Analytical Communications*, **35**, 317 (1998).
13 Hu, W., Haddad, P.R., Hasebe, K. *et al. Analytical Chemistry*, **71**, 1617 (1999).
14 Sigman, D.M., Altabet, M.A., Michener, R. *et al. Marine Chemistry*, **57**, 227 (1997).
15 Debaar, H.J.W., Vanlecuwe, M.A., Scharek, R. *et al. Deep Sea Research*, **44**, 229ff (1997).
16 Hutchins, D.A. and Bruland, K.W. *Nature (London)*, **393**, 561 (1998).
17 Ito, K. and Sunahara, H. *Busneki Kagaku*, **37**, 292 (1988).

Chapter 4

High performance liquid chromatography

This technique has been applied predominantly to organic and organo-metallic compounds with a small number of applications to cations and anions.

4.1 Organics

4.1.1 Non saline waters

4.1.1.1 Amides

Brown and Rhead [1] determined down to 0.2µg L^{-1} amides in non saline waters by high performance liquid chromatography. The procedure consists of bromination, extraction of the α,β-dibromopropionamide produced with ethyl acetate and quantification using high performance liquid chromatography with ultraviolet detection. Samples tested included river, sea and estuarine waters, sewage and china clay works effluents, and potable waters. The levels of inorganic ultraviolet absorbing impurities found in water samples did not interfere in this procedure. The experimental yields of α,β-dibromopropionamide encountered gave a mean of 70.13 ± 8.52% (95% confidence level) for acrylamide-spiked waters, over the concentration range 0.2–8.0µg L^{-1} of acrylamide monomer.

4.1.1.2 Chlorophenols

The best method for determining pentachlorophenol is conversion into the methyl ether followed by analysis using gas chromatography with an electron-capture detector, or gas chromatography coupled with mass spectrometry [2]. Both of these methods require an extensive amount of pre-treatment and highly-trained personnel for the operation of the equipment.

Ervin and McGinnis [3] attempted to overcome this problem by developing a high performance liquid chromatographic method for determining in water low concentrations of pentachlorophenol and

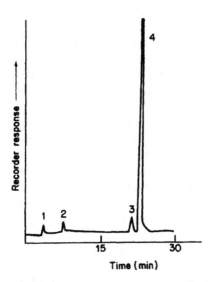

Fig. 4.1 Separation of technical pentachlorophenol using cyclohexane–acetic acid (98:2) as the eluting solvent. Peaks: 1, mixture of dioxins and other polychlorinated products; 2, mixture of chlorinated phenols including trichlorophenol; 3, tetrachlorophenol; 4, pentachlorophenol
Source: Reproduced by permission from Elsevier Science, UK [3]

chlorinated impurities that occur in technical grade material such as 2,3,4,6-tetrachlorophenol, mono-, di-, and trichlorophenols, octa-, hepta- and hexachlorodibenzo-*p*-dioxins, and a variety of other polychlorinated aromatic compounds.

The method involves chloroform extraction of acidified waste water samples and rotary evaporation without heat. After redissolving in chloroform the samples were analysed directly by high performance liquid chromatography on a microparticulate silica gel column. A number of solvent combinations are possible and 98:2 cyclohexane–acetic acid is preferred. The minimum detectable concentration is 1ppm (without sample concentration) and the coefficient of variation is 1–2%. The type of separation achieved with a microparticulate silica gel column is shown in Fig. 4.1. The first peak as determined by gas chromatographic–mass spectrometric analysis, consisted of a complex mixture of polychlorinated compounds, including octa-, hepta- and hexachlorodibenzo-*p*-dioxins as well as a mixture of products including 2,4,6–trichlorophenol. The third peak was mainly 2,3,4,6–tetrachlorophenol and the fourth peak was pentachlorophenol.

De Ruiter *et al.* [4] observed that photochemical decomposition by ultraviolet irradiation of dansyl derivatives of chlorinated phenolic compounds in methanol–water mixtures led to the formation of highly fluorescent dansyl–OH and dansyl–OH$_3$ species. The optimal irradiation time was 5.5s. This reaction was utilised in a post-column photochemical reactor in the high performance liquid chromatography determination of highly chlorinated phenols in river water. The method calibration curve (for dansylated pentachlorophenol) was linear over three orders of magnitude.

Silgouer et al. [5] have described an automated liquid chromato-graphic–mass spectrometric method for the determination of methyl, nitro and chlorophenols in river water samples. This method for determination of 19 priority phenols in water samples has been developed using on-line liquid–solid extraction followed by liquid chromatography–mass spectrometry with atmospheric pressure chemical ionisation and ion spray interfaces in the negative ionisation mode. Sixteen phenols were determined by liquid chromatography–atmospheric pressure chemical ionisation–mass spectrometry with high sensitivity. Three compounds were not detected by atmospheric pressure chemical ionisation–mass spectrometry: phenol, 4–methylphenol, and 2,4–di-methylphenol could only be determined by ion spray–mass spectrometry using a porous graphite carbon column with 100% methanol as eluent. (M–H)⁻ ion was the base peak using either liquid chromatography–atmospheric pressure chemical ionisation–mass spectrometry or liquid chromatography–ion spray–mass spectrometry. Limits of detection ranging from 5 to 0.1ng L^{-1} and 0.1–0.25ng L^{-1} were found when 50–100mL of river water was processed in full-scan and time-scheduled selective ion monitoring modes, respectively.

See also Table 4.1.

4.1.1.3 Chlorophyll pigments

The first application of high performance liquid chromatography to plant pigments was by Evans et al. [6], who separated phaeophytins a and b on Corasil(II) with a mobile phase consisting of a 1:5 (v/v) mixture of ethyl acetate and light petroleum. Eskins et al. [7] have employed two 0.62m columns of C$_{18}$–Porasil B for preparative separation of plant pigments by means of a programmed stepwise elution with methanol–water–ether. However, the method is of little value for routine application because of the time required and also because the chlorophyll degradation products, other than phaeophytin, are not separated.

Garside and Riley [8] applied high performance liquid chromatography to the determination of a range of algal pigments.

Tests carried out with the high performance liquid chromatographic technique endorsed the claim that it causes negligible degradation of both the chlorophylls and the xanthophylls. A more efficient separation of

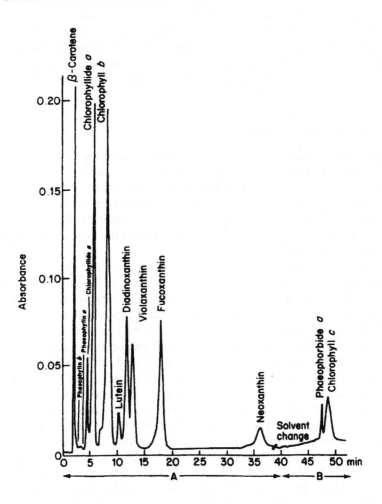

Fig. 4.2 Chromatogram of extract of pigments from a mixed culture of *Phaeondactylum tricornutum* and *Dunaliella tertiolecta*
Source: Reproduced by permission from Elsevier Science, UK [8]

plant pigments could be achieved on a silica stationary phase than on a C_{18} reversed phase medium. A 30cm column packed with Partisil 10 gave an efficient separation of the individual carotenoids, chlorophylls *a* and *b*, and many of the degradation products of the latter pair. The solvent consists of light petroleum (b.p. 60–80°C), acetone, dimethyl sulphoxide and diethylamine in the ratio 75:23.25:1.5:0.25 by volume. Unfortunately, this solvent is not sufficiently polar to elute phaeophorbide and chlorophyll *c*. With samples containing these pigments, it is necessary to carry out an additional, stepwise, elution with a more polar solvent. Further tests show that excellent resolution of these compounds could be achieved (Fig. 4.2) by means of a mixture containing light petroleum (b.p. 60–80°C), acetone, methanol and dimethyl sulphoxide in the ratio 30:40:27:3 by volume, respectively.

4.1.1.4 Carboxylic acids

Application of high performance liquid chromatography to the resolution of complex mixtures of fatty acids in water [9,10] has provided an alternative to the high temperature separation obtained by gas chromatography. Both techniques have similar limits of detection, but lack the ability to analyse environmental samples directly. Analysis requires that the fatty acids be separated from the organic and inorganic matrices, followed by concentration.

Typically, these processes can be accomplished simultaneously by the appropriate choice of methods. Initial isolation of the fatty acids is based on the relative solubility of the material of interest in an organic phase compared to the aqueous phase. Secondary separation is determined by the functional group content and affinity for a solid support.

Hullett and Eisenreich [11] used high performance liquid chromatography for the determination of free and bound fatty acids in river water samples. The technique involves sequential liquid–liquid extraction of the water sample by 0.1m hydrochloric acid, benzene–methanol (7:3) and hexane–ether (1:1). The resultant extract was concentrated and the fatty acids were separated as a class on Florasil using an ether–methanol 1:1 and 1:3 elution. Final determination of individual fatty acids was accomplished by forming the chromatophoric phenacyl ester and separating by high performance liquid chromatography. Bound fatty acids were released by base saponification or acid hydrolysis of a water sample from which the fatty acids had been removed by solvent extraction.

Kieber *et al.* [12] determined formate in non saline waters by a coupled enzymatic/high performance liquid chromatographic technique. The precision is approximately ±5% relative standard deviation. Intercalibration with an anion chromatographic technique showed an agreement of 98%. Down to 0.5μmol L^{-1} absolute of formate could be determined. Kieber *et al.* [12] found 0.2–0.8mol L^{-1} formate in seawater and 0.4–10μmol in rainwater.

The procedure involves precolumn oxidation of formate with formate dehydrogenase which is accompanied by a corresponding reduction of β-nicotinamideadenine dinucleotide (β(NAD)$^+$) to reduced β-nicotinamide dinucleotide (ie β NADH). The latter is quantified by high performance liquid chromatography. See also Table 4.1.

4.1.1.5 Non ionic surfactants

Cassidy and Niro [13] have applied high-speed liquid chromatography combined with infrared spectroscopy to the analysis of polyoxyethylene surfactants and their decomposition products in industrial process waters. Molecular sieve chromatography combined with infrared spectrometry give a selective method for the analysis of trace concentrations of these surfactants. These workers found that liquid–solid chromatography and reversed phase chromatography are useful for the characterisation and analysis of free fatty acids.

Okada [14] used indirect conductiometric detection of poly(oxyethylene) surfactants following chromatographic separation.

See also Table 4.1.

4.1.1.6 Pesticides and herbicides

Wilimott and Dolphin [15] applied their high performance liquid chromatography–electron capture detector approach to the determination of chlorinated insecticides in surface waters. Because the concentration of pesticides in most surface waters is less than $1\mu g\ L^{-1}$, some form of extraction and concentration from large volumes of water is necessary before analysis is possible. These workers applied a conventionally coated chromatographic support as a reverse liquid–liquid partitioning filter to the extraction of pesticides. They used uncoated polyether polyurethane foam; 0.5g of the flexible foam were inserted into a 10mm id quartz tube and cleaned by washing with consecutive 100mL aliquots of acetone, *n*-hexane, ethanol and distilled water.

Crathorne [16] reported a method for the determination of Pyrazon (5-amino-4-chloro-2-phenyl-3-pyridazone) in non saline waters. Pyrazon was isolated from water samples (500mL) by rotary evaporation to dryness *in vacuo*, extraction of the solid residue with methanol (2 × 25mL) and further evaporation of the methanol extract (to approx. 2mL). Final concentration (to 0.5mL) was achieved by removal under a stream of nitrogen.

The equipment used consisted of two Model 6000A solvent delivery systems and a Model 660 gradient former (Waters Associates) and a Model CE212 variable wavelength ultraviolet monitor (Cecil Instruments) operated at 270nm. Syringe injections were made through a stop–flow septumless injection port. The column (15cm × 7mm id) was packed in an upward manner with Spherisorb–ODS by a slurry procedure using acetone as slurry medium. A linear gradient was established from two solvent mixtures consisting of (a) 10% methanol 1% acetic acid in water and (b) 80% methanol in 0.1% acetic acid in water. The initial concentration was 35% (b) in (a) and the final concentration was 100% (b) with the gradient terminated after 20min. The flow rate was maintained at 2.0mL min $^{-1}$ throughout the analysis.

Miles and Moye [17] resolved several classes of pesticides by high performance liquid chromatography and detected them by fluorescence after post-column ultraviolet photolysis with or without prior conversion to o-phthalaldehyde-2-mercaptoethanol derivatives. In the presence of o-phthalaldehyde-2-mercaptoethanol fluorescence labelling reagent, most carbamates, carbamoyloximes, carbamothioic acids and substituted ureas gave sensitive responses, whereas dithiocarbamates, phenylamides and phenylcarbamates gave varied responses. In the absence of o-phthalaldehyde-2-mercaptoethanol reagent, strong fluorescence was observed following photolysis of several substituted aromatic pesticides. Detection limits for Aldicarb sulphoxide, Aldicarb, Propoxur, Thiram and Neburon, representing several classes of pesticides were 2.5, 2.3, 3.3, 3.8 and 2.0g L $^{-1}$ respectively. The relative fluorescence (compared to equimolar methylamine) of some 50 pesticides from 10 different classes, in three solvents, and in the presence of o-phthalaldehyde-2-mercaptoethanol were recorded. Fluorescence responses were significantly affected by the choice of solvent.

Carpenter *et al.* [18] have described a method for the assay of two metabolites of the herbicides dimethyl tetrachloroterephthalate, monomethyl tetrachoroterephthalate and tetrachloroterephthalic acid via high performance liquid chromatography with ion pairing. Samples are analysed via direct injection, without preparation, and analyte detection is accomplished with an ultraviolet photodiode array detector. The metabolises are extracted from positive samples with a petroleum ether–diethyl ether mixture, derivatised with *N,O-bis*-(trimethylsilyl) trifluoroacetamide, and confirmed by gas chromatography–mass spectrometry. The high performance liquid chromatographic analysis of spiked drinking water samples yielded a recovery range of 92–106% with a mean recovery of 101% for tetrachloroterephthalate acid and a recovery range of 92–101% with a mean recovery of 96% for monomethyl tetrachloroterephthalate. The minimum detection limits for these two metabolites were 2.4 and 2.7µg L $^{-1}$, respectively. In addition the gas chromatography–mass spectrometry analysis of spiked reagent water yielded mean recoveries of 91% for monomethyl tetrachlorophthalate and 86% for tetrachlorophthalic acid. Twenty drinking water samples were split and analysed by the high performance liquid chromatographic and the gas chromatographic–mass spectrometric methods and by US EPA method 515.1. Comparable results were obtained. The high performance liquid chromatographic method, which is amenable to automation, typically allows for the analysis of up to 40 samples overnight.

See also Table 4.1.

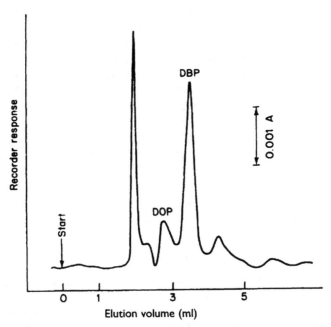

Fig. 4.3 Chromatogram of an extract of river water measured with system A. Sample volume injected, 100μL; detector, UV at 224nm; attenuation, 0.01 AUFS. DPB, di-*n*-butyl phthalate; DOP, di-2-ethylhexylphthalate
Source: Reproduced by permission from Elsevier Science, UK [20]

4.1.1.7 Phenols

Ratanathanawongs and Crouch [19] have described an on-line post-column reaction based on air–segmented continuous flow for the determination of phenol in natural waters by high performance liquid chromatography. The reaction used was the coupling of diazotised sulphanilic acid with the phenol to form high coloured azo dyes. The detection limit for phenol was 17μg L^{-1} which represents a 16-fold improvement over determination of phenol with ultraviolet detection.
See also Table 4.1.

4.1.1.8 Phthalate esters

Mori [20] has identified and determined very low levels of phthalate esters in river water using reversed phase high performance liquid chromatography using an ultraviolet detector. Phthalates were extracted with *n*-hexane and the uncleaned or concentrated extracts were injected into three chromatographic systems, these being cross-linked porous beads (Shodex HP–225, Showa Penko Co.), porous polymer beads and polystyrene GPC gel. The eluants were respectively n-hexane (system A) and methanol (system B), and chloroform (system C).

An example of a chromatogram of the extract of a river water is shown in Fig. 4.3. The presence of n-dibutylphthalate and di-2-ethylhexyl-phthalate was observed. The concentrations of phthlates in the extract were 450ppb of n-dibutylphthalate and 100ppb of di-2-ethylhexyl-phthalate and their concentrations in river water were 45 and 100ppb, respectively. The first peak in Fig. 4.3 is contaminant(s) in n-hexane.

Schouten et al. [21,22] used high performance liquid chromatography to determine very low levels of di-2-ethylhexylphthalic and di-n-butyl-phthalate in Dutch river waters and compared results with those obtained by gas liquid chromatography. Good agreement was obtained between the two techniques, although high performance liquid chromatography was shown to be the less time-consuming technique.

4.1.1.9 Polyaromatic hydrocarbons

Various workers [23–32] have studied the application of high performance liquid chromatography to the determination of PAHs in water samples. Hagenmaier et al. [31] used a reversed phase high-pressure liquid chromatography procedure for the determination of trace amounts of polycyclic aromatic hydrocarbons in water. Different column packing materials were tested, in conjunction with non polar stationary phases of various polarities, for separation efficiency, detection limits and long-term stability. The method was suitable for concentrations as low as 2ng L^{-1} in a IL sample. Compounds studied included fluoranthene, benzofluoranthene isomers, benzopyrene and perylene derivatives.

Schönmann and Kern [33] have used on-line enrichment for microgram per litre analysis of PAHs in water by high performance liquid chromatography. The trace enrichment method is based on the affinity of non polar pollutants for reversed phase chromatography supports.

When aqueous samples are passed through a reversed phase column these compounds and any other non polar organic compounds present in the sample are immobilised at the head of the column. When detectable quantities of pollutants have been accumulated on the column, they can be analysed by introducing a mobile phase of the desired eluent strength.

They used a Varian Model 5500 liquid chromatograph with three reservoir capability for on-line enrichment methods: one reservoir can be used to quantitatively transfer large sample volumes into the column and a binary gradient can then be introduced from the remaining reservoirs.

Workers at Perkin–Elmer [34] have studied the high-speed separation of PAHs using C$_{18}$-bonded phase packings (5μm particles) using both isocratic and gradient elution. The analysis of several PAH standards was performed using the 5μm bonded phase column with gradient elution

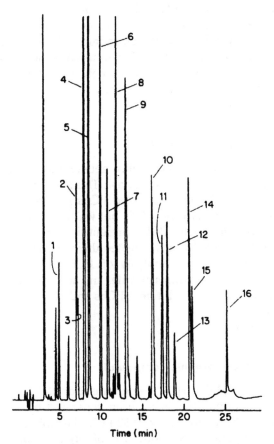

Fig. 4.4 Analysis of polyaromatic hydrocarbons. Column: two 100 × 4.6mm id in series, C₁₈ bonded–phase packing; 3μm particles. Mobile phase: acetonitrile–water, linear gradient from 65 to 90% in 20min, at 1.8mL min⁻¹; inlet pressure: 4500psig (31.0mPa) initial, 2700psig (18.6MPa) final; ambient temperature; UV detector at 254nm. Peaks: I, naphthalene; 2, fluorene; 3, acenaphthalene; 4, phenanthrene; 5, anthracene; 6, fluoranthene; 7, pyrene; 8, benzo(b)fluoranthene; 9, chrysene + benzo(a)anthracene; 10, benzo(e)pyrene + perylene; 11, benzo(a)pyrylene; 15, indeno(1,2,3–cd)pyrene; 16, coronene

Source: Reproduced by permission from Perkin–Elmer, USA [34]

from 60 to 100% acetonitrile in 5min at a flow rate of 4mL min⁻¹. Although not completely resolving all of the components, this method offers a rapid analysis which is adequate for many of these compounds. These high-speed columns can also be used in a very high resolution mode. At the expense of analysis time, very high resolution can be obtained by connecting two columns in series, eg two 3μm columns coupled result in an efficiency of about 38,000 plates under isocratic conditions (Fig 4.4). These workers utilised these coupled columns, in this case, with gradient elution, to analyse the same PAH standards. The analysis time is increased to 2min, but the extremely high resolution achieved is quite apparent. This very high resolution is extremely useful in analysing complex samples containing many components.

See also Table 4.1.

4.1.1.10 Polychlorobiphenyls

Brinkman et al. [35,36] used a silica gel column which elutes the higher chlorinated PCBs in the normal phase. This system produced a reasonable separation of the lower chlorinated PCBs present predominantly in the commercial mixture Arochlor 1221 but was less efficient in separating the more highly-chlorinated PCBs present in Arochlors 1254 and 2160). Kaminsky and Fasco [37] investigated the potential reversed phase liquid chromatography to the analysis of PCB mixture in environmental samples. They used mixtures of water and acetonitrile as the mobile phase to achieve analysis of 49 different PCBs and of samples of Arochlor 1221, 1016, 1254 and 1260.

See also Table 4.1.

4.1.1.11 Volatile organic compounds

Bruckner et al. [38] used a flame ionisation detector to detect volatile organic compounds that have been separated by water-only reversed phase liquid chromatography (WRPLC). The mobile phase is 100% water at room temperature without use of organic solvent modifiers. An interface between the liquid chromatograph and detector is presented, whereby a helium stream samples the vapour of volatile components front individual drops of the liquid chromatographic eluent, and the vapour-enriched gas stream is sent to the flame ionisation detector. The design of the drop headspace cell is simple because the water-only nature of the liquid chromatographic separation obviates the need to do any organic solvent removal prior to as phase detection. Despite the absence of organic modifier, hydrophobic compounds can be separated in a reasonable time due to the low phase volume ratio of the WRPLC columns. The drop headspace interface easily handles liquid chromatographic column flows of 1mL min $^{-1}$.

4.1.1.12 Ethylene diamine tetraacetate

Nowack et al. [39] have described a procedure for the determination of dissolved EDTA in rivers and waste water treatment plants. The

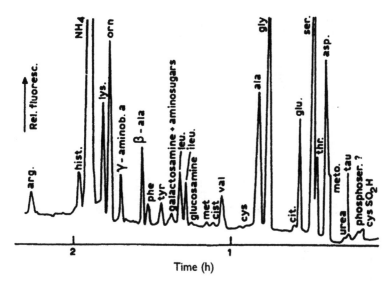

Fig. 4.5 Chromatogram of a saline extract (20mL) sample for amino acids collected at 6m in the Kiel Fjord. The concentrations of the individual acids were quantified as follows (nmol L^{-1}): meto, 11; asp, 34.4; thr, 23.2; ser, 88; glu, 36; gly, 100; ala, 56; val, 16; ileu, 9.6; leu, 12; galactosamine and aminosugars, 4; tyr, 6.8; ph, 7.2; β-ala, 20.8; γ-aminoba, 14.4; orn, 44; lys, 12; hist, 7.2; arg. 8.6; cysSO$_2$H, 4; cit, trace; tan, cys, trace; glucosamine, trace; met, trace; urea, trace; phosphoser, trace. OH–lys, trace. The total concentration of amino acid in the sample lies around 51µg L^{-1} assuming a mean molecular weight of 100
Source: Reproduced by permission from Elsevier Science, UK [40]

procedure is based on high performance liquid chromatography. A distinction can be made between Fe(III)–EDTA and all other species. Nickel–EDTA can be detected semi-quantitatively. The fraction of EDTA adsorbed on to suspended particles or to sediment can be determined after desorption with phosphate. After complexation with Fe(III)–EDTA, the EDTA is detected by reversed phase ion pair liquid chromatography as the Fe(III)–EDTA complex at a wavelength of 258mn. The behaviour of a number of metal–EDTA complexes during analysis was checked. Fe(III)–EDTA was found to be the main species (60–70%) Ni–EDTA was less than 10% in most samples.

4.1.1.13 Other applications

Other applications of high performance liquid chromatography to the determination of organic compounds in non saline waters are reviewed in Table 4.1.

Table 4.1 High performance liquid chromatography of organic compounds in non saline waters

Organic	Stationary phase	Elution solvent	Comments	Detection	LD	Ref.
Aromatic hydrocarbons	–	–	–	UV	–	[43]
Polyaromatic hydrocarbons	Glass columns	–	Application of high efficiency glass columns	–	–	[44]
Fluoranthrene, benzopyrene	Reversed phase C column	MeOH:H₂O 1:1 & pure methanol	–	–	0.018mg L⁻¹	[45]
Polyaromatic hydrocarbons	–	–	–	Time residual fluorometry	1.8fg	[46]
Polyaromatic hydrocarbons	–	–	–	UV at 254nm	–	[47]
Polyaromatic hydrocarbons	Silica gel	–	–	Fluorescence	–	[48]
Polyaromatic hydrocarbons	Chemically bonded phthalimidopropyl trichlorsilane	–	–	–	–	[49]
Polyaromatic hydrocarbons	–	–	–	Spectrofluoro-metric	0.1μg L⁻¹	[50]
Polyaromatic hydrocarbons	–	–	–	Spectrofluor-escence	–	[51]
Benzo(c) pyrene, benzo(b) fluor-anthrene, benzo(b) flouranthrene, fluoranthrene benzo(ghi), perylene indeno (1,23–cd) pyrene	Sephadex Ltd	–	–	–	–	[52]

Table 4.1 continued

Organic	Stationary phase	Elution solvent	Comments	Detection	LD	Ref.
Chlorinated polyaromatic hydrocarbons	–	–	–	–	$\mu g\ L^{-1}$	[53]
Phenols	–	–	–	Electrochemical	–	[54]
Phenols	–	–	–	Photothermal interferometry	70ppb	[55]
Aminophenols and chloroanilines	–	–	On-line method, preconcentration on porous graphite	–	–	[56]
EDA priority phenolic pollutants	–	–	–	Electrochemical	–	[57]
EPA priority phenolic pollutants	30mn NH_4O $OCCH_3$, pH5.0: acetonitrile: NeOH (50:34:10) or MeCN:MEOH:H_2O: CH_3COOH 33.3:33.3.:33.3;0.1	–	Isocratic HPLC	Dual UV	–	[58,59]
Phenols	–	–	Reverse phase HPLC	Electrochemical	mgL^{-1}	[60]
Phenol, hydro-quinone, pyrocatechol, hydroxyhydroquinone, resorcinol, phloro-glucinol	–	–	–	–	–	[61]
Chlorinated phenols	(1) silica, (2) amino propylsilica, (3) octa-decyl or silica	–	–	–	–	[62]

Table 4.1 continued

Organic	Stationary phase	Elution solvent	Comments	Detection	LD	Ref.
Nitrophenols	–	–	Reversed phase ion pair	UV at 280nm	1mg L^{-1}	[63]
Phenols	–	2-fluorensulphonylchloride	–	UV	0.05ng	[64]
Phenols	–	–	Reverse phase HPLC	UV	–	[65,66]
Phenols	–	–	On-line enrichment	–	mg L^{-1}	[67]
Phenols cresols	–	–	Reverse phase microcolumn HPLC	–	0.5µg L^{-1}	[68]
Phenols	–	6.4 MeOH.H$_2$O	Conversion to nitrophenazo derivative, range 10^{-6}–10^{-4} mole L^{-1}	UV at 345nm	1.6ng	[69]
Phenols	–	–	Preconcentrated by solid phase extraction	Spectrofluorometric	Trace	[70]
Chlorophenols	–	–	–	On-line electron capture	–	[71]
Acrylic acid	–	–	–	UV	–	[72]
Carboxylic acids	–	–	Derivitivisation–HPLC linear calibration 0.5–10nm	–	–	[73]
Aliphatic acids	–	–	Detected as 2.4 dinitrophenyl hydrazones	Spectrometric	1µg L^{-1}	[74]
Formaldehyde	–	–	RSD 2.5%	–	–	[75]
Formaldehyde, acetaldehyde	–	–	–	HCHO	0.2µg L^{-1}	[76]
Propionaldahyde, butyraldehyde, benzaldehyde	Low capacity anion exchange resin	Acetonitrile	Separated as 2,4-dinitrophenyl hydrazones	–	0.06µg L^{-1}	[77]
Alkylbenzene sulphonate	–	–	–	–	–	[78–81]

Table 4.1 continued

Organic	Stationary phase	Elution solvent	Comments	Detection	LD	Ref.
Alkylbenzene sulphonate	YWG–CH	Aqueous MeOH–NaClO₄	–	–	–	[82]
Alkylbenzene sulphonates	–	–	–	Fluorometric	1µg L⁻¹	[83]
Linear alkyl benzene sulphonates	Li Chrosorb	0.05M NaClO₄ acetonitrile (2:3)	–	Fluorometric excite, 225nm emission 275nm	10ng 0.1mg L⁻¹	[84]
Sodium alkyl benzene subphonates	–	–	–	–	–	[85]
Cationic surfactants	–	–	–	–	–	[86]
Non ionic surfactants	–	–	–	–	–	[78,79, 87]
Aromatic amines	–	–	–	Amperometric	–	[88]
Amino acids	–	–	–	Spectrofluoro metric	–	[89]
Benzidine	–	–	–	–	–	[90]
C₂–C₄ aliphatic diamines	Sep–Pak C₁₈ cartridge	Aqueous medium 1–4% K₂HPO₄	Derivitivisation with acetyl acetone. Reverse phase HPLC Precision 3.2%	–	0.14– 1.8µg L⁻¹ 50pm	[91]
Amino acids	–	–	–	–	–	[92]
p-cresidine	–	–	–	–	–	[93]
Nitro compounds	–	–	–	–	–	[94]
Haloforms	–	–	–	–	1µg L⁻¹	[95]
Tetrachloroethylene	–	–	Direct aqueous injection	–	0.06µM	[96]
Chlorinated degreasing solvents	–	–	–	–	–	[97]

118

Table 4.1 continued

Organic	Stationary phase	Elution solvent	Comments	Detection	LD	Ref.
Mono, di, tri and tetra PCBs	–	–	Determination of retention indices	–	–	[98]
PCBs	–	–	Analysis of hexane extract of water sample	–	–	[37,99,100]
PCBs	Porous graphitic carbon	–	Fractionation of PCBs	–	–	[101]
Polychlorodibenzofurans	–	–	Reverse phase HPLC	–	–	[102]
Polychlorodibenzo-p-dioxins	–	–	–	–	–	[103]
Polychlorinated dibenzofurans	–	–	–	UV, fluorescence and sensitised room temperature fluorescence	–	[104]
Insecticides and herbicides						
Phenoxy acetic acid herbicides	–	–	–	UV at 280 and 230nm	–	[105]
Phenoxyacetic acid, MCPA, dichloropro-pmecoprop 2,4-D, 2,4-dichlorophen-oxyacetic acid	–	–	–	–	5µg L^{-1}	[106]
2-(2,4,5-trichloro-phenoxy)propionic acid	–	–	–	–	–	[107]

Table 4.1 continued

Organic	Stationary phase	Elution solvent	Comments	Detection	LD	Ref.
2,4–dichlorophenoxyacetic acid, 2,4,5–trichlorophenoxyacetic acid, 2-(2,4,5–trichlorophenoxy)propionic acid	Baker 16SPF	–	–	UV at 280nm	10µg L^{-1}	[108]
Phenoxy acetic acid herbicides	–	–	Liquid–liquid extraction of of water with methylene dichloride	UV at 230 and 280nm	–	[105]
Chlorophenoxy herbicides	–	–	–	UV at 280nm	0.05µg L^{-1}	[109]
N-methyl carbamate	–	–	–	–	ng L^{-1}	[110]
Carbamate insecticides	–	–	Range 2–7ppm	Electrochemical	<2mg L^{-1}	[111, 112]
Phenyl urea herbicides	–	–	–	Mass spectrometry	–	[113]
Phenyl urea herbicides	Platinum loaded filter	–	C$_{18}$ pre-column to platinum loaded filter	–	–	[114]
Phenyl urea herbicides	–	–	–	–	–	[115–123]
30 pesticides, including atrazine, simazine, alachlor, molinate	–	–	On-line solid phase extraction	–	0.01–0.5ppb	[124]
Simazine, atrazine herbicides	–	–	Concentration on graphitised carbon black	UV at 220nm	ppt range	[125]

Table 4.1 continued

Organic	Stationary phase	Elution solvent	Comments	Detection	LD	Ref.
Aldicarb residues, aldicarb sulphoxide, aldicarb sulphone	–	–	–	Mass spectrometry	–	[126]
Aldicarb, aldicarb sulphoxide, aldicarb sulphone	–	–	–	–	–	[127]
Organophosphorus type, triethylphosphate, trimethyl-phosphate, dimethyl phosphonothioate	–	–	–	Molecular emission cavity detection	–	[128]
Organochlorine type	–	–	–	Radioactivity flow detection	$ng\ L^{-1}$	[129]
Aldicarb and degradation products	Cyanopyrol bonded	Acetonitrile: H_2O	Isocratic reverse phase HPLC	UV at 254	$<1\mu g\ L^{-1}$	[130]
Organophosphorus type	–	–	On-line method	–	–	[131]
Miscellaneous organic compounds including pesticides and herbicides						
Alkylphenols, chloro-phenols, dihydroxy-phenols, trihydroxy-benzene, benzenes, biphenols	–	–	–	–	–	[132]
Pesticides	–	–	–	–	–	[133]
5 ethoxydin and metabolites	–	–	Reverse phase HPLC	–	–	[134]

Table 4.1 continued

Organic	Stationary phase	Elution solvent	Comments	Detection	LD	Ref.
Bromacil, diuron	–	–	Methylene dichloride extract of water	UV at 254nm	–	[135]
27 polar pesticides	–	–	–	–	1ppb	[136]
Fluazifopbutyl, fluazifop	–	–	–	UV	10ng	[137]
166 pesticides and related compounds	–	–	Study of conditions; type of column, packing column, dimensions and mobile phase composition	UV spectroscopic	–	[138]
Pyrazon (5-amino-4-chloro-2-phenyl-3-pyridazone)	–	–	–	–	<1.0µg L^{-1}	[16]
Propoxur, carbofuran, propham, ceptan, chloropropham, barban, butyrate	–	–	Automated HPLC, solid phase extraction	–	–	[139]
Oxamyl methomyl Phoxan, 2,4,5 tri-chlorophenoxy acidic acid 2,4DB, NCPB	–	–	Removal of pesticides from water on graphitised carbon block	–	0.003–0.07µg L^{-1}	[140]
52 pesticides	–	–	–	Mass spectrometry	–	[141]
Paraquat diquat	–	–	With or without post column reaction with NaHSO$_3$	UV parquat 257nm diquat 310nm	7,5ppb	[142]
Rotenone	Acetonitrile:H$_2$O (70:30)	–	–	UV at 200nm	15µg L^{-1}	[143]

Table 4.1 continued

Organic	Stationary phase	Elution solvent	Comments	Detection	LD	Ref.
Diquat	-	-		-	-	[144]
Benomyl carbendazin	-	-	Benomyl converted to 3-butyl 2,4-dioxo-5-triazine (1,2 alpha) bendimidazole	-	0.03μg L^{-1}	[145]
Cobalamin	-	-	Benzyl alcohol extract of water	-	1pg L^{-1}	[146]
Dyes, azo, diazo and anthra-quinone Cl disperse yellow 42	-	-		Thermospray	0.01μg L^{-1}	[147]
Chlorophyll a and b	Partisil PXS 1025	Aqu 95% MeOH	-	-	-	[148]
Chlorophylls and degradation products (chlorophylls a, b, c, chlorophyllide a, phaeophorebide a, phaeophytins a,b)	-	-	-	Spectrofluoro-metric	-	[149] [150]
Methylated and unmethylated humic substances	-	!		-	-	[152]
Acrylamide	-	-		-	0.1μg L^{-1}	[153, 154]
Adenosine triphosphate	-	-	Derivitivised with 0.3m chloroacetaldehyde at pH4–5. Reverse phase column	Spectrofluoro-scence	-	[154]

Table 4.1 continued

Organic	Stationary phase	Elution solvent	Comments	Detection	LD	Ref.
Carbonyl compounds, halogenated anilines, benzidines, nitro-compounds, phenols, polychlorinated compounds, alkyl-benzene, sulphonates, alkyl sulphonates, alkylbenzene sulphonates, carbamates	–	–	–	–	–	[155]
Benzothiazoles, aromatic amines, chlorinated phenols	–	–	–	–	–	[156]
Phenol, acetophenone nitrobenzene, toluene	–	–	–	–	–	[157]
Misc. organics	–	–	–	UV, spectroscopic flame ionisation, electrochemical conductivity, mass spectrometry	–	[158–161]

Source: Own files

4.1.2 Seawater

4.1.2.1 Amino acids

The classical work of Dawson and Pritchard [40] on the determination of α-amino acids uses a standard amino acid analyser modified to incorporate a fluorometric detection system. In this method the samples are desalinated on cation exchange resins and concentrated prior to analysis. The output of the fluorometer is fed through a potential divider and low-pass filter to a compensation recorder.

Dawson and Pritchard [40] point out that all procedures used for concentrating organic components from estuary or sea water, however mild and uncontaminating, are open to criticism, simply because of the ignorance as to the nature of these components in the sample. It is for instance feasible that during the process of desalting on ion exchange resins under weakly acidic conditions, labile peptide linkages are disrupted or metal chelates dissociated and thereby larger quantities of 'free' components are released and analysed.

An example of a chromatogram obtained from a saline sample and the mole percentage of each amino acid in the sample is depicted in Fig. 4.5.

Mopper [41] has discussed developments in the reversed phase high performance liquid chromatographic determination of amino acids in estuarine and sea water. He describes the development of a simple, highly sensitive procedure based on the conversion of dissolved free amino acids to highly fluorescent, moderately hydrophobic isoindoles by a derivatisation reaction with excess o-phthaldehyde and a thiol, directly in sea water. Reacted samples were injected without further treatment into a reversed phase high performance liquid chromatography column, followed by a gradient elution. The eluted amino acid derivatives were detected fluorometrically. Detection limit for most amino acids was 0.1–0.2nmol per 500µL injection. Problems of inadequacies with the method itself, sample handling and whether chromatographically determined concentrations might be considered as biologically available concentrations in sea water are discussed.

4.1.2.2 Aromatic amines

Varney and Preston [42] discussed the measurement of trace aromatic amines in estuary and sea water using high performance liquid chromatography. Aniline, methyl aniline, 1–naphthylamine and diphenylamine at trace levels were determined using this technique and electrochemical detection. Two electrochemical detectors (a thin layer, dual glassy carbon electrode cell and a dual porous electrode system) were compared. The electrochemical behaviour of the compounds was investigated using hydrodynamic and cyclic voltammetry. Detection limits of 15 and 1.5nM were achieved using coulometric and amperometric cells respectively when using an in-line preconcentration step.

4.1.2.3 Other applications

Amides, polyaromatic hydrocarbons, trichlorfon, fluorines and organo-copper complexes have been determined in seawater (Table 4.2).

4.1.3 Aqueous precipitation

4.1.3.1 Phenols

Cruezwa *et al.* [168] reported a method in which traces of phenols and cresols in rain waters were determined by continuous liquid–liquid extraction and normal phase high performance liquid chromatography with ultraviolet fluorescence detection.

4.1.3.2 Formate

A coupled enzymatic high performance liquid chromatographic method has been described for the determination of 4–10µm of formate in rain [169].

4.1.4 Ground and surface waters

4.1.4.1 Pesticides

Ruepert and Ploeger [170] have described a procedure which allows the simultaneous determination of 22 pesticides (eg metamitron, simazine, diuron chlorpham) which is applicable to potable, ground and surface waters. The method relied on high performance liquid chromatography with a diode array detector monitoring adsorption at 220, 230 and 245nm. Reported detection limits are 5µg L^{-1} for all the pesticides studied.

Miles and Moye [171] have shown that several classes of nitrogen containing pesticides responded to a high performance liquid chromatography post-column reaction detector that employed ultraviolet photolysis with optional reaction with *o*-phthalicdicarboxaldehyde-2-mercaptoethanol followed by fluorescence detection. It was applied to the determination of N-methylcarbamates, carbamoyl oximes, carbamethoic acids, dithiocarbamates and phenyl ureas, phenyl amides and phenyl carbamates in groundwater. See also Table 4.3.

4.1.4.2 Other applications

Other types of organic compounds that have been determined in ground and surface waters include polyaromatic hydrocarbons, amino acids,

Table 4.2 High performance liquid chromatography of organics in seawater

Organic	Stationary phase	Eluent	Comments	Detection	LD	Ref.
Amides	–	–	–	–	–	[162]
PCBs	–	–	–	–	–	[163]
Polyaromatic hydrocarbons	–	–	–	UV	–	[164,165]
Trichlorophon	–	MeCN–NaH$_2$PO$_4$ 0.1M, pH5.5 (20:80)	–	UV at 205nm	–	[167]
Organocopper complexes and organic matter	–	–	Estuarine water	–	–	[167]

Source: Own files

nitrobenzene, parathion ethyl, parathion methyl, Aldicarb and phenyl urea herbicides, see Table 4.3.

4.1.5 Potable waters

4.1.5.1 Polyaromatic hydrocarbons

Sorrell and Reding [179] present an extension of this technique, for analysing 1–3ng L $^{-1}$ amounts of 15 polynuclear aromatic hydrocarbons in environmental water samples using high performance liquid chromatography, preceded by a clean-up using alumina, with ultraviolet monitoring and fluorescence emission–excitation spectra for identification. The analytical procedure is outlined in Fig. 4.6. Reproducible retention times (which varied less than 2% over a nine-month period) eliminated the need for column re-equilibrium. Although some coeluting PAHs (eg anthracene and phenanthrene) had been placed into different fractions, it was clear that no single ultraviolet wavelength was capable of resolving all of the PAHs within a fraction. The sensitivities of the ultraviolet detectors, defined as a signal-to-noise ration of 2, ranged from 0.25 to 1ng L $^{-1}$.

In earlier work using this procedure, Sorrell [180] had shown the recoveries of six PAHs from distilled water to range from 61 to 91% and to average 78 ± 8%. Subsequent measurements of the recoveries of 11 PAHs from seven raw or finished water samples taken over a nine-month period ranged from 53 to 116% and averaged 86 ± 12%.

Workers at Perkin–Elmer have studied the high speed separation of PAHs using a C_{18}-bonded phase packing (5µm particles) using both isocratic and gradient elution [181].

Morizane [182] has applied high performance liquid chromatography with fluorescence detection to the determination of benzo(a) pyrene in methylene dichloride extracts of potable waters in amounts down to 0.28pg.

An early reference to the use of high performance liquid chromatography in the analysis of PAHs is the work of Jentoft and Gouw [183] and Vaughan et al. [184]. These workers used ultraviolet, visible fluorescence detection and were able to detect 0.4ng of anthracene and 15ng of acenaphthalene.

Sorrell et al. [185] have listed 15 PAHs commonly found in water and have described a high performance liquid chromatographic method for determining 13 trace compounds. Limits of detection are as low as 1ng L $^{-1}$, well below the collective limit of 200ng L $^{-1}$ for six PAHs recommended by the World Health Organization.

Hunt et al. [186] have shown that all six representative PAHs tested in the World Health Organization standards for potable water (ie fluoranthene, 3, 4-benzfluoranthene, benzo(b)fluoranthene), 11, 12-benzo-

128

Table 4.3 High performance liquid chromatography of organics in ground and surface waters

Organic	Stationary phase	Eluent	Comments	Detection	LD	Ref.
Polyaromatic hydrocarbons	–	–	Ground waters	–	–	[172]
Amino acids	–	–	Surface waters	Spectrofluorescence	–	[173]
Nitrobenzene	–	–	Surface waters	–	–	[174]
Parathion ethyl, parathion ethyl	–	–	Surface waters	–	–	[175]
Aldicarb	–	–	Ground waters	–	UV at 203nm	[176]
Phenylurea herbicides	–	–	Surface waters	–	–	[177]
Phenyl urea herbicides	–	–	Surface waters	Electrochemical	–	[178]

Source: Own files

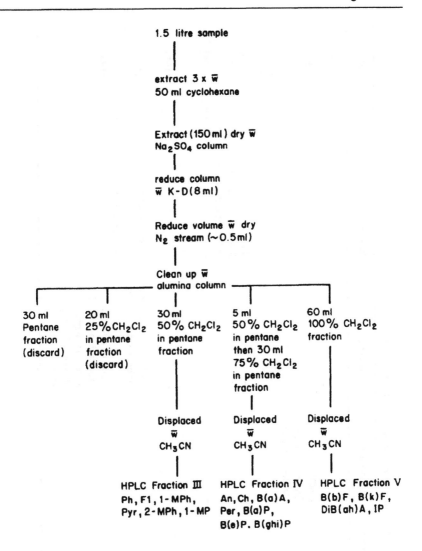

1.5 litre sample

extract 3 x w̄
50 ml cyclohexane

Extract (150 ml) dry w̄
Na₂SO₄ column

reduce column
w̄ K-D (8 ml)

Reduce volume w̄ dry
N₂ stream (~0.5 ml)

Clean up w̄
alumina column

| 30 ml Pentane fraction (discard) | 20 ml 25% CH₂Cl₂ in pentane fraction (discard) | 30 ml 50% CH₂Cl₂ in pentane fraction | 5 ml 50% CH₂Cl₂ in pentane then 30 ml 75% CH₂Cl₂ in pentane fraction | 60 ml 100% CH₂Cl₂ fraction |

Displaced w̄ CH₃CN

Displaced w̄ CH₃CN

Displaced w̄ CH₃CN

HPLC Fraction III
Ph, Fl, 1-MPh,
Pyr, 2-MPh, 1-MP

HPLC Fraction IV
An, Ch, B(a)A,
Per, B(a)P,
B(e)P. B(ghi)P

HPLC Fraction V
B(b)F, B(k)F,
DiB(ah)A, IP

Fig. 4.6 Analytical scheme for PAHs
Source: Reproduced by permission from the American Water Works Association,
Colorado, USA [180]

fluoranthene (benzo(k)fluoranthene), 3, 4-benzpyrene (benz(a) pyrene), 1,
12-benzperylene (benzo(ghi)perylene) and indeno(1,2,3-cd) pyrene), can
be separated successfully by high performance liquid chromatography,
using as a packing material phthalimidopropyltrichlorosilane bonded to
microparticulate silica gel with a non polar mobile phase of toluene–

Table 4.4 High performance liquid chromatography of organics in potable waters

Organic	Stationary phase	Elution solution	Comments	Detection	LD	Ref.
Pentachlorophenol	–	–	–	–	–	[187]
Phthalate esters	–	–	–	–	–	[188]
Diglycidyl ether	C_{18} bonded silica gel	MeOH Cl_2CH_2(1,1) then MeOH:Cl_2CH_2 (1:3)	–	–	–	[189]
4-amino–phenol	–	–	–	–	–	[190]
Glycol dinitrate	–	–	–	–	–	[191]
Acrylamide	–	–	–	UV	–	[163,192]
Thiafluorin	C_{18} Sep–Pak	–	–	Spectrofluorescence and electrochemical	µg L^{-1}	[193]
Desethylatrazine, simazine, atrazine, terbethylazine	Tenax TA	–	–	–	15–32ppb	[194]
Non volatile chlorinated hydrocarbons	–	–	–	–	–	[195]
Volatile hydrocarbons	–	–	Analysis of petroleum ether extract of water	Electron capture	–	[196]
Phenols	–	–	–	Electrochemical	0.03µg	[197]
Ozonisation products	–	–	Study of effect of ozonisation rate of water on degradation of impurities in potable water	–	–	[198]

Source: Own files

hexane. They used a non polar phase of toluene–hexane (1:10) and achieved good separations.

4.1.5.2 Other applications

Other organic compounds that have been determined in potable waters include pentachlorophenol, phthalic acid esters, diglycidyl ether, phenols, 4-aminophenol, ethylene glycol dinitrate, acrylamide, thiafluron, azine herbicides and ozonisation products (Table 4.4).

4.1.6 Waste waters

4.1.6.1 Nitroaromatic compounds

Godjehann et al. [199] have applied on-line high performance liquid chromatography coupled to proton NMR to the determination of nitro-aromatic compounds in ammunition plant ground waters.

This was the first time these two techniques were coupled. Using a continuous flow mode at very low sample flow rates of about 0.017mL m^{-1} and large injection volumes of 400μL the confirmation of the identity of many different types of nitroaromatic compounds was achieved at the microgram per litre level after solid-phase extraction of the ground water sample at a flow rate of 0.006μL m^{-1}. Less than 29nmol (5μg) of 1:3 dinitrobenzene was determined when injected on to a 75mm × 4mm reversed phase C$_{18}$ column.

4.1.6.2 Polyaromatic hydrocarbons

Das and Thomas [200] used fluorescence detection in high performance liquid chromatography to determine nine PAHs in occupational health samples including process waters. The nine compounds studied were benzo(a)anthracene, benzo(k)fluoranthene, benzo(a)pyrene/fluoranthene, chrysene, benzo(k)fluorene, perylene, benzo(e)pyrene, deibenz(ah)–anthracene and benz(ghi)perylene.

The method involves the use of a deuterium light source and excitation wavelengths below 300nm. Limits of detection in the 0.5–1pg range were obtained for several PAHs of environmental or toxicological significance. Limits of detection close to sub-picogram levels were obtained (eg benzo(a)anthracene, 0.6pg; benzo(k)fluoranthene, 0.4pg; benzo(a)pyrene, 1.1pg). Precision studies gave a relative standard deviation from 0.32 to 2.66% (eg benzo(a)anthracene, 0.33% benzo(k) fluoranthene, 0.70%; benzo(a)pyrene, 0.50%). The system allows the use of dilute solutions, thus eliminating the usual clean-up procedures associated with the trace analysis.

High performance liquid chromatography in reversed phase mode was performed isocratically with 82% acetonitrile in water. Detection wavelengths of 254 and 280nm were used and the following three sets of excitation and emission wavelengths conditions were used in fluorescence detection with excitation prefilter Corning 7.54; (a) λ_{ex} 280nm, λ_{ex} > 389nm; (b) λ_{ex} 250nm, λ_{ex} > 370nm; (c) λ_{ex} 240nm, λ_{ex} > 470nm. Between 0.5 and 3.0pg of PAHs can be detected by this procedure.

4.1.6.3 Chlorinated polyaromatic hydrocarbons

Oyler et al. [50] have determined aqueous chlorination reaction products of polyaromatic hydrocarbons in amounts down to micrograms per litre using reversed phase high performance liquid chromatography–gas chromatography. The method involves filtration through a glass microfibre filter and concentration on a high performance liquid chromatography column. The polyaromatic hydrocarbon material is then eluted using an acetonitrile–water gradient elution technique. The fractions are injected separately on to a gas chromatographic column equipped with a photoionisation detector.

4.1.6.4 Ozonisation and chlorination products

Jolly et al. [201] characterised non volatile organics produced in the ozonisation and high chlorination of waste water. The treated effluents were analysed by high performance liquid chromatography and gas–liquid chromatography as a means of detecting both non volatile and volatile organic constituents. While chlorinated primary effluent was repeatedly found to be mutagenic towards a particular strain, neither chlorinated, ozonated nor ultraviolet–irradiated secondary effluents exhibited mutagenic activity even when tested at 10–20 fold concentrations.

4.1.6.5 Other applications

High performance liquid chromatography has been applied to the following multiorganic mixtures in waste waters: guanidines, substituted guanidines and s-triazines [202], pyridazinones [203], maleic hydrazine, ethoxyquinathibendazole [204], ametryne [205], alkylphenols, alkyl-, mono- and diethoxylates [206], chlorophenols [207], chloroacetanilides and chloranilines [208] and phenols, aromatic amines, heterocyclic bases [209], EDTA [210], organophosphorus insecticides [211], dehydroacetic acid [212], nitrocompounds [213,214], nitrosulphonic acids [215], penta, tetra and dichlorophenol [216] and miscellaneous organic compounds [217].

4.1.7 Trade effluents

4.1.7.1 Mercaptobenzthiazole

Cox [218] determined mercaptobenzthiazole in trade effluent samples by a high performance liquid chromatographic procedure. An ultraviolet detector operating at 325nm was used to monitor the column effluent. The column was constructed from stainless steel tubing (15cm × 4.6mm) and was packed with Microkosorb Sil60 (5µm silica gel) at 3500psi (24.1MPa) from a slurry in 2,2,4-trimethylpentane. The packing material was retained in the column by stainless steel wire mesh of nominal pore size 8µm (Sankey Wire Weaving, Warrington, UK) inserted into a drilled-out Swagelok coupling. A similar disc of wire mesh was pressed on the top of the column packing and was retained by a plug of silanised glass wool. Ethanol-2,2,4-trimethypentane (1:9) was used as the mobile phase with flow rate of 1mL min $^{-1}$. All solvents used were of spectroscopic quality (Fisons, Loughborough, UK).

The aqueous sample (2mL) was acidified with two drops of concentrated hydrochloric acid. This mixture was shaken with chloroform (2mL) for 1min using a flask shaker. Aliquots (2µL) of the chloroform layer were used for the chromatographic analysis.

4.1.7.2 Polyaromatic hydrocarbons

Kasiske et al. [219] have described a high performance liquid chromatographic method for six polynuclear aromatic hydrocarbons in trade effluents whose concentration in potable water is regulated by European Community standards, viz. fluoranthene, benzo(e)acetphen-anthracene benzo(k)fluoranthene, benzo(def)chrysene, indeno(1,2,3,c,d) pyrene and benzo(ghi)perylene.

In this method a 1L sample of water is extracted three times with 30mL of cyclohexane. The combined organic phase is concentrated in a vacuum rotary evaporator to a volume of about 0.5mL and filtered through alumina, activity II. The polycyclic aromatic hydrocarbons adsorbed to the alumina are eluted with 3mL of cyclohexane–benzene (1:1 v/v) and evaporated to dryness.

The residue is taken up in 200µL of methanol and 20µL of the latter are injected into the chromatograph. Separation is achieved by elution through a 250 × 4mm id column packed with Nucleosil reversed phase C_{18} (particle size 5µm).

4.1.7.3 Miscellaneous pollutants

Jones et al. [220] have reviewed a procedure involved in the sample preparation and analysis of industrial waste waters with the object of

Table 4.5 High performance liquid chromatography of organics in trade effluents

Organic	Stationary phase	Eluent	Comments	Detection	LD	Ref.
Acrylamide	–	–	–	–	–	[163,221]
Phenols	–	–	–	–	–	[222–228]
Polyoxyethylene surfactant	–	–	Molecular sieve chromatography	Infrared spectroscopy	–	[13]
Zectron (4-dimethylamine-3,5 xylyl) methyl carbamate and xylenol degradation product	–	–	–	–	–	[229]
Misc organic compounds	–	–	–	–	–	[230]

Source: Own files

providing a rational, systematic approach to pollution monitoring and control programmes. A simple sequence of analytical procedures is described as a basic approach to the detection of pollutants, followed by a more complex and detailed scheme for samples selected for further examination on the basis of the initial survey. Details of the separation achieved and operating conditions for high performance liquid chromatography and spectroscopic techniques are given followed by the use of gas–liquid chromatographic separation in conjunction with other techniques for identification and quantitation of individual compounds.

4.1.7.4 Other applications

Other applications of high performance liquid chromatography to the analysis of trade effluents are listed in Table 4.5.

4.1.8 Sewage effluents

4.1.8.1 Non ionic surfactants

Boyd-Boland and Pawliszyn [231] have used solid-phase micro-extraction coupled with high performance liquid chromatography to determine alkylphenol ethoxylate (Triton X–100) surface active agent in sewage effluent samples. These workers used normal phase gradient elution and detection by ultraviolet absorbance at 220µm to determine alkylphenol ethoxylates (Triton X–100) surface active agent in sewage effluent samples. Carbowax–template resin or Carbowax–divinylbenzene coatings on the column allowed successful analysis of the ethoxylates with a linear range of 100–0.1mg L^{-1}. Limits of detection were in the low microgram per litre range.

4.1.8.2 Cobalamins

Beck and Brink [232] have described a sensitive method for the routine assay of cobalamins in activated sewage sludge. The method involves extraction with benzyl alcohol, removal of interfering substances using a combination of gel filtration and chromatography on alumina, concentration of the extract by lyophilisation, and direct determination of total cobalamin by high performance liquid chromatography, in comparison with cobalamin standards.

4.1.8.3 Other applications

High performance liquid chromatography has also been applied to the determination of other types of organic compounds in sewage including

Table 4.6 High performance liquid chromatography of organics in sewage effluents

Organic	Stationary phase	Eluent	Comments	Detection	LD	Ref.
n-alkanes, linear alkyl-benzene sulphonates, polynuclear aromatic compounds 4-nonyl phenol	–	–	Combination of steam distillation with HPLC or gas chromatography	–	–	[236]
Aminocarb	–	–	–	–	–	[237]
Sodium N-methyl-dithiocarbamate and methyl isothio-cyanate fungicides	–	–	Miscellar mobile phase	UV	–	[238]
Polyaromatic hydro-carbons	–	–	–	–	$2\,\mu g\ L^{-1}$	[239,240]
Alkyl phenols and alkyl phenol mono and di-ethoxylates	Zorbax NH_2	–	–	Fluorometric	$0.2\,\mu g$	[241]

Source: Own files

aklylbenzene sulphonates [233], acrylamide [234] and cationic surfactants [235]. See also Table 4.6.

4.1.9 Ultrapure waters

Because the greater proportion of the organic material in water samples is non volatile and therefore not amenable to gas–liquid chromatographic analysis, the Water Research Centre (UK) [242] developed a routine high performance liquid chromatographic system that is capable of separating unknown and complex extracts with the best possible resolution of individual non volatile components. A system has been developed employing liquid–solid adsorption chromatography with gradient elution by the application of a series of solvents of rising polarity. A series of solvent reservoirs is connected via a programmable multiple port rotary valve to a solvent mixing chamber containing primary solvent. A high pressure reciprocating pump draws solvent from this system at the same time, causing a solvent gradient to be generated by pulling a secondary solvent from one of the reservoirs into the mixing chamber. Solvent combinations are changed at appropriate intervals by switching the rotary valve position. Sample introduction is effected by loop or septum injection. The column is a high efficiency 250×4.2mm prepacked column containing microparticulate silica. The detectors incorporated into the system are a combined, variable wavelength ultraviolet and fluorescence photometer and a transport ionisation detector.

There remains some doubt concerning the optimum number of solvents to be employed in the multiple solvent, gradient elution programme. Scott and Kucera [243] recommend a 12 solvent series, while Snyder [244] proposed a six solvent series to cover the entire polarity range adequately.

Reust and Meyer [245] have shown that high performance liquid chromatography on reversed phase columns with ultraviolet detection is a useful tool for the determination of organic impurities present in ultrapure water. A sufficiently sensitive analysis of different types of ultrapure water requires sample volumes of several hundred millilitres combined with a very low detection wavelength. In this work sample volumes of 480mL were analysed at a detection wavelength of 210nm. Using these experimental conditions it was possible to detect and eliminate several contamination sources.

4.2 Organometallic compounds

4.2.1 Non saline waters

4.2.1.1 Organomercury compounds

Krull et al. [246] have described a procedure for the determination of

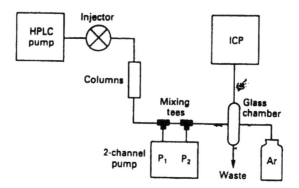

Fig. 4.7 Schematic diagram of the HPLC–cold vapour ICP instrumentation
Source: Reproduced by permission from the Royal Society of Chemistry [246]

inorganic and organomercury compounds using high performance liquid chromatography with an inductively coupled plasma emission spectrometric detector with cold vapour generation. In this method post–column cold vapour generation was used to obtain improved detection limits. The replacement of the conventional polypropylene spray chamber of the inductively coupled plasma by an all glass chamber is described. A comparison of band broadening indicates that the glass chamber is useful when a severe memory effect is observed with the polypropylene spray chamber. Detection limits ranged from 32 to 62µg L^{-1} of mercury, based on a signal to noise ratio of 2:1. This represents a three to four order of magnitude enhancement over detection limits obtained without cold vapour generation. The approach is linear over three orders of magnitude. A blind, spiked distilled water study illustrates the reproducibility and accuracy of the method.

The chromatographic system used in this method consisted of two Laboratory Data Control (LDC) (Riviere Beach, FL, USA) Constametric III pumps with a gradient controller and a Rheodyne Model 7125 injection valve (Rheodyne Corp., Cotati, CA, USA) fitted with a 200µL loop. The separation was performed on two Waters Resolve columns (Waters Assoc., Milford, MA, USA), 5µmC$_{18}$ stationary phase, 15 cm × 3.9mm id placed in series with a mobile phase consisting of 0.06M ammonium acetate and 0.005% V/V 2–mercaptoethanol with a gradient from 15 to 75% of acetonitrile. A flow–rate of 1.0ml min^{-1} was used for all analyses. The post-column reaction system has been described by Bushee *et al.* [247] and is shown schematically in Fig. 4.7. An aqueous solution of 0.5% m/V

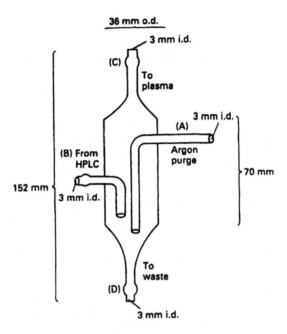

Fig. 4.8 Schematic diagram of glass sample introduction device
Source: Reproduced by permission from the Royal Society of Chemistry, London [246]

sodium tetrahydroborate(III) in 0.25M sodium hydroxide solution and a 1.2M solution of hydrochloric acid served as the two reagents. The reagent flow rates were 0.5ml min⁻¹.

An Instrumentation Laboratory Model 200 plasma (Allied Analytical Systems, Waltham, MA, USA), modified for autotune operation, was used to monitor the high performance liquid chromatography effluent at a wavelength of 253.7nm. The polypropylene nebuliser–spray chamber of the ICP showed a significant memory effect for mercury [248]. An all-glass chamber (Fig. 4.8) was developed, which replaced the polypropylene nebuliser–spray chamber completely. A short length of 0.15cm Teflon tubing was inserted into each arm. Inlet A of the chamber was used to introduce an argon purge gas at a flow rate of 0.4L min⁻¹. For optimum performance, the end of the Teflon tubing was turned upwards to direct the gas towards the plasma torch. The tubing was held in place by the walls of the chamber. The inlet B of the chamber served to introduce the high performance liquid chromatography effluent. The Teflon tube in this arm extended about 1cm below the glass tube, to prevent the gaseous effluent from travelling back up the glass arm. Just prior to the glass chamber, the effluent had passed through the post-column reactor and was in the form of a gaseous mixture. The mercury compounds, now in their cold vapour form were swept up into the plasma with the purge gas, by way of outlet C. The chamber was connected to the plasma by means of a short length of Tygon tubing. The aqueous mobile phase flowed to waste through the bottom of the chamber, outlet D. A drain trap was positioned directly below the glass chamber. The ICP peak–height response was monitored on a Honeywell Corp. (Minneapolis, MN) strip-chart recorder, and peak–area data were collected on a Radio Shack TRS–80 Model II computer (Tandy Corp., Fort Worth, TX, USA).

A series of blind spiked, distilled water samples, with known concentrations of the mercury compounds, were studied by high performance liquid chromatography–cold vapour ICP spectrometry (Table 4.7). These results indicate a general agreement between the levels of mercury species spiked and the values determined. A linear regression analysis of these results and the actual spiked values gave correlation coefficients of 0.9440, 0.9971 and 0.9869 for mercury(II) chloride, methyl-mercury chloride and ethylmercury chloride, respectively. The correlation for mercury(II) chloride was not as good as that for the two organo-mercury species. This was due to the lack of baseline resolution between the mercury(II) chloride and the methylmercury chloride, and could be improved by further optimisation of the chromatographic conditions.

4.2.1.2 Organotin compounds

For either ion exchange resolution of aqueous cations, $R_nSn^{(4-n)+}$aq [249] or their separation as ion pairs, $[R_nSn^{(4-n)+}X^-4 - n]^0$, on reverse bonded-phase columns [250], the method is restricted to 'free' tin analytes. Unlike the vigorous hydride derivatisation used in the gas chromatography–flame photometric detector method, common high performance liquid chromato-graphy solvent combinations or their ionic addends usually will not provide sufficient co-ordination strength to labilise organotin ions strongly bound to solids in environmental samples. Moreover, the high perform-ance liquid chromatography separations require that injected samples be free of particulates that may clog the column or pumping system.

On the other hand, high performance liquid chromatography if coupled with a sensitive element-specific detection system such as atomic absorption spectrometry offers a valuable tool for organotin speciation in complex fluids (especially for high organic loadings) not readily amenable to gas phase derivatisation methods. Fig. 4.9 shows a schematic of the basic high performance liquid chromatography set-up described in Brinkmann [251] coupled to a graphite furnace atomic absorption spectrometry in a manner giving automatic periodic (typically 45s intervals) sampling of the

Table 4.7 HPLC–cold vapour ICP of blind spiked distilled water

Mixture No.	Hg species	Hg spiked µg L^{-1}	Hg measured µg L^{-1}	Recovery, %*
1	HgCl$_2$	0	ND†	–‡
	CH$_3$HgCl	290	310 ± 20	107 ± 8
	CH$_3$CH$_2$HgCl	179	190 ± 20	110 ± 10
2	HgCl$_2$	461	530 ± 60	120 ± 10
	CH$_3$HgCl	0	ND	–
	CH$_3$CH$_2$HgCl	303	290 ± 30	96 ± 9
3	HgCl$_2$	0	ND	–
	CH$_3$HgCl	572	610 ± 40	107 ± 7
	CH$_3$CH$_2$HgCl	242	270 ± 20	111 ± 8
4	HgCl$_2$	248	220 ± 40	90 ± 20
	CH$_3$HgCl	1090	1040 ± 50	95 ± 5
	CH$_3$CH$_2$HgCl	446	510 ± 30	114 ± 4
5	HgCl$_2$	419	380 ± 70	90 ± 20
	CH$_3$HgCl	537	570 ± 20	106 ± 3
	CH$_3$CH$_2$HgCl	0	0	–
6	HgCl$_2$	0	ND	–
	CH$_3$HgCl	767	780 ± 20	102 ± 2
	CH$_3$CH$_2$HgCl	575	580 ± 20	101 ± 3

*Numbers represent the average ± the standard deviation of a minimum of three injections of each mixture
† ND = not detected
‡ Indicates no addition of this mercury species to the mixture

Source: Reproduced by permission from the Royal Society of Chemistry, London [246]

resolved eluents for tin-specific determination [250]. Injected sample volumes may vary from 10 to 500µL $^{-1}$. Consequently, system sensitivity is broad and samples can be very representative.

Mixtures of R$_3$Sn$^+$ compounds (R = *n*–butyl, phenyl, cyclohexyl) were separated by ion exchange–high performance liquid chromatography–graphite furnace atomic absorption spectrometry [252]. The small spread in calibration slopes in Fig. 4.10 signifies similar efficiencies for their separation and column recovery, as well as graphite furnace sensitivities.

Fig. 4.10 shows that considerably more sensitivity is possible with P/T– gas chromatography–flame photometric detector speciation of related organotin species known [253–255] to occur in environmental media. Much greater divergence in the P/T gas chromatography–flame photometric detector system calibration slopes (ratios > 25) is obtained, probably a result of different rates of hydride derivatisation during the fixed P/T purge time (10min), different partition coefficients affecting the

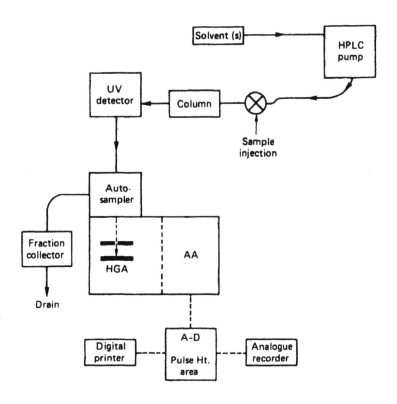

Fig. 4.9 The HPLC–GFAA system with automated peripherals
Source: Reproduced by permission from Kluwer Academic, Plenum Amsterdam [251]

rates at which end species is sparged from the solution or different retentivities on the Tenax–GC sorbent [256]. On the basis of the values obtained for the gas chromatographic method (Fig. 4.11) with 10ml sample volumes, nominal working ranges of 10–40ng L^{-1} organotin are feasible.

Both systems are capable of at least a 10-fold increase in sensitivity with only minor changes in procedure and equipment. For high performance liquid chromatography–graphite furnace atomic absorption spectrometry, this can be achieved by both increasing injected sample size and optimising flow rates with a graphite furnace–atomic absorption spectrometry thermal programme designed to give maximum atomisation efficiency for a specific organotin analyte [254,255]. For high performance liquid chromatography–graphite furnace atomic absorption spectrometry, improvements are realised by adjusting purge flow rate and time while

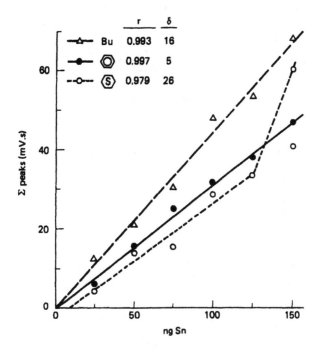

Fig. 4.10 Calibration curves for R_3Sn^+ (R = butyl, phenyl, c–hexyl) separated by HPLC–GFAA with strong cation exchange (SCX) columns using MeOH/H$_2$O/NH$_4$OAc eluents are shown with respective correlation coefficients (r) and system detection limits (δ) (95% confidence level)
Source: Reproduced by permission from Kluwer Academic Plenum, Amsterdam [251]

altering sodium borohydride additions to optimise evolution of a given organotin analyte [257]. Also, both increasing sample volumes [253, 254] and operating the Tenax–GC trap at subambient temperatures [258, 259] will yield lower working ranges.

Nygren *et al.* [260] interfaced on-line a liquid chromatograph to a continuously heated graphite furnace atomic absorption spectrometer to determine di- and tributyltin species in non saline waters with a detection limit of 0.5μg tin absolute.

Wang and Yang [261] used liquid chromatography with indirect photometric detection to determine down to 0.8μg triorganotin compounds in non saline waters.

Shen *et al.* [262] evaluated indirectly coupled plasma mass spectrometry as an element detector for the supercritical fluid chromatography of organotin compounds in water. Detection limits of 0.04 and 0.047pg absolute were obtained, respectively, for tetrabutyltin and tetraphenyltin.

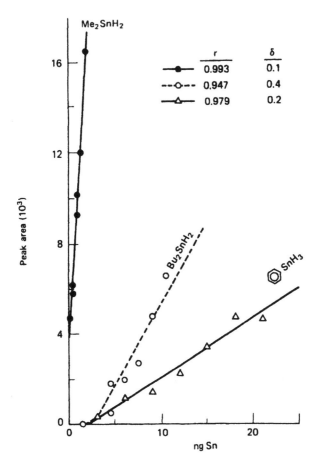

Fig. 4.11 Calibration curves for three aqueous organotins inducated separated by P/T–GC–FPD using the hydride generation mode are shown with respective r and δ estimates
Source: Reproduced by permission from Kluwer Academic Plenum, Amsterdam [251]

High performance liquid chromatography coupled with hydride generation–direct current plasma emission spectrometry has been used for trace analysis and speciation studies of methylated organotin compounds in water [263].

Total tin was determined by continuous on-line hydride generation followed by direct current plasma emission spectroscopy. Interfacing the hydride generation–DC plasma emission spectrometric system with high performance liquid chromatography allowed the determination of tin species. Detection limits, sensitivities and calibration plots were determined.

Liu [264] reported good chromatographic performance in the separation of organotin compounds including tetracyclohexyltin and tetraphenyltin regardless of the type of injection port, although electronic pressure control was needed to achieve acceptable results for tetracyclohexyltin and tetraphenyltin. Interlaboratory method precision ranged from 11–40% RSD over the concentration range tested.

4.2.1.3 Organoarsenic compounds

High performance liquid chromatography coupled with hydride generation–atomic absorption spectrometry has been used for the determination of arsenic species in non saline water samples [265].

Arsenic species were preconcentrated on Zipax, a pellicular anion exchange material and separated on a column packed with high performance liquid chromatography grade strong anion exchange resin, then continuously reduced with sodium tetrahydroborate and detected by atomic absorption spectrometry. Detection limits were 2ng for arsenite, arsenate and monomethylarsinate and 1ng for dimethylarsonate.

Arsenical herbicides (methane arsonate, dimethyl arsenite) and inorganic arsenic at the $\mu g \ L^{-1}$ level have been determined in run-off water [266].

Jyn Yynn and Wai [267] chelated organoarsenic compounds with sodium bis (trifluorethyl) dithiocarbamate prior to application of high performance liquid chromatography. He applied the technique to mixtures of arsenite, arsenate, dimethylarsonic acid, dimethylarsinic acid and other organoarsenic compounds.

Blais et al. [268] determined arsenobetaine, arsenocholine and tetra-methyl arsonium ion in non saline waters in amounts, respectively, down to 13.3, 14.5 and 7.8µg by a procedure based on liquid chromatography–thermochemical hydride generation atomic absorption spectrometry.

4.2.1.4 Organomanganese compounds

Walton et al. [269] separated organomanganese and organotin compounds by high performance liquid chromatography using laser excited atomic fluorescence in a flame as a high sensitivity detector.

The absolute detection limit for manganese was 8–22pg for various organomanganese compounds including methyl cyclopentadienyl manganese tricarbonyl.

4.2.1.5 Organocopper compounds

Brown *et al.* [270] applied high performance liquid chromatography to the determination of organocopper speciation in soil–pore waters.

4.2.2 Aqueous precipitation

4.2.2.1 Organolead compounds

Blaszkewicz *et al.* [271] determined trialkyllead species in rain water. Interfering metal ions in rain water were complexed with EDTA before adjustment of the pH to 10. Samples were pumped through an extraction column of silica gel to absorb lead compounds which were then desorbed with acetate buffer containing methanol at pH3.7. The eluate was diluted and adjusted to pH8 with borate buffer before further concentration on a Nucleosil 10–C_{18} precolumn and separated with methanoic acetate buffer. On-line detection used a post-column chemical reaction detector. Detection limits for sample volumes of 500ml were 15pg mL^{-1} and 20pg mL^{-1} for trimethyl and triethyllead respectively. Standard deviation was less than 4% for a sample containing 90pg triethyllead per millilitre. Concentrations of trimethyl- and triethyllead in rain, melt water and surface water samples were between 20 and 100pg mL^{-1}.

4.2.3 Sea and estuary water

4.2.3.1 Organotin compounds

Ebdon and Alonso [272] have determined tributyltin ions in estuarine waters by high performance liquid chromatography and fluorometric detection using morin as a micellar solution. Tributyltin ions were quantitatively retained from 100–500ml of sample on a 4cm long ODS column. After washing off the salts with 20ml of distilled water the ODS column was back-flushed with methanol–water (80 + 20) containing 0.15m ammonium acetate, on a 25cm long Partisil SCX analytical column. The eluent from the column (1ml min^{-1}) was mixed with fluorometric reagent (acetic acid, (0.01m) morin (0.0025% m/v); Triton X–100, (0.7% m/v 2.5ml min^{-1}) for detection at 524mn with excitation at 408nm. The detection limit (2σ) is 16ng of tributyltin (as Sn).

4.3 Cations

The most common detector used in the separation of metals by high performance liquid chromatography is based on monitoring the column effluent with a variable wavelength spectrometer operating in the ultraviolet or visible region. To make the metals visible before they enter the detector, they must be complexed with an organic reagent to produce complexes which absorb in the above regions of the spectrum and which, incidentally, improve the sensitivity of detection of the metals. Various metal complexing agents have been studied including 4-(2-pyridylazo) resorcinol) (PAR), dithizone [273], diethyldithiocarbamates [274,297], bis(n-butyl-2)-naphthyl methyl (dithiocarbamate) zinc(II) [276]. Since 1982 various investigators have studied the application of high performance liquid chromatography to the determination of very low concentrations of metals in water samples.

Some investigators have produced organic chelate complexes from the metals after they have been separated in the inorganic form on the column, but before entry into the detector, ie post-column derivatisation, while other investigators form the organic complexes before the sample is applied to the separation column, ie pre-column derivatisation.

In the pre- and post-complexation techniques the metals are complexed with an organic chelating agent in order to facilitate their detection with a visible or ultraviolet spectrophotometric detector, thereby improving detection limits. When polarographic detection is being employed, the electrochemical properties of the complexes rather than those of the uncomplexed metals are being relied on. In the case of an atomic absorption detection linked to a high performance liquid chromatograph there is no need to form chelates with the metals prior to or after the chromatographic separation.

4.3.1 Non saline waters

4.3.1.1 Heavy metals, mercury and aluminium

Shih and Carr [277,278] showed that metal complexes of bis(n-butyl-2-naphthyl-methyldithiocarbamate) are thermodynamically stable and chemically inert and that the nickel(II), iron(III), copper(II) and mercury (II) complexes of this dithiocarbamate can be separated by high performance liquid chromatography and detected with a variable wavelength detector.

A typical chromatogram for the separation of the bis(n-butyl-2-naphthyl methyldithiocarbamate complexes of iron(III), nickel(II), copper(II),

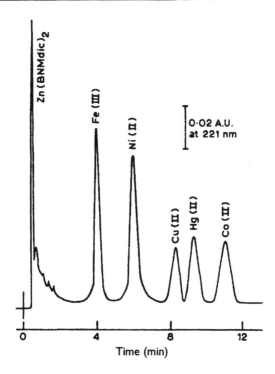

Fig. 4.12 Chromatogram of the metal (BNMDTC) complexes. Sample was 20 µl of a synthetic mixture which was 1×10^{-4} M in each complex. Flow rate 2 ml min $^{-1}$; pressure drop less than 1500 psi
Source: Reproduced by permission from Elsevier Science, UK [277]

mercury(II) and cobalt(II) is shown in Fig. 4.12. It is clear that these metals are very easily separated.

	Wavelength of detection, nm
nickel	350
cobalt	350
copper	355
iron	440

The absorptions of the different complexes at their wavelength were very different, cobalt and copper being the most sensitive.

Edward Iratami [273] applied high performance liquid chromatography to the determination in river water of mercury(II), copper(II), nickel(II), cobalt(II) and lead(II) as their dithizonates and their diethyldithiocarbamates. The metals were first complexed, then the complexes

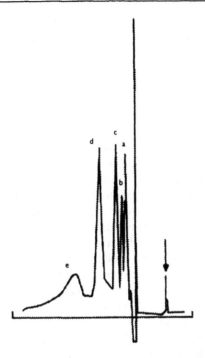

Fig. 4.13 Separation of a standard trace metal sample (extracted with dithizone solution). Divisions on baseline: 1 min per division; peaks, dithizone complexes of (a) Hg, (b) Cu, (c) Ni, (d) Co and (e) Pb.
Source: Reproduced by permission from Elsevier Science, UK [279]

extracted by chloroform extraction from the water sample prior to chromatographic separation and detection with an ultraviolet detector.

The separation of five metal dithizone complexes extracted from a standard metal solution, is illustrated in Fig. 4.13.

Jones et al. [280] used dithizone for post column derivativisation of cadmium, cobalt, copper, lead, nickel and zinc. The separation was achieved in aqueous media on a sulphonated 10% cross-linked poly-styrene resin. Cadmium(II), cobalt(II), copper(II), lead(II), nickel(II) and zinc ions were detected down to nanogram levels.

The detection used was based on monitoring the absorbance of unreacted dithizone at 590nm.

Liquid chromatography–absorption spectrophotometry was used by Vlacil and Hamplova [281] for the determination of lead and copper in natural waters. The metal diethyldithiocarbamates are extracted and concentrated by evaporation, followed by reversed phase liquid chromatography of the chelates. The copper and lead chelates can also be sequentially detected by spectrophotometry at 440 and 280nm. The detection limits for copper and lead were 8.6 and 17µg L^{-1}, respectively, when liquid chromatography was used, and were 58 and 17µg L^{-1}, respectively when spectrophotometry was used.

Nagaosa et al. [282] simultaneously determined aluminium, iron and manganese in non saline waters using high performance liquid chromatography with electrochemical and spectrophotometric detection.

In this method the water sample was directly injected as their 8-quinolinol complexes onto a Bondasphere ODS column. Chromatographic separation can be made with the mobile phase of 2:3 acetonitrile/20 mM acetate buffer solution containing 5mM 8-quinolinol reagent. Excellent sensitivity is obtained by spectrophotometric detection at 390nm. The spectrometric detection limits of these metals are at the part per billion levels of test solution. The tolerance limits of numerous other metals ions are reported. Only chromium(III) and nickel(II) interfere with the determination of aluminium, and molybdenum(VI) interferes with the determination of manganese at concentrations less than a 100-fold excess. Analytical data obtained on river and sea water samples were in agreement with expected values. Amperometric detection of iron and manganese with a thin layer flow cell and a glassy carbon working electrode is also described. This method is less sensitive but more specific than spectrophotometric detection.

Rottman and Henmann [283] determined heavy metal interactions with dissolved organic materials in natural aquatic systems by coupling a high performance liquid chromatography system with an inductively coupled plasma mass spectrometer. They employed direct coupling to get specific distribution patterns of the heavy metal complexes, and on–line isotope dilution mass spectrometry was performed to quantify heavy metals accurately on different organic fractions. With respect to the separation properties of a size exclusion column by molecular size, different distribution patterns could be found for the heavy metals depending on the type of aquatic system. Different distribution patterns in the various fractions of dissolved organic material (preferably of humic substances) could also be observed for the metals in the same natural water sample. In addition, a high resolution inductively coupled mass spectrometer was applied for the first time as an element specific detector in connection with a high performance liquid chromatographic system which also allow interference free detection of iron species.

4.3.1.2 Selenium

Nakagawa et al. [284] have described a method based on selenotrisulphide formation followed by high performance liquid chromatography with fluorometric detection for the determination of selenium(IV). The method involves precolumn reaction of selenium(IV) with penicillamine (Pen) to produce stable selenotrisulphide (Pen-SSeS-Pen) and subsequent derivatisation to a fluorophore by reaction with 7-fluoro-4-nitrobenz-2,1,3-oxidazole. The fluorophore was separated by reversed phase high performance liquid chromatography and selenium content was determined by fluorometric detection. The calibration plots showed a linear relationship in the range of 10–2000 ppb of selenium(IV) with a detection limit of 5µg L $^{-1}$ (signal to noise ratio (S/N) > 2). The method can determine total content of selenium in environmental samples after digestion of the samples and reduction of selenium(VI) to selenium(IV). The results from standard samples indicated satisfactory agreement with those obtained by other established methods and certified values with good reproducibility. This method is as sensitive as, but simpler in operation than, conventional fluorometry using diaminonaphthalene.

See also Table 4.8.

4.3.1.3 Vanadium

Mieura [285] has described a method for the determination of traces of vanadium in non saline waters with 2(-8 quinolyl azo)-s(dimethylamino) phenol by reversed phase liquid chromatography spectrometry.

In this reversed phase high performance liquid chromatographic method for neutral and cationic metal chelates with azo dyes, tetraalkyl-ammonium salts are added to an aqueous organic mobile phase. The tetra-alkylammonium in salts are dynamically coated on the reversed stationary support. As a result of the addition of tetraalkylammonium salts, the retention of the chelates is remarkably reduced. Tetrabutylammonium bromide permits rapid separation and sensitive spectrophotometric detection of the vanadium(V) chelate with 2-(8-quinolylazo)-5-(dimethyl-amino)-phenol, making it possible to determine trace vanadium(V).

Nagoasa and Kimata [286] determined down to 1ppb of vanadium in non saline waters by high performance liquid chromatography with electrochemical detection. See also Table 4.8.

4.3.1.4 Other applications

Other applications of high performance liquid chromatography to the determination of cations in non saline waters are tabulated in Table 4.8.

4.3.2 Seawater

4.3.2.1 Heavy metals

Cassidy and Elchuk [287,288] carried out trace enrichments and high performance liquid chromatography of solutions of nickel, cobalt, copper,

Table 4.8 High performance liquid chromatography of cations in non saline waters

Cation	Stationary phase	Eluent	Comments	Detection	LD	Ref.
As species	–	–	–	Wall jet cell and microsized platinum disc electrodes	–	[266]
As species	–	–	–		–	[290]
Ca, Mg	Zipex SCX	–	–	–	$<1 \times 10^{-1}$M	[291]
Se(VI), total Se	–	–	Complexed with 2,3 diamino-napthalene with EDTA–sodium fluoride	Spectrofluorometry	0.4ng	[292]
Se(IV) and total Se	–	–	Pre-column reduction to produce selenotrisulphide and derivitivisation with 7-fluoro-4-nitrobenz 2,1,3–oxidazole	Spectrofluorometry	5µg L^{-1}	[293]
Sn(II), Sn(IV)	–	–	–	–	0.2ng	[294]
Pb, Sn	–	–	–	ICPAES	–	[295]
V	–	–	–	Electrochemical	1ppb	[286]
Heavy metals Al, Fc, Mn	Bondasphere ODS	2:3 aceto-nitrile:20mm acetate buffer	Injected as 8-quinolinol complexes	Electrochemical spectro-photometric at 390nm	ppb	[282]
Cr(III), Cr(VI)	–	–	On-line analysis	AAS	30µg L^{-1}	[296]
Cu, Co, Ni, Pb, Fe	–	–	Converted to dithio-carbamates	Spectrophotometric at 320–440nm	–	[297]
Cu, Ni, Co, Cr(IV), Cr(III)	–	–	Converted to dithiocar-bamates externally before injection on column	Electrochemical	1ng	[274]

Table 4.8 continued

Cation	Stationary phase	Eluent	Comments	Detection	LD	Ref.
Fe	–	–	–	–	10ppb	[275]
Cr	–	–	Paired ion HPLC	Phosphorescence at 515nm	1.4×10^{-7}M	[298]
Cr(III), Cr(IV), total Cr	–	–	–	–	30–180ppt	[299]
Cu species	–	–	–	–	1–3µg L^{-1}	[300]
As species	Zipex preconcentration, then separation on pellicular anion exchange resin	–	–	Hydride generation AAS	1–2ng	[265]

Source: Own files

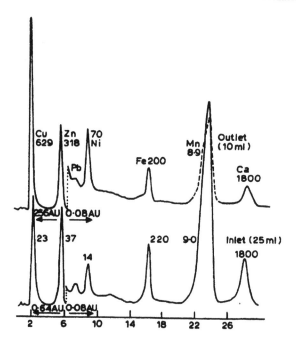

Fig. 4.14 High performance liquid chromatography analysis of fresh water coolant. Experimental conditions: samples were made 0.0001mol L^{-1} in citrate (pH4.5) and allowed to sit for 2–3 days

Reproduced by permission from Kluwer Academic Plenum, Amsterdam [287]

zinc and lead in the low µg L^{-1} range. The metal ions were enriched on a short bonded phase ion exchanger and then separated on a 13µm styrene divinylbenzene resin. The eluted metal ions were detected with a variable wavelength ultraviolet visible detector after a post-column reaction with 4-(2-pyridylazo) resorcinol monosodium salt.

Detection limits for these metal ions are in the range of 0.5µg L^{-1} (cobalt) to 15µg L^{-1} (nickel).

In Fig. 4.14 is shown a chromatogram obtained by this technique for two waste water samples. Excellent resolution is obtained for all elements examined.

See also Table 4.10.

4.3.2.2 Plutonium

Delle Site *et al.* [279] have used extraction chromatography to determine plutonium in seawater, sediments and marine organisms.

Table 4.9 Experimental check of the method with IAEA samples

Sample	Size (g)	IAEA activity (fCi ± SD)*		Present activity (fCi ± SD)†			
		$^{239,240}Pu$	^{238}Pu	$^{239,240}Pu$	Mean value	^{238}Pu	Mean value
Seawater	1000	103 ± 7	18 ± 1	113 ± 7 96 ± 7 119 ± 8	109 ± 7	17 ± 4 17 ± 3 15 ± 4	16 ± 4
Sediment	1‡	501–585	14–31	629 ± 19 516 ± 20 575 ± 19	574 ± 19	17 ± 3 23 ± 4 21 ± 4	20 ± 4

*Standard deviation for mean of 5 (^{238}plutonium) and 10 (^{239,240}plutonium) results
†Standard deviation of a single source α–counting
‡Sample dried at 105°C

Source: Reproduced by permission from Elsevier Science UK [279]

These workers used double extraction chromatography with Microthene-210 (microporous polyethylene) supporting tri-*n*-octyl-phosphine oxide (TOPO); a technique that has been used previously to isolate plutonium from other biological and environmental samples [289]. 236-plutonium and 242-plutonium were tested as the internal standards to determine the overall plutonium recovery, but 242-plutonium was generally preferred because 236-plutonium has a shorter half-life and an α-emission (5.77MeV) which interferes strongly with the 5.68MeV (95%) α-line of 224-radium, the daughter of 228–thorium. However, the 5.42MeV α-lines of 228-thorium interfere with those of 238-plutonium (5.50MeV) and so a complete purification from thorium isotopes is required.

Plutonium sources were counted by an α–spectrometer with good resolution, background and counting yield. The counting apparatus used had a resolution of 40keV. The mean (±SD) background value was 0.0004 ± 0.0003cpm in the 239- and 240-plutonium energy range and 0.0001 ± 0.0001cpm in the 238-plutonium energy range. The mean (±SD) counting yield, obtained with 239-plutonium, 240-plutonium reference sources counted in the same geometry, was found to be 25.08 ± 0.72%.

The method proposed was checked by analysing some seawater and sediment reference samples prepared by the IAEA Marine Radioactivity Laboratory (Monaco) for intercomparison programmes. The values reported by IAEA and the experimental values obtained here are compared in Table 4.9; the agreement is fairly good.

Table 4.10 High performance liquid chromatography of cations in seawater

Organic	Stationary phase	Eluent	Comments	Detection	LD	Ref.
Al, Fe, Mn	–	–	–	Spectrophotometric and electrochemical	–	[282]
Cu, Ni, V	C₁₈ column	Xylene solution of capriquat	Complexes with 2-(5,bromo-2-pyridylazo)-5-(N-propyl-sulpho–propyl-amino) phenol	Spectrophotometric	–	[301]
Transition metals	Chitosan	–	Ca and Mg do not interfere	–	–	[302]
Zn, Fe, Mg, Cu, Ni	–	–	–	Atomic fluorescence	0.5µg L⁻¹(Zn), 0.7µg L⁻¹(Fe)	[303]
Cr(III), Cr(V)	–	–	–	ICPAES	µg L⁻¹	[304]
¹⁵NH₄, ¹⁴NH₄	–	–	Post-column labelling with o-phthalaldehyde-2-mercaptoethanol	Spectrofluorometric	–	[305, 306]

Source: Own files

4.3.2.3 Other applications

Other applications of high performance liquid chromatography to the determination of cations in seawater are tabulated in Table 4.10.

4.3.3 Ground waters

4.3.3.1 Uranium

Kerr et al. [307] employed high performance liquid chromatography for the determination of uranium in groundwaters. The sample was passed through a small reversed phase enrichment cartridge, to separate the uranium from the bulk of the dissolved constituents. The uranium was then back flushed from the cartridge onto a reversed phase analytical column. The separated species were monitored spectrophotometrically after reaction with arsenazo(III). The detection limit was in the 1–2μg L^{-1} range with a precision of approximately 4%.

Cassidy and Elchuck [288,308] applied high performance liquid chromatography to the determination of uranium(IV) in ground water samples. They studied conventional cross linked and bonded phase ion exchangers, both cation and anion with aqueous mobile phases containing tartrate, citrate or α-hydroxylisobutyrate. The best chromatography was obtained on bonded phase cation exchangers with an α-hydroxylisobutyrate eluent. The metal ions were detected either by visible spectrophotometry of the arsenazo(III); VI complex at 650nm, after a post-column reaction with a complexing reagent, or with a polarographic detector. Detection after post-column reaction gave the best sensitivity, the detection limit (2 × baseline noise) was 6ng or 60μg L^{-1} for 100μL samples. In-line trace enrichment was used to decrease detection limits and linear calibration curves were observed in the range 0.5–50μg L^{-1} for ground waters.

The other metal ions that exhibited an appreciable reaction with Arsenazo(III) at 650nm under the separation and detection conditions used, were iron(III), zirconium(IV), thorium(IV) and the lanthanides. The lanthanides, iron(III) and zirconium(IV) were eluted at or near the solvent front before uranium(VI) and thorium(IV) was eluted after uranium(VI).

4.3.4 Potable water

4.3.4.1 Copper, beryllium, aluminium, gallium, palladium and iron

Ichinaka et al. [309] determined these elements in acetylacetone extracts of potable waters by high performance liquid chromatography.

Table 4.11 High performance liquid chromatography of cations in waste waters and trade effluents

Organic	Stationary phase	Eluent	Comments	Detection	LD	Ref.
Cr	–	–	Formation of red complex with 1,5 diphenyl carbazide	Spectrophotometric at 540nm	–	[310]
Ni, Co, Cu, Zn, Pb	Styrene–divinyl benzene resin	–	Post-column reaction with 4-(2-pyridyl(azo)) resorcinol	UV/visible	ng L^{-1}	[287]
Co, Cu, Hg, Ni	–	–	Conversion to ammonium-bis-(2-hydroxyethyl) dithiocarbamate	–	–	[311]
Cr	–	–	Reverse phase HPLC	UV	–	[312]
Trace metals	–	–	Reverse phase HPLC in electroplating solutions	–	–	[313]
NH$_4$	–	–	Conversion to m-toluyl derivative and dichloromethane extraction prior to chromatography	–	–	[314]

Source: Own files

4.3.5 Waste waters and trade effluents

4.3.5.1 Heavy metals and ammonium

Some applications of high performance liquid chromatography to the determination of heavy metals and ammonium in these types of samples are reviewed in Table 4.11.

4.4 Anions

4.4.1 Non saline waters

4.4.1.1 Nitrate and nitrite

Alawi [315] has described a simple and specific high performance liquid chromatographic method for the determination of nitrite and nitrate in non saline water based on the nitration of an excess of phenol by nitrate or oxidised nitrite ions; the orthonitrophenol produced is extracted, separated on a reversed phase column and quantified using an amperometric detector in the reduction mode. Nitrite, if present, is first oxidised to nitrate by addition of 1% hydrogen peroxide. The method has been successfully applied to waters containing nitrate plus nitrite at the 5μg L^{-1} level. By using a larger sample volume (500ml) and injecting a larger aliquot (100μL) onto the column the sensitivity could be improved, giving a lower limit of about 1μg L^{-1}.

Recoveries were 82.6 ± 2.37% for nitrate and 76.9 ± 1.94% for nitrite in the 0.01–1μg mL^{-1} concentration range.

This method is free from interferences, which when coupled with the specificity and the high sensitivity of the electrochemical detection mode, renders it suitable for the determination of trace levels of nitrate and nitrite in surface, ground and main water.

This technique has been applied by Iskandarni and Petrzyk [316] to the determination of nitrate and nitrite in water. Chemically bonded amine materials have been used to remove interference by humic substances prior to the ion chromatographic determination of nitrate and sulphate in non saline waters.

High performance liquid chromatography on a small bore column packed with microparticulate silica based ion exchange material has been used [317] to determine down to 0.01mg L^{-1} nitrate in water without interference from other ions associated with potable, pond, river and stream water.

Detection of the separated peaks was achieved using a Pye LC–UV variable wavelength ultraviolet detector fitted with a 8μL flow cell and set at 265nm. The chromatographic system consisted of a Partisil 10μm SAX column, and a mobile phase consisting of 10^{-3} mol L^{-1} potassium hydrogen phthalate (pH 3.95) in deionised water.

Lee and Field [318] have discussed a technique of post-column fluorescence detection of nitrite, nitrate, thiosulphate and iodide anions by high performance liquid chromatography. These anions react with cerium(IV) to produce fluorescent species in a post-column packed bed reactor.

Alawi [319] has discussed an indirect method for the determination of nitrite and nitrate in surface, ground and rain water by reaction with excess phenol (nitrite ions first being oxidised to nitrate) and extraction of the o-nitrophenol produced, followed by separation on a reversed phase high performance liquid chromatography column with amperometric detection in the reduction mode. Recoveries were 82% for nitrate and 77% for nitrite in the concentration range 10–1000µg L^{-1}. The method is claimed to be free of interferences from other ions.

Using a radial compression C_{18} column and a mobile phase of aqueous tetramethylammonium phosphate Kok et $al.$ [320] analysed mixtures of nitrate and nitrite at the 0.1mg L^{-1} level (3α) and compared the results obtained with those found by ultraviolet spectroscopic screening methods.

Schroeder [321] has determined nitrate in non saline waters and wastewaters by high performance liquid chromatography using aqueous phosphoric acid–sodium dihydrogen phosphate as the mobile phase and ultraviolet detection. The optimum nitrate concentration was 0.3–3mg L^{-1} as nitrate with a linear response for <3mg L^{-1} as nitrogen. Relative standard deviations in the optimum range were <1%, and the detection limit was 0.0007mg L^{-1} as nitrogen. Various potential interferences, including nitrite, are separated from the nitrate on the high performance liquid chromatography column.

See also Table 4.13.

4.4.1.2 Phosphate

Sakurai et $al.$ [322] have described a high performance liquid chromatographic procedure for the determination of down to 0.5mg L^{-1} phosphate in waste water and river waters. The method is based on the solvent extraction of molybdoheteropoly yellow with methyl propionate. The corresponding silicon compound is not extracted into this solvent. Thus, the interference by silicon is excluded even at concentrations as high as 10mg L^{-1}. A wavelength of 251nm at the ultraviolet detector gave an absorption maximum of phosphorus molybdoheteropoly yellow in methyl propionate.

In Fig. 4.15 is shown a typical chromatogram, obtained by this technique for a mixture of orthophosphate, pyrophosphate and $P_3O_{10}^{5-}$. Morgan and Danielson [323] have described the development of an enzymatic method of determining phosphate eluted from a reverse phase

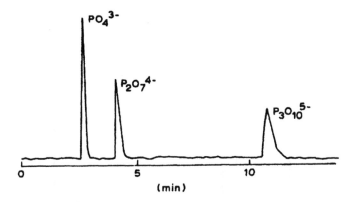

Fig. 4.15 Chromatogram of orthophosphate, diphosphate and triphosphate. Column is µ–Gondapak–NH$_2$ and the eluant was oxalate–Mg buffer at the flow rate of 1.0ml min $^{-1}$, 50µL of mixed solution which contains 10–3M each phosphate
Source: Reproduced by permission from Springer Verlag, Heidelberg [322]

high performance liquid chromatographic column. The method is applicable to sewage effluents and non saline water samples. It is based on the nucleoside phosphorylase catalysed conversion of inosine and orthophosphate to hypoanthine. This method is claimed to give better results than standard molybdenum blue spectrophotometric methods.

$$\text{Phosphate} + \text{inosin} \underset{Mg^{2+}}{\overset{NP}{\rightarrow}} \text{ribose-1-phosphate} + \text{hypoxanthine}$$

The inosine and hypoxanthine were separated by reversed phase high performance liquid chromatography and the amount of hypoxanthine produced was related to the phosphate concentration. Quantitation of the hypoxanthine peak was found to be linear with orthophosphate up to about 30mg L $^{-1}$. A detection limit of 0.75mg L $^{-1}$ could be obtained after dialysis of the commercial enzyme. Interference studies showed that the enzymatic assay unlike the colorimetric molybdate blue technique was essentially unaffected by complex matrices such as polyphosphates and phosphoesters.

Fig. 4.16 shows a typical chromatogram for a standard 42µg L $^{-1}$ phosphate solution and hypoxanthine peaks resulting from various phosphate samples after reaction with the enzyme. The calibration curve had a slope of 0.043 ± 0.002 and an intercept of 0.124 ± 0.033 with a correlation coefficient of 0.998. Linearity up to 30mg L $^{-1}$ was observed. Relative standard deviation of triplicate runs was 10% or less. The detection limit, twice the signal of the blank, was determined to be 1.5mg L $^{-1}$.

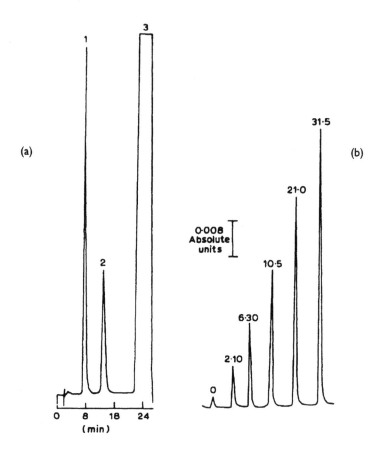

Fig. 4.16 (a) Sample chromatogram of a 42mg L^{-1} phosphate sample. Absorbance units full scale, su.f.s. = 0.08. (a) Peaks: 1 = hypoxanthine; 2 = *m*–aminophenol; 3 = inosine. (b) Hypoxanthine peak for various phosphate concentrations in mg L^{-1} (indicated above each peak)
Source: Reproduced by permission from Springer Verlag Heidelberg [322]

Fig. 4.17 is a plot of phosphate concentration given by the enzymatic method vs that found by the molybdate blue method. The line fit to these points had a slope of 0.999 ± 0.046 and an intercept of 0.421 ± 1.222 with a correlation coefficient of 0.996. High performance liquid chromatography has also been linked to an inductively coupled plasma atomic spectrometric detector [322,323] to determine orthophosphate, pyrophosphate and tripolyphosphate.

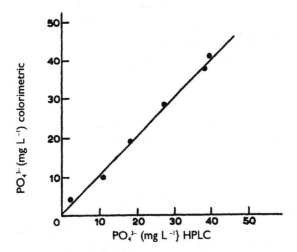

Fig. 4.17 Concentration of phosphate obtained by the colorimetric method versus the concentration obtained by the enzymatic high performance liquid chromatographic method
Source: Reproduced by permission from Springer Verlag Heidelberg [322]

Speciation and quantitative analysis of orthophosphate, pyrophosphate and tripolyphosphate are performed by using high performance liquid chromatography with inductively coupled argon plasma emission spectrometric detection. High performance liquid chromatography is used to separate mixtures of phosphates on an anion exchange column using tartrate magnesium buffer. The ICP is used as a selective detector by observing P II emissions at 214.9nm. The detection limit is 0.5, 1 and 3μg respectively for ortho-, pyro- and tripolyphosphate respectively.

In this procedure a Waters Associates liquid chromatograph was connected to anion exchange columns. Column packings used were μ–Bondapak–NH$_2$ and Nagel SA. The outlet of the liquid chromatograph was connected to the crossflow nebuliser of the ICP spectrometer (Jarrell–Ash–Atom–Comp 750) with Teflon tubing (0.15 × 13cm). Eluant flow rate was set at 1ml min^{-1}.

Phosphorus was monitored at 214.9nm (second order). The analog signal was taken out through the profile mode of a computer program. In order to filter noise, a 50kΩ resistor and 20μF condenser were installed before the recording, giving an approximate time constant of 1s.

The ICP operational parameters were RF power 1.3kw, coolant gas 18L min^{-1}, sample gas 0.5L min^{-1}, plasma gas 0.1min^{-1} and observation height in plasma 15mm.

Table 4.12 High performance liquid chromatography variables

Column	μ–Bondapak–NH$_2$ 1/8in × 1ft
	Nucleosil N(CH$_3$)$_3$ 1/8in × 1ft
Eluant	Tartaric acid (0.1mol L^{-1}) + MgSO$_4$ (0.01mol $^{-1}$) + NaOH (0.1mol $^{-1}$)
Buffers	Oxalic acid (0.1 mol $^{-1}$) + MgSO$_4$ (0.01mol $^{-1}$) + NaOH (0.1mol $^{-1}$)
	Ammonium formate (0.2mol $^{-1}$) + MgSO$_4$ (0.01mol $^{-1}$) + HCl (0.1mol $^{-1}$)
	Acetic acid (0.2 mol $^{-1}$) + NaOH (0.1mol $^{-1}$)
	Tris(hydroxymethyl)amine (0.3mol $^{-1}$) + H$_2$SO$_4$ (0.075mol $^{-1}$)

Source: Reproduced by permission from Springer Verlag Heidelberg [322]

Under these conditions the detection response is substantially independent of the chemical form of the phosphorus compounds.

Optimum chromatographic conditions are summarised in Table 4.12. The methods of Sakurai *et al.* [322] and Morgan and Danielson [323] are both applicable to the analysis of waste waters and sewage effluents.

4.4.1.3 Halides

Akaiwa *et al.* [324] have used ion exchange chromatography on hydrous zirconium oxide, combined with detection based on direct potentiometry with an ion selective electrode, for the simultaneous determination of chloride and bromide in non saline waters.

The silver chloride electrode gave poor response to iodide and bromide, and so did the silver bromide electrode to iodide. Although the silver iodide electrode responded to all three halides, the peaks are not sufficiently resolved and they are asymmetric. Further, there was a drift of the base line after detection of a halide ion which was not a component of the electrode and this drift caused disturbance in the following peak. This difficulty is eliminated by using hydrous zirconium oxide instead of the anion exchange resin for the chromatography since it reverses the elution order for halide ions. The silver bromide electrode is then the most suitable as the detector for both bromide.

4.4.1.4 Chloride, bromide, iodide, nitrate and thiocyanate

High performance liquid chromatography with a variable, wavelength ultraviolet detector is a sensitive and precise method for the simultaneous measurement of chlorides, bromides, iodides, nitrates and thiocyanates in non saline water samples. Stetzenbach and Thompson [325] applied the method employing anion exchange columns to ground water samples achieving detection limits of 50μg L^{-1} for these ions.

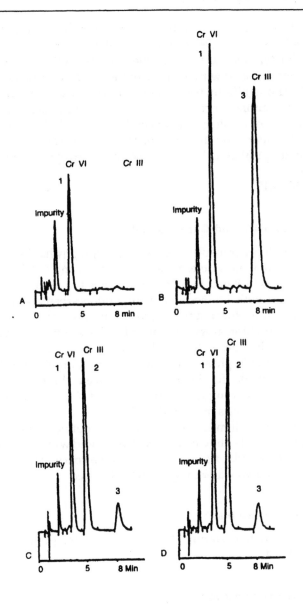

Fig. 4.18 Chromatograms of reaction mixtures of Cr(III) and Cr(VI) Conditions for high performance liquid chromatography: column RP-8, mobile phase; methanol–water (65:35); flow rate, 3cm³ min⁻¹ detection at 254nm: (A) 10.0µg Cr(III), pH 4.0; (B) 10µ.g Cr(III), pH 5.8; (C) 10.0µg Cr(IV), pH 4.0; (D) 10.0µg Cr(VI) + 10µg Cr(III) pH 4.0. 1, unidentified disulphide; 2, $Cr(S_2CH (C_2H_5)_2)_2$ $(OS_2 CN(C_2H_5)_2)$, $(Cr(VI))$; 3 $Cr(S_2CN(C_2H_5)_2)_3$, $(CR(III))$
Source: Reproduced with permission from Frederick Wieweg and Son, Wiesbaden [326]

Thiocyanate, bromide, iodide, nitrate and nitrite have large enough extinction coefficients at 194–215nm to make ultraviolet absorption detection useful for trace analysis at these wavelengths. An eluant of phosphate buffer prepared with high performance liquid chromatography grade water has no significant background absorption to 190nm and could be used over the entire functional pH range of silica based stationary phases. The choice of a suitable wavelength is generally limited only by the purity of the eluant and the limits of the detector. The practical limit of detection is about 50mg L $^{-1}$ for all of the anions except chloride which is about 1mg L $^{-1}$.

See also Table 4.13.

4.4.1.5 Chromium(III) and Chromium(VI)

Reverse phase high performance liquid chromatography has been used to carry out the simultaneous determination of Cr(III) and Cr(VI) in water [326]. The sample is buffered to pH5.8 and allowed to react overnight at room temperature with 5% sodium diethyldithiocarbamate. It is then extracted with chloroform and analysed by reversed phase high performance liquid chromatography, after the residue from evaporation of the chloroform has been dissolved in acetonitrile, using methanol: water (65:35) as mobile phase. Detection was by ultraviolet at 254nm. Corrections were made for interference between chromium(III) and chromium(VI) and concentrations of 2–10mg L $^{-1}$ of each ion were determined with accuracies of the order of ±5%.

In this method chromium(II) chelates at pH 5.8 but not at pH 4.0.

Evaluation of the peaks shown in Fig. 4.18 produced with chromium(III) and chromium(VI) standard solutions enables the concentration of these two species in unknown water samples to be deduced.

4.4.1.6 Other applications

Other applications of the determination of anions in non saline waters are reviewed in Table 4.13.

4.4.2 Aqueous precipitation

4.4.2.1 Nitrite and nitrate

Alawi [319] has discussed the determination of nitrite and nitrate in rain.

4.4.3 Seawater

4.4.3.1 Carboxylic acids

As a consequence of the above considerations Kieber *et al.* [12] have developed an enzymatic method to quantify formic acid in non saline water samples at sub-micromolar concentrations. The method is based on the oxidation of formate by formate dehyrogenase with corresponding reduction of β-nicotinamide adenine dinucleotide (β-NAD⁺) to reduced β-NAD⁺ (β-NADH); β-NADH is quantified by reversed phase high performance liquid chromatography with fluorometric detection. An important feature of this method is that the enzymatic reaction occurs directly in aqueous media, even seawater, and does not require sample pre-treatment other than simple filtration. The reaction proceeds at room temperature at a slightly alkaline pH (7.5–8.5) and is specific for formate with a detection limit of 0.5μM ($S/N = 4$) for a 200μL injection. The precision of the method was 4.6% relative standard deviation ($n = 6$) for a 0.6μM standard addition of formate to Sargasso seawater. Average recoveries of 2μM additions of formate to seawater were 103%. Intercalibration with a Dionex ion chromatographic system showed an excellent agreement of 98%. Concentrations of formate present in natural samples ranged from 0.2 to 0.8μM for Biscayne Bay seawater.

This method is also applicable to non saline waters and sediment porewaters.

Goncharova and Khomenkô [339] have described a column chromatographic method for the determination of acetic, propionic and butyric acids in seawater and thin layer chromatographic methods for determining lactic, aconitic, malonic, oxalic, tartaric, citric and malic acids. The pH of the sample is adjusted to 8–9 with sodium hydroxide solution. It is then evaporated almost to dryness at 50–60°C and the residue washed on a filter paper with water acidified with hydrochloric acid. The pH of the resulting solution is adjusted to 2–3 with hydrochloric acid (1:1), the organic acids are extracted into butanol, then back-extracted into sodium hydroxide solution; this solution is concentrated to 0.5–0.7ml, acidified, and the acids separated on a chromatographic column.

4.4.3.2 Iodide

Varma *et al.* [340] determined iodide in seawater in amounts down to 2ng L⁻¹ by a method based on precolumn derivativisation of the iodide into 4-iodo-2,6 dimethylphenol. An ultraviolet detector was employed.

Table 4.13 High performance liquid chromatography of anions in non saline waters

Organic	Stationary phase	Eluent	Comments	Detection	LD	Ref.
BO_3	Ion exchange complex (TSK Aelic–Anion DW)	–	Separated as boric acid–chromotropic and complexes interference by MoO_4, VO_3	–	–	[327]
Halides						
I	Ion exchange resin column	–	On-line preconcentration electrode	Iodide selective electrode	1nM	[328]
I, Br, Cl	–	–	Conversion to alkyl Hg(II) halide	–	–	[329]
Cl, Br, NO_2, CNS	Anion exchange column	–	–	–	–	[325]
Organic anions						
Ascorbate	–	–	Converted to 2,4-dinitrophenyl hydrazone	–	–	[330]
Formate	–	–	Enzymatic method	–	4–10µm	[12]
Palmitate, pyrenyl palmitate, lactate, propionate, formate	–	–	1-pyrenyldiazo methane used as labelling reagent	Spectrofluoro-metric at 340nm	20–30fg M	[170]

Table 4.13 continued

Organic	Stationary phase	Eluent	Comments	Detection	LD	Ref.
Fluoride, perchlorate, chloride, nitrite, phosphite, arsenate, nitrate, chlorate, sulphate, chromate, borofluoride, glycollate, acetate, lactate, propionate, acetoacetate, hydroxybutyrate, chloroacetate, isobutyrate and butyrate	–	Ruthenium (II–I), 10 phenanthroline and ruthenium (II)–2,2'bipyridine mobile phases	Ruthenium in 1:1 complexes used as ion interaction reagents	Indirect spectrofluorometric	–	[331]
NO$_3$	–	–	Reverse phase ion interaction	–	0.5mg L^{-1}	[332]
PO$_4$ (polyphosphate, monofluoro phosphates)	–	–	–	Photodiode array detection	–	[333]
SeO$_3$, SeO$_4$	–	–	Post-column derivitivisation with 2,3–diaminonaphthalene	–	–	[334]
SeO$_3$, SeO$_4$	Li Chrosorb RP–18	–	–	UV spectroscopic ICPAES	–	[335]
SO$_3$	–	–	–		–	[336]
SiO$_3$	–	–	–		–	[337]
NO$_3$, Br, NO$_2$, SO$_4$, I, CNS	–	–	Reaction with sodium naphthalene trisulphonate	Indirect photometric	1–4ng	[338]

Source: Own files

4.4.4 Potable water

4.4.4.1 Iodide, chloride, bromide, chlorate, bromate and iodate

Salov et al. [341] determined iodide, chloride, bromide, chlorate, bromate and iodate in potable water by high performance liquid chromatography with an inductively coupled argon plasma mass spectrometric detector.

4.4.4.2 Nitrate

Schroeder [342] determined nitrate in potable water using high performance liquid chromatography with a reversed phase octadeyl C_{18} column using aqueous phosphoric acid and dihydrogen phosphate as a mobile phase and ultraviolet detection. Detection limits were 7µg L $^{-1}$.

4.4.4.3 Chloride, nitrate and sulphate

Takeuchi et al. [343] described an indirect photometric detection method for anions in micro high performance liquid chromatography. Micro anion exchange octadecylsilica columns permanently coated with the hydrophobic cetyltrimethylammonium ion (cetrimide) were used in the high performance liquid chromatography of chloride, nitrate and sulphate contained in potable water. Indirect detection of analyte anions was achieved with an ultraviolet detector. The breakthrough volume of the anion exchange column was measured with aqueous sodium salicylate (1mM). Retention times of analyte anions in an artificial mixture were longer when aqueous sodium salicylate (1mM) containing acetonitrile (5%) was used as the mobile phase.

4.4.5 Trade effluents

4.4.5.1 Thiosulphate and polythionates

Takana et al. [344] used high performance liquid chromatography on an anion exchange column with differential pulse polarographic detection to determine thiosulphate and tri, tetra-, penta- and hexathionates in trade effluents. The method is accurate to within 10% at the 0.001 $^{-1}$ mM concentration range.

4.4.5.2 Fluoride

Hannah [346] has described a high performance liquid chromatographic anion exclusion method for the determination of fluorides in complex trade effluents. Pharmaceutical industrial effluents were examined for fluoride using an Ion–100 anion exchange column coupled with a GA–100 guard column. Organics were removed from the samples at pH 8.2–8.3 by passing through disposable G18 extraction columns. Conductivity detection allowed determination of fluoride at low mg L^{-1} levels in the presence of high concentrations of chloride and sulphate through an ion exclusion mechanism in the polymeric liquid chromatographic column. The limit of detection was 0.2mg L^{-1} fluoride L^{-1} with a working linear dynamic range of 0.2–100mg L^{-1}.

4.4.6 Sewage effluents

See section 4.4.1.2 [322,323].

References

1 Brown, L. and Rhead, M. *Analyst (London)*, **104**, 391 (1979).
2 Hoben, H.J., Ching, S.A., Casarette, L.J. and Young, R.A. *Bulletin of Environmental Contamination and Toxicology*, **15**, 78 (1976).
3 Ervin, H.E. and McGinnis, G.D. *Journal of Chromatography*, **190**, 203 (1980).
4 De Ruiter, C., Bohle, J.F., De Jong, G., Brinkman, U.A.T. and Frei, R.W. *Analytical Chemistry*, **60**, 666 (1988).
5 Silgouer, D.P., Grasserbauer, M. and Barcelo, D. *Analytical Chemistry*, **69**, 2756 (1997).
6 Evans, N., James, D.E., Jackson, A.H. and Matlin, S.A. *Journal of Chromatography*, **115**, 325 (1975).
7 Eskins, K., Scholfield, C.R. and Dutton, H.H. *Journal of Chromatography*, **135**, 217 (1977).
8 Garside, C. and Riley, J.P. *Analytica Chimica Acta*, **46**, 179 (1969).
9 Borch, R.F. *Analytical Chemistry*, **47**, 2437 (1975).
10 Hoffman, N.W. and Liao, J.C. Analytical Chemistry, 48, 1104 (1976).
11 Hullett, D.A. and Eisenreich, S.J. Analytical Chemistry, 51, 1953 (1979).
12 Kieber, D.J., Vaughan, G.M. and Hopper, K. *Analytical Chemistry*, **60**, 1654 (1988).
13 Cassidy, R.M. and Niro, O.M. *Journal of Chromatography*, **126**, 787 (1976).
14 Okada, T. *Analytical Chemistry*, **62**, 734 (1990).
15 Willmott, F.W. and Dolphin, R.J. *Journal of Chromatographic Science*, **12**, 695 (1974).
16 Crathorne, B. and Watts, C.D. *Journal of Chromatography*, **169**, 436 (1979).
17 Miles, C.J. and Moye, H.A. *Analytical Chemistry*, **60**, 220 (1988).
18 Carpenter, R.A., Hollowell, R.H. and Hill, K.M. *Analytical Chemistry*, **69**, 3314 (1997).
19 Ratanathanawongs, S.K. and Crouch, S.R. *Analytica Chimica Acta*, **192**, 277 (1987).
20 Mori, S. *Journal of Chromatography*, **129**, 53 (1976).
21 Schouten, M.J., Copius Peereboom, J.M. and Brinkman, U.A.T. *International Journal of Environmental Analytical Chemistry*, **7**, 13 (1979).
22 Schouten, M.J., Copius Peereboom, J.M., Brinkman, U.A.T. et al. *International Journal of Environmental Analytical Chemistry*, **6**, 133 (1979).
23 Lewis, W.M. *Water Treat. Exam.*, **24**, 243 (1975).
24 Crathorne, B. and Field, M. *Pract. Anal. Div. Chem. Soc.*, **11**, 155 (1978).

25 Crane, R.I., Crathorne, B. and Fielding, N. In *Hydrocarbons and Halogenated Hydrocarbons in Aquatic Environments* (Environmental Science Research Series, Vol. 16), (Ed. Asghan, B.K.) Plenum, New York (1976).

26 Dong, M., Locke, D.C. and Ferrand, E. *Analytical Chemistry*, **48**, 368 (1976).

27 Krstulovic, A.M., Rosie, D.M. and Brown, P.R. *Analytical Chemistry*, **48**, 1383 (1976).

28 Lankmayr, E.P. and Muller, K. *Journal of Chromatography*, **170**, 139 (1979).

29 Nielson, T.J. *Journal of Chromatography*, **170**, 147 (1979).

30 Crosby, N.T., Hunt, D.C., Philip, L.A. and Patel, I. *Analyst (London)*, **106**, 135 (1981).

31 Hagenmaier, H., Feirabend, R. and Jager, W.W. *Zeitschrift für Wasser und Abwasser Forschung*, **10**, 99 (1971).

32 Black, J.J., Dymerski, P.P. and Zapisek, W.F. *Bulletin of Environmental Contamination and Toxicology*, **322**, 278 (1979).

33 Schönmann, M. and Kern, H. *Varian Instrument News*, **15**, 6 (1981).

34 Perkin-Elmer Ltd., Beaconsfield. *The Analytical Bulletin*, No. 1 p. 4. 2/2 (1981).

35 Brinkman, A.T., Seetz, J.W.F.L. and Reymer, H.G.M. *Journal of Chromatography*, **116**, 353 (1976).

36 Brinkman, A.T., Dokok, A., De Vries, G. and Reymer, H.G.M. *Journal of Chromatography*, **128**, 101 (1976).

37 Kaminsky, L.S. and Fasco, M.J. *Journal of Chromatography*, **155**, 363 (1978).

38 Bruckner, C.A., Ecker, S.T. and Synovec, R.E. *Analytical Chemistry*, **69**, 3465 (1997).

39 Nowack, A., Kari, F.G. Hilger, S.U. and Sigg, L. *Analytical Chemistry*, **68**, 561 (1996).

40 Dawson, R. and Pritchard, R.G. *Marine Chemistry*, **6**, 27 (1978).

41 Mopper, K. *Science of the Total Environment*, **49**, 115 (1986).

42 Varney, M.S. and Preston, M.R. *Journal of Chromatography*, **348**, 265 (1980).

43 Galassi, S. Committee of European Communities. Report EUR 10388 Organic Pollutants in the Aquatic Environment, pp. 71–76 (1988).

44 Saxena, J., Basu, D.K. and Kozuchowski, J. Method Development and Monitoring of Polynuclear Aromatic Hydrocarbons in Selected US Water, Health Effects Research Laboratory, TR–77–563, No. 24, Cincinnati, Ohio (1977).

45 O'Donnell, M. *Journal of Environmental Science*, **1**, 77 (1980).

46 Furuta, A. and Otsuki, A. *Analytical Chemistry*, **55**, 2407 (1983).

47 Black, J.J., Hart, T.F. and Black, P.J. *Environmental Science and Technology*, **16**, 247 (1982).

48 Ho, S.S.J., Butler, H.T. and Poole, C.J. *Journal of Chromatography*, **281**, 330 (1983).

49 Crosby, N.T. and Hunt, D.C. *Analytical Proceedings (London)*, **17**, 381 (1980).

50 Cartoni, G., Coccioli, F., Ronchetti, M., Simonetti, L. and Zoccolillo, L. *Journal of Chromatography*, **370**, 157 (1986).

51 Xia, L., Fan, C. and Hu, Z. *Fenxi Cshi. Tonghao*, **6**, 1 (1987).

52 Zawada, H. and Zerbe, J. *Probi. Fonovogo. Monit. Sostoyaniya Prir. Sredy.*, **4**, 264 (1988).

53 Oyler, A.R., Bodenner, D.L., Welsh, K.J., Liukkomen, R.J. and Carlson, R.M. *Analytical Chemistry*, **50**, 837 (1978).

54 Bianco, L., Marucchi, M. and Gasparini, G. *Int. Allment. (Pinerolo, Italy)*, **26**, 1020 (1987).

55 Walsh, J.E., MacCraith, B.D., Meaney, M. *et al. Analyst (London)*, **121**, 789 (1996).

56 Guenu, S. and Hennion, M.C. *Journal of Chromatography*, **665**, 243 (1994).

57 Whang, C.W. *Journal of the Chinese Chemical Society (Taipei)*, **34**, 81 (1987).
58 Alacon, P., Bustos, A., Canas, B., Andres, M.D. and Polo, M.L. *Chromatographia*, **24**, 613 (1987).
59 Lee, H.K., Li, S.F.Y. and Tay, Y.H. *Journal of Chromatography*, **438**, 429 (1988).
60 Armentrout, D.N., McLean, J.D. and Long, M.W. *Analytical Chemistry*, **51**, 1039 (1979).
61 Hashimoto, S., Miyata, T., Washino, M. and Kawakamis, W. *Environmental Science and Technology*, **13**, 71 (1979).
62 Ugland, K., Lundances, E., Griebrokle, E. and Bjorseth, A. *Journal of Chromatography*, **213**, 83 (1981).
63 Roseboom, H., Berkhoff, C.J., Wammes, J.T. and Wegman, R.C.C. *Journal of Chromatography*, **208**, 331 (1981).
64 Carlson, R.M., Swanson, T.A., Dyler, A.R. *et al. Journal of Chromatographic Science*, **22**, 272 (1984).
65 Sharp, R.E. and Meyer, G.S. *Analytical Chemistry*, **54**, 1164 (1982).
66 Rennie, P.S. *Analytical Proceedings (London)*, **24**, 295 (1987).
67 Schönmann, M. and Kern, H. *Varian Instrument Applications*, **15**, 6 (1981).
68 Ruban, V.F. and Belen'skii, G.B. *Zhur Analit Khim*, **43**, 1307 (1988).
69 Yushenko, U.V., Verpovskii, N.S., Zul'figarov, O.S. and Pilipenko, A.T. *Zhur Anal. Khim.*, **42**, 2033 (1987).
70 Takami, K., Mochizuki, K., Kamo, T., Sugimae, A. and Nakamoto, M. *Bunseki Kagaku*, **36**, 601 (1987).
71 Maris, F.A., Stab, J.A., De Jong, G. and Brinkman, U.A.T. *Journal of Chromatography*, **445**, 129 (1988).
72 Brown, L. *Analyst (London)*, **104**, 1165 (1979).
73 Mueller-Harvey, I. and Parkes, R.J. *Journal of Estuarine and Coastal Shelf Science*, **25**, 567 (1987).
74 Van Hoof, F., Wittock, A., Van Buggenhart, E. and Janssens, J. *Analytica Chimica Acta*, **169**, 149 (1985).
75 Whittle, P.J. and Rennie, P.J. *Analyst (London)*, **113**, 665 (1988).
76 Matsumoto, M., Nishikawa, Y., Murano, K. and Fukuyama, T. *Bunseki Kagaku*, **36**, 179 (1987).
77 Takani, K., Kuwata, K., Sigimae, A. and Nakamoto, M. *Analytical Chemistry*, **57**, 243 (1985).
78 Marcomin, A., Capri, S. and Giger, W. *Journal of Chromatography*, **403**, 243 (1987).
79 Wu, J., Li, J., Li, Y. and Song, H. Kexue. *Tongbao* (Foreign Language edition), **32**, 863 (1987).
80 Zhan, J., Wang, J. and Zhang, W. *Huanjing Huaxue*, **5**, 58 (1988).
81 Di Corcia, A., Marchett, A. and Mariomini, A. *Analytical Chemistry*, **63**, 1779 (1991).
82 Tan, Y. and Zhang, Y. *Huanjing Huaxue*, **5**, 50 (1988).
83 Nahai, A., Tsuji, K. and Yamanka, M. *Analytical Chemistry*, **52**, 2775 (1980).
84 Takami, K., Kamo, T., Mochizuki, K., Sugimas, A. and Nakamoto, K. *Bunseki Kagaku*, **36**, 278 (1987).
85 Saito, T., Higashi, K. and Hagiwar, K. *Fresenius Zeitschrift für Analytische Chemie*, **313**, 21 (1982).
86 Wee, V.T. *Water Research*, **18**, 223 (1984).
87 Varma, M.M., Patel, D., Wan, L. *et al. Proceedings of Water Quality Technical Conference* (1986) (Adv. Water. Anal. Treat.), **14**, 723–726 (1987).
88 Rice, J.R. and Kissinger, P.T. *Environmental Science and Technology*, **16**, 263 (1982).

89 Kim-Heang, S. *International Journal of Environmental Analytical Chemistry*, **15**, 309 (1983).
90 Wanatabe, T., Hongu, A., Honda, K. *et al. Analytical Chemistry*, **56**, 251 (1984).
91 Nishikawa, Y. *Journal of Chromatography*, **392**, 349 (1987).
92 Kano, H. and Ogura, N. *Sulshitsu Okadu Kenkyu*, **108**, 495 (1987).
93 Myoshi, H., Nagai, T. and Ishikawa, M. *Shizuoka-ken Eisel Kankyo Senta Hokoku*, **26**, 29 (1988).
94 Leggett, D.C. *Analytical Chemistry*, **49**, 880 (1977).
95 Janda, V. *Vodni Hospodarstvi, Series B*, **31**, 137 (1981).
96 Kummert, R., Molnar-Kbuica, E. and Giger, W. *Analytical Chemistry*, **50**, 1637 (1978).
97 Stozek, A. and Beumer, W. *Korrespondenz Abswasser*, **26**, 632 (1979).
98 Albro, P.W. and Fishbein, O. *Journal of Chromatography*, **69**, 273 (1972).
99 Needham, L.L., Surek, A.L., Head, S.L., Burse, V.W. and Liddle, J.A. *Analytical Chemistry*, **52**, 2227 (1980).
100 Dong, M.N. and DiCesare, J.L. *Journal of Chromatographic Science*, **20**, 517 (1982).
101 Cresser, C.S. and Al Haddad, A. *Analytical Chemistry*, **61**, 1300 (1989).
102 Gonzalez, E.B., Baumann, R.A., Cooijer, C., Velthorst, W.H. and Frei, R.U. *Chemosphere*, **16**, 1123 (1987).
103 Kimata, K., Hosoya, K., Araki, T. *et al. Analytical Chemistry*, **65**, 2502 (1993).
104 Blanco, G.E., Baumann, R.A., Gooljer, C., Velhorst, N.H. and Frei, R.W. *Chemosphere*, **16**, 1133 (1987).
105 Hamann, R. and Kettrup, A. *Chemosphere*, **16**, 527 (1987).
106 Akerblom, A. *Journal of Chromatography*, **319**, 427 (1985).
107 Vaughan, C.N. *Analytica Chimica Acta*, **131**, 307 (1981).
108 Hoke, S.H., Bryeagemann, E.E., Baxter, L.J. and Trybus, T. *Journal of Chromatography*, **357**, 429 (1986).
109 Thakker, N. and Alone, B.Z. *Indian Journal of Environmental Health*, **29**, 215 (1987).
110 Hill, K.K., Holowell, R.H. and Dal, L.A. *Analytical Chemistry*, **56**, 2465 (1984).
111 Anderson, J.L. and Chesney, D.J. *Analytical Chemistry*, **52**, 2156 (1980).
112 Anderson, J.L., Whiten, K.K., Brewster, J.D., Ou, T.K. and Nonidez, W.K. *Analytical Chemistry*, **57**, 1366 (1985).
113 Lewson, K., Schafter, L.J. and Freudenthal, J. *Journal of Chromatography Review*, **271**, 51 (1983).
114 Gowie, C.E., Kwakman, P., Frei, R.W. *et al. Journal of Chromatography*, **289**, 73 (1984).
115 Farrington, D.S., Hopkins, R.G. and Ruzicka, J.H.A. *Analyst (London)*, **102**, 377 (1977).
116 Schwarz, F.P. and Wasik, S.P. *Analytical Chemistry*, **48**, 525 (1976).
117 Monarca, S., Causey, B.S. and Kirkbright, F.G. *Water Research*, **13**, 503 (1979).
118 Gotz, R. *Facum Stadt Hygiene*, **29**, 10 (1978).
119 Shtivel, N.K., Gipikova, L.I. and Smirnova, Z.S. *Soviet Journal of Water Chemistry and Technology*, **7**, 66 (1985).
120 Mallevialle, J., Lefebvre, M., Roussau, C. and Sagbier, C. *Rev. Fronc. Sci., L'eau*, **1**, 25 (1982).
121 Hunter, O. *Environmental Science and Technology*, **9**, 241 (1975).
122 Stottmeister, E., Hendel, P. and Heemenau, H. *Chemosphere*, **17**, 801 (1988).
123 Gomez-Belinchou, J.T. and Albaiges, J. *International Journal of Environmental Analytical Chemistry*, **30**, 183 (1987).
124 Chiron, S., Fernandez Alba, A. and Barcelo, D. *Envirnomental Science & Technology*, **27**, 2352 (1993).

125 Di Corcia, A., Vlarchetti, M. and Samperi, R. *Journal of Chromatography*, **405**, 357 (1987).
126 Wright, L.H., Jackson, M.D. and Lewis, R.G. *Bulletin of Environmental Contamination and Toxicology*, **28**, 740 (1982).
127 Cochran, W.P., Languettern, M. and Trudeau, S. *Journal of Chromatography*, **243**, 307 (1982).
128 Cope, M.J. *Analytical Proceedings (London)*, **17**, 273 (1980).
129 Podowski, A.A., Freaz, M., Mertens, P. and Khan, M.A.Q. *Bulletin of Environmental Contamination and Toxicology*, **32**, 301 (1984).
130 Liu, L.Y. and Cooper, W.T. *Journal of Chromatography*, **390**, 285 (1987).
131 Farren, A., DePablo, J. and Hernandez, S. *Analytica Chimica Acta*, **212**, 123 (1988).
132 Tesarova, E. and Pakova, V. *Chromatography*, **17**, 269 (1983).
133 Huen, J.M., Gillard, R., Mayer, A.G., Baltensperger, B. and Keen, H. *Fresenius Zeitschrift für Analytische Chemie*, **348**, 606 (1994).
134 Shoaf, A.R. and Carison, W.C. *Weed Science*, **34**, 745 (1988).
135 Goewle, C.E. and Hogendoorn, E.A. *Journal of Chromatography*, **410**, 211 (1987).
136 Slobodnik, J., Groenwegen, M.G.M., Brouer, E.R., Lingeman, G. and Brinkman, U.A.T. *Journal of Chromatography*, **642**, 359 (1993).
137 Negre, M., Gennari, M. and Cignetti, A. *Journal of Chromatography*, **387**, 541 (1987).
138 Lawrence, J.F. and Turton, D. *Journal of Chromatography*, **159**, 207 (1978).
139 Marvin, C.H., Brindle, I.D., Hall, C.D. and Chilia, M. *Analytical Chemistry*, **62**, 1495 (1990).
140 Jones, E.O. *Analytical Chemistry*, **63**, 580 (1991).
141 Bellar, T.A. and Buddevc, T.A. *Analytical Chemistry*, **60**, 2076 (1988).
142 Simon, V.A. *Liquid Chromatography–Gas Chromatography*, **5**, 899 (1987).
143 Bushway, R.J. *Journal of Chromatography*, **303**, 263 (1984).
144 Lauren, D.R. and Agnew, M.P. *Journal of Chromatography*, **303**, 206 (1984).
145 Chiba, M. and Singh, R.P. *Journal of Agriculture and Food Chemistry*, **34**, 108 (1986).
146 Beck, R.A. *Analytical Chemistry*, **50**, 200 (1978).
147 Voyksner, P.D. *Analytical Chemistry*, **57**, 2600 (1985).
148 Breitura, V., Packenbusch, B. and Hutzinger, O. *International Journal of Environmental Analytical Chemistry*, **32**, 135 (1987).
149 Shoaf, W.T. *Journal of Chromatography*, **152**, 247 (1978).
150 Falkawski, P.G. and Sucher, J. *Journal of Chromatography*, **213**, 349 (1981).
151 Miles, C.J. and Brenzonik, P.L. *Journal of Chromatography*, **259**, 499 (1983).
152 Husser, E.R., Stehl, R.L., Price, D.R. and DeLap, R.A. *Analytical Chemistry*, **49**, 154 (1977).
153 Ludwig, F.J. and Besand, M.F. *Analytical Chemistry*, **50**, 185 (1978).
154 Walker, G.S., Coverey, M.J., Klug, M. and Wetzier, R.G. *Journal of Microbiological Methods*, **5**, 255 (1988).
155 Smith, W.F. and Healey, J.A. *Science of the Total Environment*, **37**, 71 (1984).
156 Games, D.E. *Analytical Proceedings*, **21**, 174 (1984).
157 Clark, J. *Journal of Chromatography International*, **22**, 12 (1987).
158 Reust, J.B. and Meyer, V.R. *Analyst (London)*, **107**, 673 (1982).
159 Afghan, B.K. and Batley, G.E. *Eau du Quebec*, **14**, 204 (1981).
160 Haky, J.E. and Young, A.M. *Journal of Liquid Chromatography*, **7**, 675 (1984).
161 Gillyon, E.C.D. *Laboratory Practice*, **28**, 1194 (1979).
162 Brown, L. and Rhead, M. *Analyst (London)*, **104**, 391 (1979).

163 Petrick, G., Schulz, D.E. and Duinker, J.C. *Journal of Chromatography*, **435**, 241 (1988).
164 Dai, S., Huang, G. and He, P. *Huanjing Kexue*, **9**, 44 (1988).
165 Marcomini, A., Striso, A. and Pavoni, B. *Marine Chemistry*, **21**, 15 (1987).
166 Samuelson, O.B. *Agriculture*, **60**, 161 (1988).
167 Mills, G.L., McFadden, E. and Quinn, J.G. *Marine Chemistry*, **20**, 313 (1987).
168 Cruezwa, J., Leuenberger, C., Tromp, J., Giger, W. and Abel, M.J. *Journal of Chromatography*, **403**, 233 (1987).
169 Nimura, N., Kinoshita, T., Yoshida, T., Uetake, A. and Nakai, C. *Analytical Chemistry*, **60**, 2067 (1988).
170 Ruepert, R. and Ploeger, E. *Fresenius Zeitschrift für Analytische Chemie*, **331**, 503 (1988).
171 Miles, C.J. and Moye, H.A. *Chromatographia*, **24**, 628 (1987).
172 Knopp, D., Vaeaenaenen, V. and Niesser, R. *Proceedings SPIE Int. Society Optical Engineering*, **2504**, 531 (1995).
173 LeCloirec, C., LeCloirec, P., Morvan, J. and Martin, G. *Rev. France Sic. L'Eau*, **2**, 25 (1983).
174 Golkiewicz, W., Werkhoven, G., Goewie, C.G. *et al. Journal of Chromatographic Science*, **21**, 27 (1983).
175 Clark, G.J., Goodwin, R.R. and Smiley, J.V. *Analytical Chemistry*, **57**, 2223 (1985).
176 Miles, C.J. and Defino, J.J. *Journal of Chromatography*, **299**, 275 (1984).
177 Staber, I. and Schulten, H.R. *Science of the Total Environment*, **16**, 249 (1980).
178 Nielen, H.W.K., Koomen, G., Frei, R.W. and Brinkman, U.A.T. *Journal of Liquid Chromatography*, **8**, 315 (1985).
179 Sorrell, R.K. and Reding, R.J. *Journal of Chromatography*, **185**, 655 (1979).
180 Sorrell, R.K., Dressman, R.C. and McFarren, E.F. AWWA Water Quality Technology Conference, Kansas City, MO, December 5–7, 1977, American Water Works Association, Denver, Colo, P.3A–3 (1978).
181 Perkin–Elmer Ltd. *The Analytical Bulletin*, No. 1, p.4 20 January (1982).
182 Morizane, K. Osaka-shi Sudokyoku-Komucbu Subshitsu. Shikensho Chosa Hakaku Narabini Shiken Senseki (37) 135 (1988).
183 Jentoft, R.E. and Gouw, T.H. *Analytical Chemistry*, **40**, 1787 (1968).
184 Vaughan, G.C., Wheals, B.B. and Whitehouse, J.J. *Journal of Chromatography*, **78**, 203 (1973).
185 Sorrell, R.K., Dressman, R.C. and McFarren, E.F. *Environment Protection Agency, Cincinnati, Ohio. High Pressure Liquid Chromatography for the Measurement of PAHs in Water*. Report No. 29344 p. 39 (1977), Paper presented at Water Quality Technology Conference, Kansas City, Missouri, 5–6 Dec., 1977
186 Hunt, D.C., Wild, P.J. and Crosby, N.T. *Water Research*, **23**, 643 (1978).
187 McMurtrey, K.C., Holcomb, K.E., Ekwenchi, A.U. and Fawcett, N.C. *Journal of Liquid Chromatography*, **7**, 953 (1984).
188 Takeuchi, T., Jin, Y. and Fshii, D. *Journal of Chromatography*, **321**, 159 (1985).
189 Clathorne, B., Palmer, C.P. and Stanley, J.A. *Journal of Chromatography*, **360**, 266 (1986).
190 Schulz, B. *Journal of Chromatography*, **299**, 484 (1984).
191 Fan, T.Y., Ross, R., Fine, D.H. and Keith, L.H. *Environmental Science and Technology*, **12**, 692 (1978).
192 Standing Committee of Analysts, HM Stationery Office, London. *Methods for the Examination of Waters and Associated Materials 1988. Determination of Acrylamide Monomer in Waters and Polymers* (1987).
193 Putzien, J.Z. *Wasser und Abwasser Forschung*, **19**, 228 (1986).

194 Lintelman, J., Mengel, C. and Kettrup, A. *Fresenius Zeitschrift für Analytische Chemie*, **346**, 752 (1993).
195 Watts, C.D. Water Research Centre Technical Report TR 110. *Mass spectrometric identification of non-volatile organic compounds* (1979).
196 Reunanen, M. and Kronfeld, R. *Journal of Chromatographic Science*, **20**, 449 (1982).
197 Rennie, P. and Mitchell, S.F. *Chromatographia*, **24**, 319 (1987).
198 Brunet, R., Bourdigot, M.M., Legube, B. and Dore, M. *Aqua No. 4 Scientific and Technical Review*, **7**, 6 (1980).
199 Godjehann, M. and Priess, W.H. *Analytical Chemistry*, **69**, 3832 (1977).
200 Das, B.S. and Thomas, G.H. *Analytical Chemistry*, **50**, 967 (1978).
201 Jolly, R.L., Cummings, R.B., Harman, S.J. *et al.* US National Technical Information Services, Springfield, VA Report No. PRNL/TM–6555, p. 89 (32710) (1979).
202 Burrows, E.P. Brueggewan, E. and Hoke, S.H. *Journal of Chromatography*, **294**, 494 (1984).
203 Lchotay, J., Matisova, E., Garaj, J. and Violova, A. *Journal of Chromatography*, **246**, 323 (1982).
204 Victor, P.M., Hall, R.E., Shamis, J.D. and Whitlock, S.A. *Journal of Chromatography*, **283**, 383 (1984).
205 Cook, A.M., Beilstein, P. and Hutter, R. *International Journal of Environmental Analytical Chemistry*, **14**, 91 (1983).
206 Ahel, M. and Giger, W. *Analytical Chemistry*, **57**, 1577 (1985).
207 Nielsen, P.B. *Chromatographia*, **18**, 323 (1984).
208 Thyseen, K. *Journal of Chromatography*, **319**, 99 (1985).
209 Batley, G.E. *Journal of Chromatography*, **389**, 409 (1987).
210 Dai, J. and Helz, G.R. *Analytical Chemistry*, **60**, 301 (1988).
211 Farren, A., Figueralo, E. and De Pablo, J. *International Journal of Environmental Analytical Chemistry*, **33**, 245 (1988).
212 Richardson, P.E., O'Grady, B.V. and Bremner, J.B. *Journal of Chromatography*, **268**, 341 (1983).
213 Bauer, F., Grant, C.L. and Jenkins, F.F. *Analytical Chemistry*, **58**, 176 (1986).
214 Li, W., Yin, P. and Yang, Y. *Welshhengwuxue Zazhi*, **7**, 50 (1987).
215 Crowley, T.O. and Larson, R.A. *Journal of Chromatographic Science*, **32**, 57 (1994).
216 Melcher, R.G., Dakker, D.W. and Hughes, G.H. *Analytical Chemistry*, **64**, 2258 (1992).
217 Nielson, M.W.E., Brinkman, U.A.T. and Frei, R.W. *Analytical Chemistry*, **57**, 806 (1985).
218 Cox, G.B. *Journal of Chromatography*, **116**, 244 (1976).
219 Kasiske, D., Klinkmuller, K.D. and Sonneborn, M.J. *Journal of Chromatography*, **149**, 703 (1978).
220 Jones, P.W., Graffeo, A.P., Detrick, R., Clarke, P.A. and Jakobsen, R.J. US National Technical Information Service, Springfield, Virginia. Report No. PB–259299 Technical manual of organic materials in process streams (1976).
221 Brown, L., Rhead, H.H. and Bancroft, K.G.C. *Analyst (London)*, **107**, 749 (1982).
222 Kishan, B. *Analytical Chemistry*, **47**, 1344 (1975).
223 Bhatia, K. *Analytical Chemistry*, **51**, 1344 (1979).
224 Olsson, L., Nicolas, R. and Samuelson, O. *Journal of Chromatography*, **123**, 355 (1976).
225 Putrizyky, D.J. and Chu, C.H. *Analytical Chemistry*, **49**, 860 (1977).
226 Sparacino, C.M. and Minick, D. *Journal of Environmental Science and Technology*, **14**, 880 (1980).

227 Ogan, K. and Stavin, W.J. *Journal of Chromatographic Science*, **16**, 517 (1978).
228 Legana, A., Petronia, B.M. and Rotatori, M. *Journal of Chromatography*, **198**, 143 (1980).
229 Hasler, C.F. *Bulletin of Environmental Contamination and Toxicology*, **12**, 599 (1974).
230 Burkhordt, L.P., Kueli, D.W. and Veith, G.D. *Chemosphere*, **14**, 1551 (1985).
231 Boyd-Boland, A.A. and Pawliszyn, J.B. *Analytical Chemistry*, **68**, 1621 (1996).
232 Beck, R.A. and Brink, J.J. *Journal of Environmental Science and Technology*, **10**, 173 (1976).
233 Giger, W., Brunner, P.H., Abel, M., Mariomini, A. and Schaffner, C. *Gas Wasser und Abwasser*, **67**, 111 (1987).
234 Brown, L., Rhead, M. *Analyst (London)*, **104**, 391 (1979).
235 Fernandez, P., Alder, A.C. and Giger, W. *National Meeting American Chemical Society, Division of Environmental Chemistry*, **33**, 303 (1993).
236 Sweetman, A.J. *Water Research*, **28**, 343 (1994).
237 Brun, G.L. *Bulletin of Environmental Contamination and Toxicology*, **24**, 886 (1980).
238 Mullins, F.G.P. and Kirkbright, G.F. *Analyst (London)*, **112**, 701 (1987).
239 Grzybowski, J., Radecki, A. and Rewkowska, G. *Environmental Science and Technology*, **17**, 44 (1983).
240 Brown, L., Rhead, H.M. and Braven, J. *Water Research*, **18**, 995 (1984).
241 Holt, M.S., Perry, J., McKerrell, E.H. and Watkinson, R. *Journal of Chromatography*, **362**, 419 (1986).
242 Water Research Centre UK. Notes on Water Pollution No. 70. *Analytical Methods for Organic Compounds in Sewage Effluents* (1980).
243 Scott, R.P.W. and Kucera, P. *Analytical Chemistry*, **45**, 749 (1973).
244 Snyder, L.R. *Analytical Chemistry*, **46**, 1384 (1974).
245 Reust, J.B. and Meyer, V.R. *Analyst (London)*, **107**, 673 (1982).
246 Krull, I.S., Bushee, D.S., Schleicher, R.G. and Smith, S.B. *Analyst (London)*, **111**, 345 (1986).
247 Bushee, D., Krull, I.S., Demko, P. and Smith, S.B. *Liquid Chromatography*, **7**, 861 (1984).
248 Kuldvere, A. *Analyst (London)*, **107**, 179 (1982).
249 Jewett, K.L. and Brinkman, F. *Journal of Chromatographic Science*, **19**, 583 (1981).
250 Brinkman, F.E., Blair, W.R., Jewett, K.L. and Iverson, W.P. *Journal of Chromatographic Science*, **15**, 493 (1977).
251 Brinkman, F.E. Trace Metals in Seawater. In *Proceedings of a NATO Advanced Research Institute on Trace Metals in Seawater* 30/3–4/4/81. Sicily (eds. C.S. Wong *et al.*) Plenum Press, New York (1981).
252 Committee on Medical and Biologic Effects on Environmental Pollutants, Selenium, National Academy of Sciences, Washington, DC (1976).
253 Brama, R.S. and Tomkins, M.A. *Analytical Chemistry*, **51**, 12 (1979).
254 Hodge, V.F., Seidel, G.L. and Goldberg, F.D. *Analytical Chemistry*, **51**, 1256 (1979).
255 Brinkmann, F.E. *Journal of Organometallic Chemistry*, **12**, 343 (1981).
256 Michael, L.C., Erickson, M.P., Parks, S.P. and Pellizzari, B.D. *Analytical Chemistry*, **52**, 1836 (1980).
257 Jackson, J.A., Blair, W.R., Brinkmann, F.E. and Iverson, W.P. Environmental Science and Technology, 16, 110 (1982).
258 Hallas, L.E. PhD Dissertation, University of Maryland (1978).
259 Parris, G.E., Blair, W.E. and Brinckmann, F.E. *Analytical Chemistry*, **49**, 378 (1977).

260 Nygren, O., Nilsson, C.A. and Frech, W. *Analytical Chemistry*, **40**, 2204 (1988).
261 Wang, C.N. and Yang, L.L. *Analyst (London)*, **113**, 1393 (1988).
262 Shen, L., Vela, N.P., Sheppard, B.S. and Carns, J.S. *Analytical Chemistry*, **63**, 1491 (1991).
263 Krull, I.S. and Pararo, K.W. *Applied Spectroscopy*, **39**, 960 (1985).
264 Liu, Y., Lopez-Avila, V., Alcaraz, M. and Beckert, W.F. *Journal of the American Association of Analytical Chemists International*, **78**, 1275 (1995).
265 Tye, C.T., Haswell, S.J., O'Neill, P. and Bancroft, K.C.C. *Analytica Chimica Acta*, **169**, 195 (1985).
266 Wauchope, R.D. and Yamamoto, M. *Journal of Environmental Quality*, **9**, 957 (1980).
267 Jyn Yynn Yu and Wai, C.M. *Analytical Chemistry*, **63**, 842 (1991).
268 Blais, J.S., Monplasir, G.M. and Marshall, W.P. *Analytical Chemistry*, **62**, 1611 (1990).
269 Walton, A.P., Wei Guro Izo, Liang, Z., Michel, R.G. and Morris, I.B. *Analytical Chemistry*, **63**, 222 (1991).
270 Brown, L., Haswell, S.J., Rhead, M.M., O'Neill, P. and Bancroft, K.C.C. *Analyst (London)*, **108**, 1511 (1983).
271 Blaszkewicz, M., Baumhoer, G. and Neidhart, B. *International Journal of Analytical Chemistry*, **28**, 207 (1987).
272 Ebdon, L. and Alonso, G. *Analyst (London)*, **112**, 1551 (1987).
273 Edward Iratimi, E.B. *Journal of Chromatography*, **256**, 253 (1983).
274 Bond, A.M. and Wallace, G.G. *Analytical Chemistry*, **54**, 1706 (1982).
275 Inoue, H. and Ito, K. *Microchemical Journal*, **49**, 249 (1994).
276 Gulens, J., Leeson, P.K. and Segiun, I. *Analytica Chimica Acta*, **156**, 99 (1982).
277 Shih, Y.T. and Carr, P.W. *Analytica Chimica Acta*, **142**, 55 (1982).
278 Shih, Y.T. and Carr, P.W. *Talanta*, **24**, 411 (1981).
279 Delle Site, A., Marchionni, V. and Testa, C. *Analytica Chimica Acta*, **117**, 217 (1980).
280 Jones, P., Hobbs, P.K. and Ebdon, L. *Analytica Chimica Acta*, **149**, 39 (1983).
281 Vlacil, F. and Hamplova, V. *Sb. Yns. Sk. Chem. Technol. Praze. Anal. Chem. H₂O*, 115 (1985).
282 Nagaosa, Y., Kawabe, H. and Bond, A.M. *Analytical Chemistry*, **63**, 28 (1991).
283 Rottman, L., Henmann, K.G. *Analytical Chemistry*, **16**, 3709 (1994).
284 Nakagawa, T., Aoyama, E., Hasegawa, N., Kobayashi, N. and Tanaka, H. *Analytical Chemistry*, **6**, 233 (1989).
285 Mieura, J. *Analytical Chemistry*, **62**, 1424 (1990).
286 Nagoasa, Y. and Kimata, Y. *Analytica Chimica Acta*, **327**, 203 (1996).
287 Cassidy, R.M. and Elchuk, S. *Journal of Chromatographic Science*, **19**, 503 (1981).
288 Cassidy, R.M. and Elchuk, S. *Journal of Chromatographic Science*, **18**, 217 (1980).
289 Testa, C. and Delle Site, A. *Journal of Radioanalytical Chemistry*, **34**, 121 (1976).
290 Stosanovic, R.S., Bond, A.M. and Bulter, E.C.V. *Analytical Chemistry*, **62**, 2692 (1990).
291 Rho, Y.S. and Choi, S.G. *Archives of Pharmacological Research*, **9**, 211 (1986).
292 Islikawa, T. and Hashimoto, Y. *Bunseki Kagaku*, **37**, 344 (1988).
293 Nakagawa, T., Aoyama, E., Hawegawa, N., Kobayashi, N. and Tanaka, H. *Analytical Chemistry*, **61**, 233 (1989).
294 Ebdon, L., Hill, S.J. and Jones, P. *Analyst (London)*, **110**, 515 (1985).
295 Hill, S.J., Brown, A., Kivas, C., Sparkes, S. and Ebdon, L. *Technical Instrumentation Analytical Chemistry*, **17**, 411 (1995).
296 Posta, J., Berndt, H., Luo, S.K. and Schaldach, G. *Analytical Chemistry*, **65**, 2590 (1993).

297 Smith, R.M. and Yankey, L.E. Analyst (London), 107, 744 (1982).
298 Baumann, R.A., Schreurs, M., Cooljer, C., Velhorst, M.H. and Frei, R.W. Canadian Journal of Chemistry, 65, 965 (1987).
299 Powell, M.J., Boomer, D.W. and Wiederin, D.R. Analytical Chemistry, 67, 2474 (1995).
300 Becker, G., Oestvold, G., Paul, P. and Seip, P.M. Chemosphere, 12, 1209 (1983).
301 Shijo, Y., Sato, H., Uehara, N. and Aratake, S. Analyst (London), 121, 325 (1996).
302 Riccardo, A.A., Muzzarelli, G.R. and Tubertini, O. Journal of Chromatography, 47, 414 (1970).
303 Mackey, D.J. and Higgins, H.W.J. Journal of Chromatography, 436, 243 (1987).
304 Posta, J., Alimonti, A., Petrucci, F. and Caroli, S. Analytica Chimica Acta, 325, 185 (1996).
305 Gardner, W.S. and St John, P.A. Analytical Chemistry, 63, 537 (1991).
306 Gardner, W.S., Herche, L.R., St John, P. and Seitzinger, S.P. Analytical Chemistry, 63, 1838 (1991).
307 Kerr, A., Kuperterschmidt, W. and Atlas, M. Analytical Chemistry, 60, 2729 (1988).
308 Cassidy, R.M., Elchuk, S. International Journal of Environmental Analytical Chemistry, 10, 1876 (1981).
309 Ichinaka, S., Hongo, N. and Yamazaki, M. Analytical Chemistry, 60, 2099 (1988).
310 Ruter, J., Fislage, U.P. and Neidhard, B. Chromatographia, 19, 62 (1984).
311 King, J.N. and Fritz, J.S. Analytical Chemistry, 59, 703 (1987).
312 Andrei, C.M. and Broekaert, J.A.C. Fresenius Journal of Analytical Chemistry, 346, 653 (1993).
313 Jen, J. and Chen, C. Analytica Chimica Acta, 1, 55 (1992).
314 Chau, E.C.M. and Farquharson, R.A. Journal of Chromatography, 178, 358 (1979).
315 Alawi, M.A. Fresenius Zeitschrift für Analytische Chemie, 313, 239 (1982).
316 Iskandarani, Z. and Petrzyk, J.D. Analytical Chemistry, 54, 2601 (1982).
317 Cooke, M. Journal of High Resolution Chromatography and Chromatography Communications, 6, 383 (1983).
318 Lee, S.H. and Field, L.R. Analytical Chemistry, 56, 2647 (1984).
319 Alawi, N.A. Fresenius Zeitschrift für Analytische Chemie, 317, 372 (1984).
320 Kok, S.H., Buckle, K.A. and Wootton, M. Journal of Chromatography, 160, 189 (1983).
321 Schroeder, D.C. Journal of Chromatographic Science, 25, 405 (1987).
322 Sakurai, N., Kadohata, K. and Ichinose, N. Fresenius Zeitschrift für Analytische Chemie, 314, 634 (1983).
323 Morgan, D.K. and Danielson, N.D. Journal of Chromatography, 262, 265 (1983).
324 Akaiwa, H., Kawamoto, H. and Osumi, M. Talanta, 29, 689 (1982).
325 Stetzenbach, K.J. and Thompson, G.M. Groundwater, 21, 36 (1983).
326 Tande, T., Pettersen, J.E. and Torgrimsen, T. Chromatographia, 13, 607 (1980).
327 Zun, J., Oshima, M. and Motomuzu, S. Analyst (London), 113, 1631 (1988).
328 Butler, E.C.V. and Gershev, R. Analytica Chimica Acta, 164, 153 (1984).
329 Moss, P.E. and Stephen, M.A. Analytical Proceedings (London), 22, 5 (1985).
330 Kishida, E., Nishimoto, Y. and Kojo, S. Analytical Chemistry, 64, 1505 (1992).
331 Rigas, P.G. and Pietrzyk, P.K. Analytical Chemistry, 60, 160 (1988).
332 Marengo, E., Gennaro, H.C. and Abrigo, C. Analytical Chemistry, 64, 1885 (1992).
333 Yaza, N., Nakashima, S. and Nakazato, V.T. Analytical Chemistry, 64, 1499 (1992).
334 Shibata, Y., Morita, M. and Fuwa, K. Analyst (London), 110, 1269 (1985).

335 Cartoni, G.P. and Coccioli, F. *Journal of Chromatography*, **360**, 225 (1986).
336 Migneault, D.R. *Analytical Chemistry*, **61**, 272 (1989).
337 Lin, C.I. and Huber, C.O. *Analytical Chemistry*, **44**, 2200 (1972).
338 Maki, S.A. and Danielson, N.C. *Analytical Chemistry*, **63**, 699 (1991).
339 Goncharova, I.A. and Khomenkô, A.N. Gidrokhim Mater., 53, 36 (1970). Ref. Zhur Khim. 199D Abstract No. 59339 (1971).
340 Verma, K., Jain, A. and Vernia, A. *Analytical Chemistry*, **64**, 1484 (1992).
341 Salov, V.V., Voshinaga, J., Shibita, Y. and Moritana, M. *Analytical Chemistry*, **64**, 2425 (1992).
342 Schroder, D.C. *Journal of Chromatography*, **25**, 405 (1987).
343 Takeuchi, T., Suzuki, E. and Ishii, D. *Journal of Chromatography*, **447**, 221 (1988).
344 Takana, K. and Ishizuki, T. *Water Research*, **16**, 719 (1982).
345 Hannah, R.E. *Journal of Chromatographic Science*, **24**, 336 (1986).

Chapter 5

High performance liquid chromatography–mass spectrometry

5.1 Organic compounds

5.1.1 Non saline waters

5.1.1.1 Sulphonylurea herbicides

Sulphonylureas form a class of herbicides introduced in the 1980s. From a chemical point of view sulphonylureas are labile and weakly acidic compounds. Compared to older herbicides, sulphonylurea herbicides have much lower use rates and are more rapidly degraded in soil. When present, very low concentrations (low parts per trillion region) of these herbicides in environmental waters are then to be expected. For this reason and because of the chemical characteristics cited above, monitoring of these herbicides in water is a particularly challenging problem. Although various separation techniques, such as gas chromatography–mass spectrometry of sulphonylurea derivatives [1], supercritical fluid chromatography [2] and capillary electrophoresis [3,4] have been proposed to analyse sulphonylurea in various matrices, liquid chromatography is the technique of choice [5–10]. Liquid chromatographic methods have become even more attractive since the introduction of robust and sensitive devices, such as thermospray and electrospray, to interface liquid chromatography to mass spectrometry.

Volmer *et al.* [11] evaluated the performances of both thermospray and electrospray ion sources for determining trace levels of sulphonylurea herbicides in water. In terms of specificity, they concluded that the latter source, in combination with a tandem mass spectrometer, was superior to the former one.

Further examples of the application of high performance liquid chromatography to the analysis of non saline waters are reviewed in Table 5.1.

5.1.1.2 Organophosphorus insecticides

Barcelo *et al.* [12] characterised 10 organophosphorus insecticides by high performance liquid chromatography–mass spectrometry using acetonitrile:water:chloroacetonitrile (69:30:1) as the mobile phase. The negative ion chemical ionisation spectra were dominated by the [M–R]⁻ ion with R being a methyl or ethyl group. At lower source temperatures (M + Cl)⁻ ions were formed. The negative ion chemical ionisation sensitivity was approximately an order of magnitude higher than in the positive ion mode.

Barcelo [13] characterised selected pesticides by negative ion chemical ionisation thermospray high performance liquid chromatography–mass spectrometry. Ions observed in the negative ion chemical ionisation spectra corresponded to mechanisms of anion attachment ([M + acetate]⁻, electron capture ([M]⁻) and dissociative electron capture ([M–R]⁻). Sensitivity was lower in the negative ion chemical ionisation mode than in the positive ion mode.

See also Table 5.1.

5.1.1.3 Ozonisation products

Jolly *et al.* [33] have conducted a study into the non volatile products produced during the chlorination, ozonisation and ultraviolet irradiation of supply water and water produced in waste water treatment plants. High performance liquid chromatography and mass spectrometry were used to separate organic constituents from concentrates of samples before and after treatment. Chromatographic profiles reveal the differences due to disinfection of the treated sample. High performance liquid chromatographic analysis of chlorinated, ozonated and ultraviolet irradiated secondary effluents indicates that both chlorination and ozonisation destroy chromatographic constituents and also produce chromatographic constituents. Ultraviolet irradiation at disinfection levels appears to have little chemical effect on the non volatile constituents separated by high performance liquid chromatography.

5.1.1.4 Other applications

Other applications of high performance liquid chromatography mass spectrometry to the detection of organic compounds are reviewed in Table 5.1.

5.1.2 Potable waters

5.1.2.1 Sulphonylurea herbicides

Di Corcia *et al.* [34] have discussed methods for the trace analysis of sulphonylurea herbicides in potable waters.

Table 5.1 High performance liquid chromatography–mass spectrometry of organic compounds in non saline waters

Organic	Stationary phase	Eluent	Comments	Detection	LD	Ref.
Alkylated ammonium ion	–	–	–	–	–	[14]
Polyethylene surfactants	–	–	Graphitised carbon black preconcentration	LC–MS with electrospray interface	ppt	[15]
Sulphonated Azo dyes	–	–	Optimisation of pH, ion source parameters and buffer composition.	Thermospray HPLC–MS	–	[16]
			Extraction from water with C_{18} resin and SAX	Spectrofluorescence and mass spectrometric	–	[17]
Methyl, nitro and chlorophenols (EPA priority pollutants)	–	–	–	LC–MS (atmospheric pressure chemical ionisation and ion spray interfaces)	ppt	[18]
Alkyl sulphonates and alkylethyl sulphonates	–	–	–	Ion spray mass spectrometry	–	[19]
DNA	–	–	–	LC–MS with thermospray interface	–	[20]
168 pollutants	–	–	–	LC–GC–MS	–	[21]
52 pesticides	–	–	–	Quadruple MS through a thermospray interface	–	[22]
Acidic, basic and neutral pesticides	–	–	–	LC–mass particle beam interface	–	[23]
20 acidic pesticides	Reverse phase C_{18} column	Mobile phase 0.1M K_2HPO_4 –0.2M BuNF	–	Electrospray LC–MS	2.5–200ng	[24]

Table 5.1 continued

Organic	Stationary phase	Eluent	Comments	Detection	LD	Ref.
Carbamate, methyl urea and oxime pesticides and herbicides	–	Ag 0.1M NH_4O CCH_3	Large dynamic range of 10^4	Thermospray HPLC–MS	1ppb	[25]
Organophosphorus and carbamate insecticides	–	–	–	LC–MS with thermospray or electrospray interface	sub ppb	[26]
Organophosphorus, carbamate, phenylurea and triazine insecticides	–	–	–	LC–MS or LC–MS–MS	ppb to ppt	[27]
Carbamate insecticides	–	–	–	Off-line LC–MS with thermospray interface	sub ppb	[28]
Polar pesticides	–	–	–	Thermospray LC–MS	1–100ppt	[29]
Acidic pesticides	–	–	Teflon-coated electron impact ion source	LC–MS	0.1ppb	[30]
Polar pesticides	–	–	–	LC–MS atmospheric pressure chemical ionisation interface	sub ppb	[31]
Tamephos	–	–	–	Off-line LC–MS thermospray interface	sub ppb	[32]

Source: Own files

Seven commonly used sulphonylureas, *viz.* thifensulphuron methyl, metsulphuron methyl trisulphuron, chlorosulphuron, rimusulphuron, tribenzuron methyl and bensulphuron methyl were extracted from water by off-line solid-phase extraction with a Carbograph 4 cartridge. Sulphonylurea herbicides were then isolated from both humic acids and neutral contaminants by differential elution. Analyte fractionation and quantification were performed by liquid chromatography with ultra-violet detection. Recoveries of sulphonylurea herbicides extracted from 4L of potable water (10ng L^{-1} spike level), 2L of ground water (50ng L^{-1} spike level) and 0.2L of river water (250ng L^{-1} spike level) were not lower than 94%. Depending on the particular sulphonylurea herbicides, method detection limits were 0.6–2ng L^{-1} in drinking water, 2–9ng L^{-1} in ground water and 13–40ng L^{-1} in river water.

5.1.2.2 Non volatile organic compounds

Crathorne *et al.* [35] have described a procedure for solvent extracting, separating and identifying trace levels of non volatile organic compounds in potable water. A method was developed for analysing four selected chlorine–containing compounds and this was then extended using high performance liquid chromatography linked to mass spectrometry for the analysis of a wider range of non volatile organics. These workers showed that either vacuum evaporation or freeze drying can be successfully used to isolate and concentrate non volatile organic matter from potable water.

This technique was applied to several potable water samples. Crathorne *et al.* [35] used the above technique to identify and determine the following four non volatile chlorinated organic compounds in potable water.

Fig. 5.1 shows a high performance liquid chromatogram obtained for the four compounds present in a sample of potable water. The large peak at the start of each chromatogram is essentially unretained and is probably

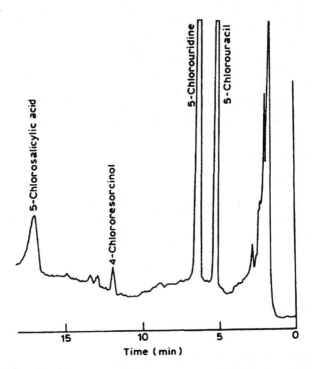

Fig. 5.1 High performance liquid chromatogram of four components of potable water sample. Conditions (200ng each component): column 15cm × 7mm id; packing 5...m Spherisorb ODS. Diluent, linear gradient from 1% methanol in 0.1% aqueous acetic acid to 90% methanol in 0.1% aqueous acetic acid over 30min. Detections: UV absorption at 280nm for 5-chlorouracil, 5-chlorouridine and 4-chlororescorcinol and at 315nm for 5–chlorosalicyclic acid
Source: Reproduced by permission from Elsevier Science, UK [35]

equivalent to an excluded peak containing much of the higher molecular weight material. A large peak in this position was always observed when analysing potable water extracts. Monitoring the four compounds was carried out using ultraviolet detection at two wavelengths. The absolute detection limit is defined as the amount injected onto the column which results in a signal-to-noise ratio of 2:1.

5.1.3 Ground waters

5.1.3.1 Organophosphorus insecticides

Lacorte and Barcelo [36] have described a procedure for the determination of nanogram per litre levels of organophosphorus pesticides in ground waters based on automated on-line liquid–solid extraction followed by liquid chromatography. The detection used on the liquid chromatograph is an atmospheric pressure chemical ionisation mass spectrometer using negative and positive ion modes of operation.

The technique was used to determine several organophosphorus pesticides (E)– and (Z)–Mevinphos, Dichlorvos, Azinphos–methyl, Azinphosethyl, Parathion–methyl, Parathion–ethyl, Malathion, Fenitrothion, Fenthion, Chlorfenvinphos and Diazinon, in groundwater.

5.1.4 Waste waters

5.1.4.1 Azo dyes

High performance liquid chromatography–mass spectrometry has been applied to the determination of down to 1mg L^{-1} of azo dyes in waste waters [37]. Games et al. [38] used high performance liquid chromatography for solute focusing prior to superfluid chromatography–mass spectrometry in the analysis of waste streams.

References

1 Klaffenbach, P. and Holland, P.T. Journal of Agriculture and Food Chemistry, 41, 388 (1993).
2 Berger, A. Chromatographia, 41, 133 (1995).
3 Garcia, F. and Henion, J. Journal of Chromatography, 606, 237 (1992).
4 Dinelli, G., Vicari, A. and Bonetti, A. Journal of Chromatography, 700, 195 (1995).
5 Raiser, R.W., Barefoot, A.C., Dietrich, R.F. et al. Journal of Chromatography, 554, 91 (1991).
6 Howard, A.L. and Taylor, L.T. Journal of Chromatographic Science, 30, 374 (1992).
7 Schneider, G.E., Koeppe, M.K., Naidu, M.V. et al. Journal of Agriculture and Food Chemistry, 41, 2404 (1993).
8 Nilve, G., Knutsson, M. and Jonsson, J.A. Journal of Chromatography, 688, 75 (1994).
9 Galletti, G.C., Bonetti, A. and Dinelli, G. Journal of Chromatography, 692, 27 (1995).
10 Cambon, J.P. and Bastide, J. Journal of Agriculture and Food Chemistry, 41, 333 (1996).
11 Volmer, D., Wilkes, J.G. and Leveson, K. Rapid Communications Mass Spectrometry, 9, 767 (1995).
12 Barcelo, D., Maris, F.A., Geerdink, R.B. et al. Journal of Chromatography, 394, 65 (1987).
13 Barcelo, D. Liquid Chromatography Gas Chromatography, 6, 324 (1988).
14 Stein, M. Vom Wasser, 69, 39 (1987).
15 Crescenzi, C., Di Corcia, A., Samperi, R. and Mariomini, A. Analytical Chemistry, 67, 1797 (1995).
16 Flory, D.A., McLean, M.M., Vestai, M. and Belowski, L.D. Rapid Communications in Mass Spectrometry, 1, 48 (1987).
17 Scullion, S.D., Clench, M.R., Cooke, M. and Ashcroft, A.E. Journal of Chromatography, 732, 207 (1996).

18 Puig, D., Silgance, I., Grasserbauer, M. and Barcelo, D. *Analytical Chemistry*, **69**, 2756 (1997).
19 Popenoe, D.D., Morris, S.J., Horn, P.S. and Norwood, K.T. *Analytical Chemistry*, **66**, 1620 (1994).
20 Kuchl, D.W., Serrano, J. and Naumann, S. *Journal of Chromatography, A*, **684**, 113 (1994).
21 Bulterman, A.J., Vreuls, J.J., Ghissen, R.T. and Brinkman, U.A.T. *Journal of Chromatography*, **16**, 397 (1993).
22 Bellar, T.A. and Budde, W.L. *Analytical Chemistry*, **60**, 2076 (1988).
23 Cappiello, A., Famiglini, G. and Bruner, F. *Analytical Chemistry*, **66**, 1416 (1994).
24 Cresenzi, C., Di Corcia, A., Marchese, S. and Sampori, R. *Analytical Chemistry*, **68**, 1968 (1996).
25 Yoyksner, R.D. Chemical Analysis (N.Y.), 91 (Appl. New Mass Spectrum Tech. Pest. Chem. pp.146–160) (1987).
26 Barcelo, D., Chiron, S., Honing, M., Molina, C. and Abian, J. *Advanced Mass Spectrometry*, **13**, 465 (1995).
27 Hobanboom, A., Slobodnick, J., Vreuls, J.J. *et al. Chromatographia*, **42**, 506 (1996).
28 Honing, M., Riu, J., Barcelo, D., Van Baar, B.L.M. and Brinkman, U.A.T. *Journal of Chromatography*, **733**, 283 (1996).
29 Sennert, S., Volmer, D., Levsen, K. and Weunsch, G. *Fresenius Journal of Analytical Chemistry*, **351**, 642 (1995).
30 Capielo, A., Famiglini, G., Palma, A., Berloni, A. and Bruner, F. *Environmental Science and Technology*, **29**, 2295 (1995).
31 Spliid, N.H. and Koppen, B. *Journal of Chromatography*, **736**, 105 (1996).
32 Lacorte, S., Ehresmann, N. and Barcelo, D. *Environmental Science and Technology*, **30**, 917 (1996).
33 Jolly, R.L., Cummings, R.B., Denton, M.S. *et al.* Oak Ridge National Laboratory, Union Carbide Corp., Report ORNL/TM6555 (1985).
34 Di Corcia, A., Cresenzi, C., Samperi, R. and Scappaticco, L. Analytical Chemistry, 69, 2819 (1997).
35 Crathorne, B., Watts, C.D. and Fielding, M. *Journal of Chromatography*, **185**, 671 (1979).
36 Lacorte, S. and Barcelo, D. *Analytical Chemistry*, **68**, 2464 (1996).
37 Belowski, L.D., Pyle, S.D., Ballard, J.M. and Shaul, G.M. *Biomedical Environmental Mass Spectrometry*, **14**, 343 (1987).
38 Games, D.E., Rontree, J.A. and Fowlis, I.A. *Journal of High Resolution Chromatography*, **17**, 68 (1994).

Chapter 6

Conventional column chromatography

Conventional column chromatography which predates more modern techniques such as high performance liquid chromatography is still of value in the analysis of waters. Applications of column chromatography are similar to those of high performance liquid chromatography. This technique is discussed in this chapter.

6.1 Organic compounds

6.1.1 Non saline waters

6.1.1.1 Hydrocarbons

Bundt et al. [1] separated low-boiling petroleum hydrocarbons from the aliphatic, mono-, di- and polyaromatics by column chromatography. This separation was simplified by removal of non volatile polar components using a silica gel–aluminium oxide column. See also Table 6.3.

6.1.1.2 Non ionic detergents

Huber et al. [2] have described a column chromatographic method for the rapid separation and determination of non ionic detergents of the polyoxyethylene glycol monoalkylphenyl type and applied it to the analysis of water samples. Such adducts ranging in chain length from 1 to 20 ethylene oxide units were separated on 23cm columns of PEG–400 on Spherosil with three mobile phases, viz, 2,2,4-trimethylpentane, 2,2,4-trimethylpentane-CCl₄ (2:1) and 2,2,4-trimethylpentane-CCl₄ (1:2) representing the extremes in polarity for separating the oligomers in a reasonable time. Use of the first mobile phase permitted separation of the oligomers produced by the reaction between 1mol of phenol and from 3 to 7mol of ethylene oxide, and of the second mobile phase those produced similarly with 7 to 14mol of ethylene oxide (but with no separation on the shorter oligomers). With use of the third mobile phase there was no separation of oligomers, but for each sample a peak of Poisson distribution was obtained that represented the oligomer distribution of the sample. Detection was by ultraviolet absorption; the relative precision was 0.4% and the detection was 0.2µg.

Otsuki and Shiraishi [3] used reversed phase absorption liquid chromatography and field desorption mass spectrometry to determine polyoxyethylene alkylphenyl ether non ionic surfactants in water. In the separation of polyoxyethylene octylphenyl, nonylphenyl and dodecyl-phenyl ethers by gradient elution with a holding process by holding the mobile phase composition of 50:50 water–methanol, polyoxyethylene alkylphenyl ethers were eluted by a further initiation of the gradient in the order octylphenyl, nonylphenyl and dodecylphenyl ethers, indicating that the retention time was related to the number of alkyl carbon atoms.

Calibration graphs of polyoxyethylene, octylphenyl and nonylphenyl ethers with a scale range of 0.05 absorbance units were linear between 0 and 20µg.

Otsuki and Shiraishi [3] examined the separated substances by field desorption mass spectrometry in order to assess the carbon number and purity.

From the field desorption mass spectra of standard samples, a table for identification of poly(oxyethylene) alkylphenyl ethers and determination of the degree of polymerisation of ethylene oxide was constructed as shown in Table 6.1; n is the number of alkyl carbon atoms and m is the degree of polymerisation of ethylene oxide. When the field desorption mass spectrum having a peak pattern with the difference of $44m/z$ was obtained such as the peaks at 484, 528, 572, 616 and $660m/z$, Table 6.1 would show that those peaks are due to poly(oxyethylene) nonylphenyl ethers with the degree of polymerisation of 6–10 of ethylene oxide. Table 6.2 also shows the identification of poly(oxyethylene) dialkylphenyl ethers and determination of the degree of polymerisation of ethylene oxide based on calculations of the molecular weight.

Evans et al. [4] carried out a quantitative determination of linear primary alcohol ethyoxylate surfactants in environmental waters by thermospray liquid chromatography–mass spectrometry.

6.1.1.3 Other applications

Other applications of this technique to the determination of organic compounds in non saline waters are reviewed in Table 6.3.

6.1.2 Seawater

6.1.2.1 Hydrocarbons

Wasik [16] has used an electrolytic stripping cell to determine hydro-carbons in seawater. Dissolved hydrocarbons in a known quantity of

Table 6.1 Table for identification of poly(oxyethylene)alkylphenyl ethers from field desorption mass spectra (m/z values)

$$C_nH_{2n+1} \cdot C_6H_4 \cdot (C_2H_4O)_mH$$

n	4	5	6	7	8	9	10	11	12	13	14	15	16
6	354	398	442	486	530	574	618	662	706	750	794	838	882
7	368	412	456	500	544	588	632	676	720	764	808	852	896
8	383	426	470	514	558	602	646	690	734	778	822	866	910
9	396	440	484	528	572	616	660	704	748	792	836	880	924
10	410	454	498	542	586	630	674	718	762	806	850	894	938
11	424	468	512	556	600	644	688	732	776	820	864	908	952
12	438	482	526	570	614	658	702	746	790	834	878	922	966
13	452	496	540	584	628	672	716	760	804	846	892	936	980

Source: Reproduced by permission from the American Chemical Society [3]

Table 6.2 Table for identification of poly(oxyethylene)dialkylphenyl ethers from field desorption mass spectra (m/z values)

n	3	4	5	6	7	8	9	10	11	12	13	14	15
6	394	438	526	526	570	614	658	702	746	790	834	878	922
7	422	466	510	554	598	642	686	730	774	818	862	906	950
8	450	494	538	582	626	670	714	758	802	846	890	934	978
9	478	522	566	610	654	698	742	786	830	874	918	962	1006
10	506	550	594	638	682	726	770	814	858	902	946	990	1024
11	534	578	622	666	710	754	798	842	886	930	974	1018	1062
12	562	606	650	694	738	782	826	870	914	958	1002	1046	1090
13	590	634	678	722	766	810	854	898	942	986	1030	1074	1118

Source: Reproduced by permission from the American Chemical Society [3]

seawater were equilibrated with hydrogen bubbles, evolved electrolytically from a gold electrode, rising through a cylindrical cell. In an upper headspace compartment of the cell the hydrocarbon concentration is determined by gas chromatography. The major advantages of this cell are that the hydrocarbons in the upper compartment are in equilibrium with the hydrocarbons in solution, and that the hydrogen used as an extracting solvent does not introduce impurities into samples.

Wasik [16] used this method to successfully determine ppb of gasoline in seawater. He found that a convenient method for concentration of the hydrocarbons is to recycle the hydrogen stream containing the hydrocarbons many times over a small amount of charcoal (2.3mg). The

Table 6.3 Conventional column chromatography of organic compounds in non saline waters

Organic compound	Stationary phase	Eluent	Comments	Detection	LD	Ref.
Aliphatic hydrocarbons	Kierelgel, Florasil and alumina	–	Removal of fats and fatty acids discussed	–	–	[5]
Benzene, toluene, xylene	Liquid coated furzed silica film	–	–	–	–	[6]
Polyaromatic hydrocarbons	–	–	Comparison with gas chromatography–mass spectrometry results	–	–	[7]
Carboxylic acids	–	–	1–pyrenyl diazomethane used as labelling reagent	Excitation of fluorescent reagent at 340nm (emission at 395nm)	20–30fmole	[8]
Phenols	–	–	–	Electrochemical detection	low ppb	[9]
Aromatic sulphonates	–	–	Solid phase extraction then LC	–	–	[10]
Triaryl, trialkyl phosphates	–	–	Solid phase extraction then LC or gas chromatography	–	40–100pg	[11]
Hydroxytriazines	–	Acetonitrile	Mass spectrometry	–	100ppt	[12]
Chloroaniline	–	–	–	–	0.1µg L^{-1}	[13]
Nitrophenols	–	–	–	Spectrophotometric	–	[14]
Aliphatic diamines	–	–	–	–	14–20µg L^{-1}	[15]

Source: Own files

charcoal, while still in the filter tube, was extracted three times with 5μL of carbon disulphide. A 2–3μL aliquot of this solution was then injected into a SCOT capillary column.

6.1.2.2 Organoboron compounds

Mills *et al.* [17] have reported that tetraphenylboron and diphenyboric acid in seawater samples can be preconcentration on C_{18} reverse phase resin. Naturally occurring organic matter in a black water coastal plain stream did not interfere.

6.1.3 Sewage effluents

6.1.3.1 Carboxylic acids

Methods have been described for the determination of total fatty acids in raw sewage sludge. These methods [18–20] require concentration steps such as simple distillation, steam distillation, evaporation, or extraction [21–23] which resulted in greater losses of the volatile matter [24].

Straight distillation or steam distillation of volatile acids and their chromatographic separation have been proposed in determining organic acids in sludge. In this method an acidified aqueous sample, containing relatively high concentrations of organic acids, is adsorbed on a column of silicic acid and the acids are eluted with *n*-butanol in chloroform. The eluate is collected and titrated with standard base. All short-chain (1–6 carbon) organic acids are eluted, but so are crotonic, adipic, pyruvic, phthalic, fumaric, lactic, succinic, malonic, aconitic and oxalic acids, as well as alkyl sulphates and alkyl–aryl sulphonates. No information on the individual volatile acids is obtained by this method and the results are reported collectively as total organic acids. Various chromatographic methods, such as paper [25,26] and gel chromatography [20], have been used for the analysis of sludge digester liquor. Mueller *et al.* [24] have modified the indirect chromatographic method for samples of raw sewage and river water involving tedious concentration steps, leading to losses. In paper chromatography individual volatile acid concentrations should be higher than 600mg L $^{-1}$, while in other methods the minimum detectable level is 1000mg L $^{-1}$ [25,26].

6.1.4 Swimming pool waters

6.1.4.1 Organochlorine compounds

Maierski *et al.* [27] have studied the differential determination of volatile and non volatile organochlorine compounds in swimming pool waters. They describe methods for fractionating organochlorine compounds into groups of differing polar properties during the analysis of swimming pool waters. The volatile chlorine compounds are vaporised, mineralised at 1100°C and determined by a microcoulometer; for non volatile compounds an enrichment technique is used based on the principle of reversed phase liquid chromatography.

6.1.5 Waste waters and trade effluents

6.1.5.1 Miscellaneous organic compounds

See Table 6.4.

6.2 Organometallic compounds

6.2.1 Non saline waters

6.2.1.1 Organogermanium compounds

Sazaki *et al.* [41] have studied liquid–liquid extraction of methylated and inorganic germanium ($(CH_3)_n Ge(OH)_{4-n}$ $n = 0$, 1, 2 and 3) in aqueous solution (pH 1–12) with organic ligands to develop a separation method for germanium compounds. Ligands containing a negatively charged oxygen donor were proved to be the most powerful extractants for germanium compounds. Using benzoic acid, trimethylgermanium is extracted into carbon tetrachloride, while monomethylgermanium, dimethylgermanium and inorganic germanium are not extracted into the organic phase. The extracted species is trimethylgermanium–benzoic acid, of which the benzoate ion is monodentate. Catechol and mandelic acid produce monoanionic complexes of germanium and monomethylgermanium ($[(CH_3)_n Ge-(OH)_{1-n} L_2]^-$; $n = 0$ and 1), of which the coordination number about the central germanium atom is five or six. These compounds are extracted into the nitrobenzene phase accompanied by tetrabutylammonium as a counter cation. Dimethylgermanium and trimethylgermanium are not extracted as the catecholate and mandelate complexes, because they have low affinity for higher coordinated states. This study demonstrates for the first time that germanium compounds can be separated on the basis of their stereochemistry in solution.

6.2.1.2 Organomercury compounds

On-line preconcentration followed by liquid chromatographic separation and cold vapour atomic absorption spectrometry have been used to detect 0.5ppt methylmercury in non saline water [43].

Table 6.4 Conventional column chromatography of cations in non saline waters, waste waters and trade effluents

Organic compound	Stationary phase	Eluent	Comments	Detection	LD	Ref.
Ethylene diamine tetraacetic acid and nitroacetic acid	Reverse phase column	Ag trichloro acetic acid (pH<2)	Amino acids, citric acid and fulvic acid do not interfere	Carbon paste electrode	–	[28]
Chlorotriazines	–	–	–	–	–	[29]
Abate	Reverse phase adsorption LC	–	–	–	50–15µg	[30]
Chlorotoluron (3-(3 chloro-toluyl-dimethyl urea)	–	–	–	–	–	[31]
Organophosphates	–	–	–	–	–	[32,33]
Polychlorobiphenyls	–	–	–	GC	–	[34]
Parathion-ethyl	–	–	–	–	–	[35]
Urea herbicides	–	–	–	Electrochemical Kelgraft composite electrode	62–410µg	[36]
			Waste waters			
Nitriloacetate	–	–	–	–	0.1mg	[37]

Table 6.4 continued

Organic compound	Stationary phase	Eluent	Comments	Detection	LD	Ref.
			Trade effluents			
Nitriloacetate	–	–	–	–	0.1mg	[37]
Polychlorobiphenyls and chlorinated insecticides	Silica gel	–	–	–	0.02–0.9 (organochlorine insecticides) 0.5–0.9μg L^{-1} (polychloro-biphenyls)	[38]
Arochlor 1242 (PCB)	–	–	Study of effect of suspended solids	–	–	[39,40]

Source: Own files

6.2.1.3 Organotin compounds

Yang *et al.* [42] have speciated organotin compounds by reversed phase liquid chromatography with inductively coupled plasma atomic emission spectrometric detection. The separation took less than 6min with detection limits in the range 2.8–16pg of tin for the various species.

6.2.2 Aqueous precipitation

6.2.2.1 Organolead compounds

To preconcentrate trialkyllead species in rainwater Blaszkewicz *et al.*[51] complexed interfering metal ions in rain water with EDTA before adjustment of the pH to 10. Samples were pumped through an extraction column of silica gel to adsorb lead compounds which were then desorbed with acetate buffer containing methanol at pH3.7. The eluate was diluted and adjusted to pH8 with borate buffer before further concentration on a Nucleosil 10–C_{18} pre-column. Adsorbed trialkyllead compounds were eluted by back-flushing on to a RP–C_{18} column and separated with methanoic acetate buffer. On-line detection used a post-column chemical reaction detector. Detection limits for sample volumes of 500mL were 15µg L^{-1} and 20µg L^{-1} for trimethyl- and triethyllead, respectively. Standard deviation was less than 4% for a sample containing 90pg triethyllead per ml.

6.2.3 Seawater

6.2.3.1 Organoarsenic compounds

Yamamoto [44] separated organoarsenic compounds from seawater by column chromatography. The organoarsenic compounds were reduced to arsine with sodium borohydride and analysed by atomic absorption spectrometry.

6.2.4 Potable waters

6.2.4.1 Organoarsenic compounds

Liu *et al.* [45] separated sub ng amounts of arsenite, monomethylarsenic and dimethylarsinic acids using dodeyldimethyl–ammonium bromide vesicles for liquid chromatography coupled to an inductively coupled plasma atomic emission spectrometer.

6.3 Cations

6.3.1 Non saline waters

6.3.1.1 Calcium and magnesium

Liquid chromatography was used by Rho and Choi [46] for the simultaneous determination of calcium and magnesium. A column of 4.6mm by 25cm, containing Zipax SCX was used for the separation. Linear calibration curves were obtained in the range of 1×10^{-4} to 5×10^{-4}M and the correlation coefficient of the calibration curves was in the range of 0.9952–0.9996.

6.3.1.2 Cobalt

Boyle et al. [47] have described a method for determining cobalt in non saline waters using cation exchange liquid chromatography. Cobalt was determined directly in 500μL fresh water samples with a detection limit of 20pM per kg.

6.3.2 Seawater

6.3.2.1 Transition metals

Riccardo et al. [48] showed that chitosan is promising as a chromato-graphic column for collecting traces of transition elements from salt solution, and seawater and for recovery of trace metal ions for analytical purposes. Traces of transition elements can be separated from sodium and magnesium, which are not retained by the chitosan.

6.4 Anions

6.4.1 Non saline waters

6.4.1.1 Carboxylates

Liquid chromatography separation and indirect detection using iron(II), 1,10–phenanthroline as a mobile phase additive have been used for the separation of carboxylates (and sulphonic acids) [49]. Detection limits approach 20ng.

6.4.2 Waste waters

6.4.2.1 Sulphate

Column chromatography [50] has been used to determine sulphate in waste water. A post-column solid phase reaction detector was employed in conjunction with an anion exchange separation column to determine 5 to 40mg L $^{-1}$ sulphate.

References

1 Bundt, J., Herbel, W. and Steinhart, H. *Fresenius Zeitschrift für Analytische Chemie*, **328**, 480 (1987).
2 Huber, J.F., Kolder, F.F.M. and Miller, J.M. *Analytical Chemistry*, **44**, 105 (1972).
3 Otsuki, A. and Shiraishi, H. *Analytical Chemistry*, **51**, 2329 (1979).
4 Evans, K.A., Pubey, S.T., Kravetz, L. *et al. Analytical Chemistry*, **66**, 699 (1994).
5 Götz, R. *Facum Städt Hygiene*, **29**, 10 (1978).
6 Louch, D., Motlagh, S. and Paliszyn, J. *Analytical Chemistry*, **64**, 1187 (1992).
7 Outkiewicz, T., Ryborz, S. and Maslowski, J. *Journal of Environmental Protection and Engineering*, **4**, 263 (1978).
8 Nimura, N., Kinoshita, Y., Yoshida, T., Uetake, T. and Nakai, C. *Analytical Chemistry*, **60**, 2067 (1988).
9 Patterson, B., Cowie, C.E. and Jackson, P.E. *Journal of Chromatography, A*, **731**, 95 (1996).
10 Longe, F.T., Wenz, M. and Branch. H.J. *Journal of High Resolution Chromatography*, **18**, 243 (1995).
11 Barcelo, D., Maris, F.A., Frei, R.W., De Jong, G. and Brinkman, U.A.T. *International Journal of Environmental Analytical Chemistry*, **30**, 95 (1987).
12 Farber, H., Nick, K. and Scholer, H.F. *Fresenius Journal of Analytical Chemistry*, **350**, 145 (1994).
13 Di Corcia, A. and Samperi, R. *Analytical Chemistry*, **62**, 1490 (1990).
14 Hoffsammer, J.O., Glover, D.J. and Hazzard, C.V. *Journal of Chromatography*, **195**, 435 (1980).
15 Nishikawa, J. *Journal of Chromatography*, **349**, 349 (1987).
16 Wasik, S.L. *Journal of Chromatography*, **12**, 845 (1974).
17 Mills, G.L., Schwind, D. and Adriano, D.C. *Chemosphere*, **17**, 937 (1988).
18 Andrews, J.F. and Pearson, E.A. *International Journal of Air and Water Pollution*, **9**, 439 (1965).
19 McCarty, P.L., Jens, J.S. and Murdoch, W. *The significance of individual volatile acids in anaerobic treatment.* Proceedings, 17th Purdue Industrial Waste Conference (1962).
20 Mueller, H.F., Buswell, A.M. and Larsen, T.E. *Sew. Industrial Wastes*, **28**, 255 (1956).
21 Hunter, J.V. The organic composition of various domestic sewage fractions. Ph.D. Thesis, Rutgers University (1962).
22 Hunter, J.W. and Heukelelian, H. *Journal of Water Pollution Control Federation*, **37**, 1142 (1965).
23 Painter, H.A. and Viney, M. *Journal of Biochemical Micribiol. Technol. Engineering*, **1**, 143 (1959).
24 Mueller, H.F., Larson, T.E. and Lennarz, W. *Analytical Chemistry*, **30**, 41 (1958).
25 Buswell, A.M., Gilcreas, F.M. and Morgan, G.B. *Journal Water Pollution Control Federation*, **34**, 307 (1962).
26 Manganelli, R.M. and Brofazi, F.R. *Analytical Chemistry*, **29**, 1441 (1957).
27 Maierski, H., Eicheldorfer, D. and Quentin, K.E. *Wasser Abwasser Forschung*, **15**, 292 (1982).
28 Doei, J. and Helz, G.R. *Analytical Chemistry*, **60**, 301 (1988).

29 Onnerfjord, P., Barcelo, D., Emneus, J., Gorton, L. and Marko Varga, G. *Journal of Chromatography*, **737**, 35 (1996).
30 Otsuki, A. and Takaku, T. *Analytical Chemistry*, **51**, 833 (1979).
31 Smith, A.E. and Lord, K.A. *Journal of Chromatography*, **107**, 407 (1975).
32 Barcelo, D. *Journal of Chromatography*, **643**, 117 (1993).
33 Barcelo, D., Maris, F.A., Yeerdink, R.H., Frei, R.W. and Brinkman, U.A.T. *Journal of Chromatography*, **394**, 65 (1987).
34 Noroozian, E., Maris, E.A., Nielson, M.W.F. *et al. Journal of High Resolution Chromatography*, **10**, 17 (1987).
35 Moye, H.A. *Journal of Chromatographic Science*, **13**, 268 (1975).
36 Von Nehring, O.G., Hightower, J.W. and Anderson, J.L. *Analytical Chemistry*, **58**, 2777 (1988).
37 Dai, J. and Helz, G.R. *Analytical Chemistry*, **60**, 301 (1988).
38 Lopez-Avila, V., Schoen, S., Milanes, J. and Beckert, W.F. *Journal of the Association of Official Analytical Chemists*, **71**, 375 (1988).
39 Easty, D.B. and Wahers, B.A. *Tappi*, **61**, 71 (1978).
40 Easty, D.B. and Wahers, B.A. *Analytical Letters*, **10**, 857 (1977).
41 Sazaki, T., Sohrin, Y., Hagegawa, H. *et al. Analytical Chemistry*, **66**, 271 (1994).
42 Yang, H.J., Jiong, S.J., Yang, Y.J. and Hwang, C. *Analytica Chimica Acta*, **312**, 141 (1995).
43 Fritz, S.S. *Fresenius Journal of Analytical Chemistry*, **353**, 34 (1995).
44 Yamamoto, M. *Soil Science Society of America, Proceedings*, **39**, 859 (1975).
45 Liu, Y.M., Fernandez Sanchez, M.L., Gonzalez, E.B. and Sanz-Medel, A. *Journal of Analytical Atomic Spectrosocpy*, **8**, 815 (1993).
46 Rho, Y.S. and Choi, S.G. *Archives of Pharmagological Research*, **9**, 211 (1986).
47 Boyle, E.D., Handy, B. and Van Geen, A. *Analytical Chemistry*, **59**, 1499 (1987).
48 Riccardo, A.A., Muzzarelli, G.R. and Tubertini, O. *Journal of Chromatography*, **47**, 414 (1970).
49 Rigas, P.G. and Pietrzyk, D.J. *Analytical Chemistry*, **59**, 1388 (1987).
50 Brunt, K. *Analytical Chemistry*, **57**, 1338 (1985).
51 Blaszkiewicz, M., Baumhoer, G. and Neidbert, G. *International Journal of Environmental Analytical Chemistry*, **28**, 207 (1987).

Chapter 7

Ion pair high performance liquid chromatography

7.1 Organic compounds

7.1.1 Non saline waters

7.1.1.1 Linear alkylbenzene sulphonates and p-sulphonylcarboxylate salts

Taylor and Nickless [1] have described a paired–ion high performance liquid chromatographic technique for the separation of mixtures of linear alkylbenzene sulphonates and p-sulphophenylcarboxylate salts,

$$(-C-C-C-C-C-COOH)$$

$$|$$

$$Ph$$

in river waters. Partially biodegraded linear alkylbenzene sulphonate was analysed by the same method. Structural information on measurable intermediates formed was provided by stopped–flow ultraviolet spectra, comparison of retention behaviour with that of standards and analysis of collected fractions. Samples (1.5L) were concentrated for analysis by acidification with sulphuric acid to pH2, followed by passage through a column containing 20mL of XAD–4 resin at a flow rate of 7ml min $^{-1}$. Compounds retained by the resin were eluted with 3 × 25mL portions of methanol and the combined eluates evaporated to dryness then up to 2µl.

Fig. 7.1 shows the trace obtained from the analysis of undegraded linear alkylbenzene sulphonate (LAS) mixture (sodium dodecylbenzene sulphonate). Peak identifications were made by co-injection with pure undegraded linear alkylbenzene sulphonate compounds and confirmed by analysis of the undegraded linear alkylbenzene sulphonate mixture by desulphonation followed by gas chromatography of the resultant alkylbenzenes on a 15m OV–1 support coated open-tubular column.

7.1.1.2 Cationic surfactants

De Ruiter et al. [2] described a high performance liquid chromatographic

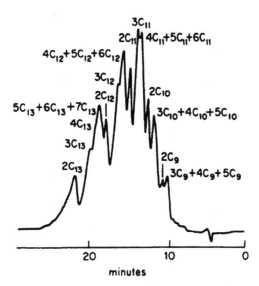

Fig. 7.1 Paired-ion high performance liquid chromatography of ungraded LAS. Column: 250 × 4.6mm, bonded C_{18} silica, d_p = 5µm. Mobile phase: 13.7 ×10⁻³M $(CTMA^+)_2SO_4^{2-}$ in 87.5% methanol, 12.6% water, pH5.4. Flow rate: 0.8ml min⁻¹; ultraviolet detection at 224nm, 0.5AUFS
Source: Reproduced by permission of Elsevier Science, UK [1]

method for cationic surfactants based on the on-line ion pair extraction of the cationic surfactant with the counter ion of methyl orange or 9,10-dimethoxyanthracene-2-sulphonate sodium salt into the organic phase. Ultraviolet or fluorescence spectrometry was used to monitor the ion pair of methyl orange and 9,10-dimethoxyanthracene-2-sulphonate sodium salt respectively. A new sandwich-type separator, as part of the extraction detector, was used successfully. Detection limits for dialkyldimethyl-ammonium chloride in river water where 2µg L⁻¹ and 10ng L⁻¹ for methyl orange and 9,10–dimethoxyanthracene-2-sulphonate sodium salt respectively.

7.1.1.3 Chlorophylls

Mantoura and Llewelyn [3] developed a reversed phase high performance liquid chromatographic system for a rapid (about 20min) separation and quantification of 14 chlorophylls and their breakdown products and 17 carotenoids from acetone extracts of non saline waters. An ion-pairing reagent is included to achieve good resolution with the acidic chloropigments (chlorophylls and phaeophorbides). Fluorescence and adsorption detectors are used to quantify chloropigments and

Table 7.1 Ion pairing high performance liquid chromatography of organic compounds in non saline waters

Organic compound	Stationary phase	Eluent	Comments	Detection	LD	Ref.
Dicamba and metabolite (3,6 dichlorosalicyclic acid)	Reverse phase column	–	–	–	ppb	[4]
Chlorinated phenoxy acids	–	–	–	Thermospray MS	–	[5]
Primary aliphatic amines	Reverse phase	–	–	–	–	[6]

Source: Own files

carotenoids respectively, within a detection limit of 0.01–0.2ng for these pigments and 200–600ng for carotenoids.

7.1.1.4 Other applications

Other applications of ion pairing high performance liquid chromatography to the analysis of non saline waters are reviewed in Table 7.1.

7.1.2 Seawater

7.1.2.1 Thiols

Shea and MacCrehan [7] determined hydrophillic thiols in marine sediment pore waters using ion pair chromatography coupled to electrochemical detection.

7.1.2.2 Flavins

Vastano *et al.* [8] have reported a method, based on solid phase extraction, with ion pair high performance liquid chromatography using fluorescence detection for the determination of flavins in seawater. Concentrations in the pm range could be determined.

7.1.3 Waste waters

7.1.3.1 Sulphonated polyphenols

Reemtsma and Jekel [9] determined sulphonated polyphenols in tannery waste waters by ion pair liquid chromatography.

7.2 Cations

7.2.1 Non saline waters

7.2.1.1 Miscellaneous metals including sodium, lithium, ammonium, potassium, magnesium, calcium, lead, copper, cadmium, cobalt, nickel, zinc, iron and 14 lanthanides

Jen and Chen [10] determined metal ions at µg L^{-1} concentrations in non saline waters using reversed phase ion pair liquid chromatography.

7.2.2 Trade effluents

7.2.2.1 Copper(II), silver(I), iron(II), iron(III), cobalt(III), nickel(II) and gold(I)

Grigorova *et al.* [11] separated and determined stable metallocyanide complexes of copper(II), silver(I), iron(II), cobalt(III), nickel(II), iron(III) and gold(I) in metallurgical plant solutions by reversed phase ion-pair partition chromatography and ultraviolet detection. The mobile phase contained 2.5mm tetrabutyl ammonium hydrogen sulphate and methanol and the stationary phase consisted of carbon–18 Novapak cartridges with 4μm packing. Precision was good with relative standard deviation ranging from 1.1 to 2.8%.

7.3 Anions

7.3.1 Non saline waters

7.3.1.1 Miscellaneous

Bidling Meyer *et al.* [12] have studied an ion pairing chromatographic method for the determination of anions using an ultraviolet absorbing co-ion in the mobile.

Bidling Meyer *et al.* [13] have also discussed the ion pair chromatographic determination of anions using an ultraviolet absorbing co-ion in the module phase. The separation and spectrophotometric determination of otherwise non ultraviolet absorbing anions is possible. Spectrophotometric detection was comparable to conductiometric detection at nanogram levels for the eight anions investigated.

References

1 Taylor, P.W. and Nickless, G. *Journal of Chromatography*, **178**, 259 (1979).
2 De Ruiter, C., Hefkens, J.C.F., Brinkman, U.A.T. *et al. International Journal of Analytical Chemistry*, **31**, 325 (1987).
3 Mantoura, R.F.C. and Llewelyn, C.A. *Analytica Chimica Acta*, **151**, 297 (1983).
4 Arjmand, K., Spittler, T.D. and Mumma, R.O. *Journal of Agricultural and Food Chemistry*, **36**, 492 (1988).
5 Hiemstra, M. and de Kok, A. *Journal of Chromatography*, **667**, 155 (1994).
6 Di Cinto, R. *Chem Ind. (Milan)*, **70**, 76 (1988).
7 Shea, D. and McCrehan, V.A. *Analytical Chemistry*, **60**, 1449 (1988).
8 Vastano, S.E., Milne, P.J., Stahovec, W.L. and Mopper, K. *Analytica Chimica Acta*, **201**, 127 (1987).
9 Reemtsma, T. and Jekel, M. *Journal of Chromatography*, **660**, 199 (1994).
10 Jen, J. and Chen, C. *Analytica Chimica Acta*, **270**, 55 (1992).
11 Grigarova, B., Wright, S.A. and Josephson, M. *Journal of Chromatography*, **410**, 1843 (1987).
12 Bidling Meyer, B.A., Sanastasia, C.T. and Warren, F.V. *Analytical Chemistry*, **60**, 192 (1988).
13 Bidling Meyer, B.A., Sanastasia, C.T. and Warren, F.V. *Analytical Chemistry*, **59**, 1843 (1987).

Chapter 8

Micelle chromatography

8.1 Organic compounds

8.1.1 Non saline waters

8.1.1.1 Polyaromatic compounds

Micellar extraction followed by liquid chromatography and fluorescence detection has been used to determine ppt of polyaromatic hydrocarbons in non saline waters [1,2].

8.1.2 Sewage effluents

8.1.2.1 Sodium-N-methyldithiocarbamate and methylisothiocyanate

A high performance liquid chromatographic procedure has been described [3] for the simultaneous determination of the fungicides sodium-N-methyldithiocarbamate and methylisothiocyanate in surface waters and in sewage, based on their separation on a reversed phase column with a miscellar mobile phase (hexadecyltrimethylammonium bromide) in 1:1 v/v methanol:water buffered to pH6.8. Detection limits were 70 and 1µg dm^{-1} when the analysis was performed with an ultraviolet detector at 247nm.

8.1.3 Trade effluents

8.1.3.1 Phthalate esters

Takeda *et al.* [4] studied the migration behaviour of phthalate esters in micellar electrokinetic chromatography with and without the addition of butyl alcohol. This procedure was applied to the determination of phthalate esters in environmental waters.

8.2 Cations

8.2.1 Non saline waters

8.2.1.1 Calcium, potassium and sodium

Jones [5] carried out a simultaneous separation of non organic cations (eg calcium, potassium and sodium) also anions (eg nitrate, thiosulphate, cyanide and thiocyanate) by ion chromatography using a single column coated with weak/strong charged three welterionic bile salt micelles.

8.3 Anions

8.3.1 Non saline waters

8.3.1.1 Iodate

Okada [6] has used micelle exclusion chromatography to determine iodate in the presence of other anions (iodide, bromide, nitrite and nitrate) in water. The method is based on partition of the anions to a cationic micelle phase and shows different selectivity from ion exchange chromatography.

8.3.1.2 Iodide, nitrate, nitrite, bromide and iodide

Dietrich et al. [7] applied micelle exclusion chromatography to the determination of iodate, nitrite, nitrate, bromide and iodide in non saline waters.

8.3.1.3 Thiocyanate, sulphite and sulphide, nitrate, thiosulphate and cyanide

Gonzalez et al. [8] have applied micelle analysis in the kinetic multi-component analysis for the simultaneous determination of thiocyanate, sulphite and sulphide in non saline waters. The method involved reaction of the anions with S,S'dithiobis(2–nitrobenzoic acid) in aqueous cetyl-trimethyl ammonium bromide micelles. Cyanide, sulphide and sulphite were determined at concentrations, respectively, down to $0.5–1.5 \times 10^{-4}$M $0.2–1 \times 10^{-4}$M and $0.2–1.5 \times 10^{-4}$M with a relative error of less than 5%.

See also section 8.2.1.1 above.

References

1 Boeckelen, A. and Niessner, R. *Fresenius Journal of Analytical Chemistry*, **346**, 435 (1993).
2 Brouwer, E.R., Hermans, A.J.J., Lingeman, H. and Brinkman, U.A.T. *Journal of Chromatography*, **669**, 45 (1994).

3 Mullins, F.G.P. and Kirkbright, G.F. *Analyst (London)*, **112**, 701 (1987).
4 Takada, S., Wakida, S., Yamari, M., Kawabara, A. and Higashi, K. *Analytical Chemistry*, **65**, 2489 (1993).
5 Jones, O. *Analytical Chemistry*, **66**, 6765 (1994).
6 Okada, T. *Analytical Chemistry*, **60**, 1511 (1988).
7 Dietrich, A.M., Ledder, J.D., Gallagher, P.L., Grabeel, M.N. and Hochn, R.C. *Analytical Chemistry*, **64**, 496 (1992).
8 Gonzalez, V., Moreno, B. and Silicia, D. *Rulios Perez–Bendito*, **65**, 1987 (1993).

Chapter 9

Ion exclusion chromatography

9.1 Organic compounds

9.1.1 Non saline waters

9.1.1.1 Carboxylates

Okada [1] has described a redox suppression for the ion exclusion chromatography of carboxylic acids with conductiometric detection. The reaction between hydriodic acid (the eluent) and hydrogen peroxide (the precolumn reagent) is used as the redox suppressor for ion exclusion chromatography of carboxylic acids. The suppressor is useful with highly acidic eluents and reduces background conductance more effectively than a conventional ion exchange suppressor.

9.1.2 Rain water

9.1.2.1 Carboxylates

Backman and Peden [2] used ion exclusion chromatography to determine weak carboxylic acids in rain water. Citrate, formate and acetate were identified.

9.2 Cations

9.2.1 Non saline waters

9.2.1.1 Arsenic

Butler [3] determined inorganic arsenic species in non saline waters by ion exclusion chromatography with electrochemical detection. Two species were separated by ion exclusion chromatography using 0.01M orthophosphoric acid eluent. Arsenic(III) was detected by its oxidation at a platinum wire electrode. Measurement of total inorganic arsenic after reduction of arsenic(V) to arsenic(III) by sulphur dioxide enabled

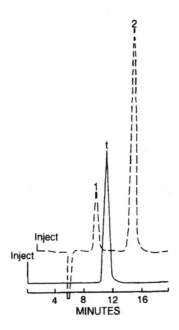

Fig. 9.1 Separation of arsenite and arsenate by ion exclusion chromatography; conductivity detector (———); UV (- - -) at 200nm, position 1; peak 1, 15mg L^{-1} AsO$_4^{2-}$; peak 2, 30mg L^{-1} AsO$_3^{2-}$
Source: Own files

arsenic(V) to be estimated. The detection limit for arsenic(III) was 0.012µM at an applied electrode potential of plus 1.0 volts.

9.2.2 Ground waters

Yokoyama *et al.* [4] used ion exclusion chromatography and continuous hydride atomic absorption spectrometry to study arsenic speciation in geothermal waters. Arsenic was determined in the range 0.01 to 10 mg L^{-1}.

9.3 Anions

9.3.1 Non saline waters

9.3.1.1 Arsenate and arsenite

Fig. 9.1 shows the separation of arsenite and arsenate by ion exclusion chromatography using 0.01mol L^{-1} hydrochloric acid as the eluent. Arsenious acid is a very weak acid and cannot be detected at low levels by the conductivity detector. However, like sulphide, it is easily detected with the ultraviolet detector. The simultaneous determination of arsenite and arsenate is possible.

Arsenite can also be separated by ion chromatography using standard conditions. It elutes early, near the carbonate dip. For these applications the ultraviolet detector should be placed between the separator and suppressor columns.

9.3.1.2 Organic anions, tartrate, maleate, malonate, citrate, glycollate, formate and fumarate

Okada [5] employed a Redox suppressor in his ion exclusion chromatographic determination of tartrate, maleate, malonate, citrate, glycollate, formate and fumarate in non saline waters. A conductiometric detector was employed.

9.3.1.3 Nitrite

Kim and Kim [6] have described an extremely sensitive determination of nitrite in drinking water (tap water and underground water) and environmental samples (rain, lake water, and soil) by ion exclusion chromatography with electrochemical detection. Potential interferences in the determination of nitrite by the standard spectrophotometric method or by the ion exchange chromatographic method with either conductivity or ultraviolet detection were eliminated. The detection limit was $0.1 \mu g \ L^{-1}$ without preconcentration.

9.3.1.4 Phosphate

Ion exclusion chromatography provides a convenient way to separate molecular acids from highly ionised substances. The separation column is packed with a cation exchange resin in the H^+ form so that salts are converted to the corresponding acid. Ionised acids pass rapidly through the column while molecular acids are held up to varying degrees. A conductivity detector is commonly used.

Tanaka et al. [7–9] have reported that the separation of phosphate from chloride, sulphate and several condensated phosphates $P_2O_7^{4-}$ (pyrophosphate) and $P_3O_{10}^{2-}$ and $P_3O_{10}^{3-}$ (tripolyphosphates) could be achieved by ion exclusion chromatography on a cation exchange resin in the H^+ form by elution with an acetone:water and dioxan:water mixture. As the separation mechanism of the ion exclusion chromatography by elution with water alone for numerous anions or their respective acids is based on the Donnan membrane equilibrium principle (ion exclusion effect), it is a highly useful technique for the separation of non electrolytes such as carbonic acid from electrolytes such as hydrochloric acid and sulphuric acid [10]. Ion exclusion chromatography has also been coupled to ion chromatography to determine simultaneously both weak and strong acids [11]. The ion exclusion chromatography separation of ortho-phosphate from the strong acid anions by elution with organic solvent: water described above is based on this ion exclusion effect and/or the partition effect between the cation exchange resin phase (water rich) and the mobile phase (organic solvent rich) owing to the hydration of the resin [7] and phosphate has been monitored as the corresponding acid (H_3PO_4) with a flow coulometric detector for the detection of H^+ ion and a conductometric detector.

9.3.2 Aqueous precipitation

9.3.2.1 Chloride, bromide, iodide, nitrate, nitrite and thiocyanate

An ion exclusion procedure described by Stetzenback and Thompson [12] has been applied to the determination of these anions in rainwater.

Moss and Stephen [13] determined chloride, bromide and iodide in rainwater by converting them to aklylmercury(II) halides and measurement by high performance liquid chromatography.

9.3.3 Seawater

9.3.3.1 Silicate

Hioki et al. [14] have described an on-line determination of dissolved silica in seawater by ion exclusion chromatography in combination with inductively coupled plasma atomic emission spectrometry.

This method was developed as a second, independent method to complement the usual colorimetric procedure in the determination of a certified concentration of dissolved silica in a planned seawater reference material. Ion exclusion affords a separation of the dissolved silica not only from the major seawater cations but also from potentially interfering anions. The detection limit, conservatively estimated at 2.3ng g^{-1} as Si (0.08µM), is superior to that achievable by direct analysis by inductively coupled plasma atomic emission spectrometry.

Other studies on the application of this technique have been carried out [15–18].

9.3.4 Potable waters

9.3.4.1 Bicarbonate and carbon dioxide

Tanaka and Fritz [19] have described a procedure for the determination of bicarbonate and carbon dioxide in potable water by ion exclusion

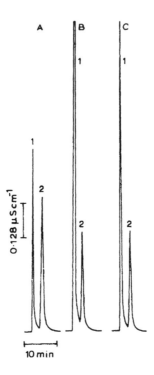

Fig. 9.2 Ion exclusion chromatograms of HCO_3^- in some tap waters: (A) raw tap water of City of Ames, IA, after 10-fold dilution (0.653mM × 10); (B) potable water after softening treatment (0.466mM); (C) potable water (La Salle, IL) after 20-fold dilution (0.470mM × 10). Peak 1 is strong acid anions and peak 2 is HCO_3^-
Source: Reproduced with permission from the American Chemical Society [19]

chromatography using water as the eluent and an electrical conductivity detector. The sensitivity of detection is improved approximately 10-fold by the use of two ion exchange enhancement columns inserted in series between the separating column and the detector. The first enhancement column converts carbonic acid to potassium bicarbonate and the second enhancement column converts the potassium bicarbonate to potassium hydroxide.

The plastic separating column was 7.5 × 100mm and was packed with a cation exchange resin in the H⁺ form TSK SCX 5µm (polystyrene divinyl benzene copolymer-based material with a high cation exchange capacity).

The first enhancement column was constructed of plastic (4.6 × 50mm) and packed with a cation exchange resin in the K⁺ form (TSK SCX Tµm; TSK IC–Cation for cation chromatographic use, 10µm silicon based material with low cation exchange capacity).

The second enhancement column was constructed of plastic (4.6 ×
50mm) and packed with an anion exchange resin in the OH⁺ form TSK
SAX (5μm). The precolumn was constructed of plastic (7.5 × 100mm) and
packed with an anion exchange resin in the OH⁺ form (TSK SAX, 5μm).

All samples were filtered with a 0.45μm polytetrafluoroethylene
(PTFE) membrane filter before injection into the column.

Tanaka and Fritz [19] applied this procedure to the determination of
bicarbonate or carbon dioxide in some potable water samples. Fig. 9.2
shows the ion exclusion chromatograms obtained before and after a
softening treatment. The results indicated that the method is useful in the
field of water quality control of water treatment facilities.

9.4.3.2 Nitrite

Kim and Kim [6] determined nitrite in potable water by ion exclusion
chromatography with electrochemical detection. See section 9.3.1.3.

References

1 Okada, T. *Analytical Chemistry*, **60**, 1666 (1988).
2 Backman, S.R. and Peden, M.E. *Water Soil Pollution*, **33**, 191 (1987).
3 Butler, E.C.V. *Journal of Chromatography*, **450**, 353 (1988).
4 Yokoyama, T., Takahashi, Y. and Tarutani, T. *Chemical Geology, Part 1*, **48**, 27 (1992).
5 Okada, T. *Analytical Chemistry*, **60**, 1668 (1988).
6 Kim, H.J. and Kim, Y.K. *Analytical Chemistry*, **61**, 1485 (1989).
7 Tanaka, K. and Ishizaka, T. *Journal of Chromatography*, **190**, 7 (1980).
8 Tanaka, K. and Sunahara, H. *Bunseki Kagaku*, **27**, 95 (1978).
9 Tanaka, K., Nakajima, K. and Sunahara, M. *Bunseki Kagaku*, **26**, 102 (1977).
10 Tanaka, K., Ishizuka, T. and Sunahara, M. *Journal of Chromatography*, **174**, 153 (1979).
11 Dionex Application Note No. 19. Ion Chromatography Systems Analysis of strong and weak acids in coffee extracts. Dionex Corporation (1979).
12 Stetzenbach, K.J. and Thompson, G.M. *Ground Water*, **21**, 36 (1983).
13 Moss, P.E. and Stephen, W.I. *Analytical Proceedings*, **22**, 5 (1985).
14 Hioki, A., Lam, J.W.H. and McLaren, J.W. *Analytical Chemistry*, **69**, 21 (1997).
15 Spencer, C.P. In *Chemical Oceanography*, 2nd edn., eds. J.P. Riley and G. Skirrow, Academic Press, New York, Vol. 2, Chapter 11, pp. 250–251 (1975).
16 Sakai, H., Fujiwara, T. and Kumamaru, T. *Bulletin of the Chemical Society of Japan*, **66**, 3401 (1993).
17 Konno, S. and Goto, R. *Kogyo Yosui*, **433**, 50 (1994).
18 Sakai, H., Fujiwara, T. and Kumamaru, T. *Analytica Chimica Acta*, **302**, 173 (1995).
19 Tanaka, K. and Fritz, J.S. *Analytical Chemistry*, **59**, 708 (1987).

Chapter 10

Size exclusion high performance liquid chromatography

Size exclusion techniques provide rapid, gentle separations with a constant sample matrix but have generally been designed to separate large organic molecules. Metal separations have been restricted to relatively high molecular weight metal organic components.

Another approach for separation is to adjust chromatographic conditions to maximize chemical or apparent molecular size differences of dissolved components. Then, chemically different groups can be fractionated by size exclusion stationary phases even if molecular weights are similar. In the absence of salts or buffers (to reduce charge effects) sample components may be separated due to factors other than molecular weight. Distilled water may increase the apparent molecular size of ionic dissolved components due to hydration layer formation or other hydrogen bonding or ionic repulsion mechanisms. Sample column interactions, causing fractionation, may result from charge attractions or repulsions in the absence of salts or buffers in the mobile phase.

10.1 Organic compounds

10.1.1 Potable waters

10.1.1.1 Ozonisation products

Since the detection of halogenated organics in potable waters much research effort has been directed towards finding water treatment processes to remove such organics and their precursors and toward finding disinfectants other than chlorine. Great interest has been focused upon ozonisation because both disinfection and organic removal can be accomplished with this process. As ozonated end-products will occur in water produced by such processes and these could be potentially toxic and would accumulate in waste water after repeated cycles of use it is necessary to ascertain what end-products occur in water that has been ozonated and subsequently chlorinated.

Glaze *et al.* [1] analysed ozonisation by-products produced before and after chlorination in potable water and waste water using size exclusion chromatography and halogen-specific microcoulometry. Gas chromatography with an electron capture detector and gas chromatography–mass spectrometry was also used. Glaze *et al.* [1] showed that the majority of the organic halogen found in chlorinated lake water is not accounted for by trihalomethanes but is present, to a large extent as 2,4,6,2′,4′,6′-hexachlorobiphenyl. They studied the ozonisation and photolytic ozonisation of this compound.

10.1.1.2 Chlorinated humic acids

Becker *et al.* [2] have discussed the application of size exclusion chromatography to the determination of chlorinated humic acids in potable waters.

10.1.2 Waste waters and trade effluents

10.1.2.1 Polyacrylamide

Leung [3] used size exclusion high performance liquid chromatography to determine down to 20ppm of polyacrylamide flocculants in waste waters. Ultraviolet detection at 205nm showed higher sensitivity than infrared detection.

Leung *et al.* [4] determined polyacrylamides in coal washery effluents by ultrafiltration size exclusion chromatography with ultraviolet detection. A column of TSK 500 PW hydrophilic and semi-rigid porous polymer gel was used with 0.05m sodium sulphate as mobile phase.

10.2 Cations

10.2.1 Non saline waters

10.2.1.1 Calcium and magnesium

Gardner *et al.* [5] recognised that because component separations on size exclusion columns with distilled water are affected by chemical physical interactions as well as component molecular size, distilled water size exclusion chromatography should also fractionate dissolved metal forms. They interfaced distilled water size exclusion chromatography with inductively coupled argon plasma detection to fractionate and detect dissolved forms of magnesium and calcium in lake and river waters. Inductively coupled argon plasma detection is ideally suited to this application because it provides continuous metal monitoring for aqueous samples and accepts sample flow rates appropriate for distilled water size exclusion chromatography.

10.2.1.2 Copper complexes

Adamic and Bartak [6] used high pressure aqueous size exclusion chromatography with reverse pulse amperometric detection to separate copper(II) complexes of poly(amino carboxylic acids), catechol and fulvic acids. The commercially available size exclusion chromatography columns were tested. Columns were eluted with copper(II) complexes of poly(aminocarboxylic acids), citric acids, catechol and water derived fulvic acid. The eluent contained copper(II) to prevent dissociation of the labile metal complexes. Reverse pulse electrochemical measurements were made to minimise oxygen interferences at the detector. Resolution of a mixture of DTPA, EDTA and NTA copper complexes was approximately the same on one size exclusion chromatography column as on Sephadex G–25 columns but with a 10-fold increase in efficiency. A linear detector response to amounts of water derived fulvic acids separated on the gel permeation chromatography 100 column was obtained enabling direct measurement of amounts of fulvic acid in water samples. The detection limits were 40ng for copper EDTA and 12 µg for fulvic acid.

10.3 Anions

10.3.1 Non saline waters

10.3.1.1 Sulphide and sulphite

The ion chromatographic determination of weak acid anions is complicated by ion exclusion in the suppressor column, resulting in faster elution and sharper peaks, directly proportional to the degree of exhaustion of the suppressor column [7]. A 10mg L^{-1} nitrite standard showed a 37% increase in peak height over an 8h period when monitored · with the conductivity detector while on only a minor 2% increase in peak height was observed over the same time period by using the ultraviolet detector after the separator column.

The ultraviolet detector can also be used in some cases to resolve overlapping peaks. Determination of the nitrite peak by using the conductivity detector is complicated by both the ion exclusion effect and the incomplete resolution between the large chloride peak and the much smaller nitrite peak.

The ion exclusion interference can be eliminated for ultraviolet active anions by placing the ultraviolet detector between the separator and suppressor columns (position 1). In addition, the problem of overlapping peaks can sometimes be resolved spectrophotometrically by proper choice of wavelength.

Sulphide cannot be detected under normal ion chromatographic conditions. It is converted into hydrogen sulphide in the suppressor column.

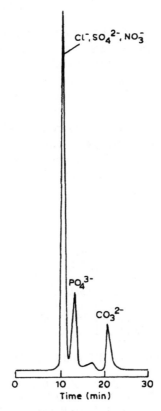

Fig. 10.1 Ion exclusion chromatogram of mixture of 10mg L^{-1} each of Cl$^-$, SO$_4^{2-}$, NO$_3^-$, PO$_4^-$ and CO$_3^{2-}$ obtained by elution with 60% acetone–water. Detected with FCD (100mV full scale)
Source: Reproduced by permission from Elsevier Science, UK [8]

Hydrogen sulphide acid is a very weak acid that does not ionise sufficiently to be detected with the conductivity detector. The ultraviolet detector, however, is able to detect sulphide at low levels.

10.3.2 Waste waters

10.3.2.1 Phosphate, chloride, carbonate and sulphate

Tanaka and Ishizuki [8] have investigated the possibility of determining orthophosphate in waste waters by ion exclusion chromatography on a cation exchange resin in the H$^+$ form by elution with acetone water. They discuss optimal conditions for the separation of phosphate from chloride, sulphate, carbonate, etc. which are always present in waste water and sewage samples.

Fig. 10.1 shows the chromatogram of a mixture of 10mg L^{-1} of chloride sulphate, nitrate, phosphate and carbonate obtained with the flow colorimetric detector by elution with 60% acetone water. As can be seen from this figure phosphate could be separated from the strong acid anions and carbonate. The RS value between the strong acid anions and phosphate was about 1.7. This RS value suffices for the quantitation of phosphate by the peak area measurement with a computing integrator.

In Fig. 10.1 an unknown peak was observed between phosphate and carbonate. This peak was due to water in the sample solution introduced into the chromatograph. However, the peak did not interfere with the quantitation of phosphate.

References

1 Glaze, W.H., Peyton, C.R., Salch, F.Y. and Huang, F.Y. *International Journal of Analytical Chemistry*, **7**, 143 (1979).
2 Becker, G., Carlberg, G.E., Gjessing, E.T., Hougslo, J.K. and Monarta, S. *Environmental Science and Technology*, **19**, 422 (1985).
3 Leung, W.M. *Process Technology Proceedings* (Hocculation, biotechnology separation systems), pp. 149–165 (1987).
4 Leung, R.W.M., Pandey, R.N. and Das, B.S. *Environmental Science and Technology*, **21**, 476 (1987).
5 Gardner, W.S., Lundrum, P.F. and Yates, D.A. *Analytical Chemistry*, **54**, 1198 (1982).
6 Adamic, M.L. and Bartak, D.E. *Analytical Chemistry*, **57**, 279 (1985).
7 Tanaka, F. and Ishizuki, T. *Journal of Chromatography*, **190**, 7 (1980).
8 Tanaka, K. and Ishizuki, T. *Water Research*, **16**, 719 (1982).

Chapter 11

Gel permeation chromatography

11.1 Organic compounds

11.1.1 Non saline waters

11.1.1.1 Haloforms

Schnoor et al. [1] have determined the apparent molecular weight range of trihalomethane precursor compounds in the Iowa river and a reservoir near Iowa City. Soluble organics were size fractionated by gel permeation chromatography and the fractions were chlorinated and analysed for trihalomethane yields by electron-capture gas chromatography. Of the trihalomethanes formed, 75% were derived from organics of molecular weight less than 3000 and 20% from those of molecular weight less than 1000.

11.1.1.2 Humic and fulvic acids

Naturally occurring organic compounds are found in significant concentrations in waters throughout the world. A survey of 27 Canadian rivers and lakes [2] revealed that the majority of dissolved organic carbon was of natural origin with fulvic acid, tannins and lignins being the major components.

Humic and fulvic acids are among the most widely distributed products of plant decomposition on the earth's surface, occurring in soil, water, sediments and a variety of other deposits. They are amorphous, yellow–brown or black, hydrophillic, acidic, polydisperse substances of wide-ranging molecular weight <10000 for the fulvic acid, 10000–3000000 for humic acid are the usual ranges although some values outside of these ranges have been reported by Rashid and King [3]. Humic and fulvic acids are responsible for the yellow–brown coloration of many lakes and rivers. By definition, humic acid is that fraction of organic material soluble at pH9 but insoluble at pH2, while fulvic acid is soluble at both pH9 and pH2 [4].

Tannins and lignins are high–molecular–weight polycyclic aromatic compounds widely distributed through the plant kingdom. Pearl [5] defined lignin as 'the incrusting material of plants which is built up of methoxy- and hydroxy–phenylpropane units'. It is not hydrolysed with acids but is readily oxidised and soluble in hot alkali and bisulphite. The exact chemical composition of lignin-like compounds commonly found dissolved in non saline waters is not known. Tannins, according to Geissman and Crout [6] are either polymers of gallic acid linked to carbohydrate residues (hydrolysable tannins) or polymeric flavonoid compounds (condensed tannins). Both tannins and lignins are highly resistant to chemical and biological degradation. How closely these compounds resemble their laboratory counterparts, tannic acid and lignosulphonic acid, is not really known.

Chian and De Walle [7] used gel permeation gas chromatography on Sephadex G–75 and G–200, ultrafiltration and gas chromatography to characterise soluble organic matter in heavily polluted ground water samples and leachates from landfills. The largest organic fraction consisted of three volatile fatty acids, and the next largest was a fulvic-like material with a relatively high carboxyl and aromatic hydroxyl group density. There was also a small percentage of a high molecular weight humic carbohydrate–like complex with which liquids were associated.

Gel filtration chromatography using Sephadex G100 as column packing and ultraviolet detectors, have been used in studies carried out on the elution of humic acid [8] and in characterisation studies on secondary sewage effluents [9] and in organic substances in river waters [10].

11.1.1.3 Polyaromatic hydrocarbons

McKay and Lathan [11] used gel permeation chromatography to determine polyaromatic hydrocarbons in non saline waters.

11.1.2 Seawater

11.1.2.1 Crude oils

Done and Reid [12] applied gel permeation chromatography to the identification of crude oils and products isolated from estuary and sea water, The technique, which appears more suited to the analysis of crude oils, is based on the separation of oil components in order of their molecular size, for practical purposes their molecular weight.

Oils were dissolved in tetrahydrofuran or toluene and each solvent resulted in a different output profile, and therefore more information for identification. The solutions were pumped through a 0.6m × 9.5mm column packed with 6nm (60Å) Styragel and the eluent monitored by a differential refractometer. Elution time was 1h, although this was improved by staggered injection, using 6mg samples. 'Fingerprints' of crude oils investigated, fall into several discrete groups. Crude oil residues (bp 525°C) were excluded by the Styragel employed, while crude oil and weathered samples were found to give similar traces, some changes were introduced by evaporation processes, and it could be difficult to differentiate between similar but weathered crude oils. Heavy fuel oils were more difficult to identify, owing to the variety of fuel oils in use.

11.1.3 Trade effluents

11.1.3.1 Carbohydrates

Collins and Webb [13] used gel chromatography to detect and determine carbohydrates in pulp mill effluents. The phenol–sulphuric acid method was used to monitor column effluents.

11.1.3.2 Calcium sulphonate

Sagfors and Starck [14] used gel permeation chromatography to study calcium lignosulphate of high molecular weight in acid and alkaline kraft pulp bleaching effluents.

11.2 Cations

11.2.1 Seawater

11.2.1.1 Lithium

Rona and Schmuckler [15] used gel-permeation chromatography to separate lithium from Dead Sea brine. The elements emerged from the column in the order potassium, sodium, lithium, magnesium and calcium and it was possible to separate a lithium-rich fraction also containing some potassium and sodium but no calcium and magnesium.

References

1 Schnoor, J.L., Nitzschke, J.L., Lucas, R.D. and Veenstra, J.N. *Environmental Science and Technology*, **113**, 1134 (1979).
2 Lawrence, J. *National Inventory of Natural Organic Compounds. An Interim Report.* Canada Centre for Inland Waters. Unpublished Report, Burlington, Ontario.
3 Rashid, M.A. and King, L.H. *Chemical Geology*, **7**, 37 (1971).
4 Schnitzer, M. and Khaus, I.A. *Humic Substances in the Environment.* Marcel Dekker, New York (1972).
5 Pearl, I.A. *The Chemistry of Lignin*, Marcel Dekker, New York (1967).

6 Geissman, T.A. and Crout, D.H.G. *Organic Chemistry of Secondary Plant Metabolism.* Freeman Cooper, San Francisco (1969).
7 Chian, E.S.K. and De Walle, F.B. *Environmental Science and Technology*, **11**, 158 (1977).
8 Aho, J. and Lehto, A. *Archiv. Hydrobiol.*, **101**, 21 (1984).
9 Manka, M., Mandelbaum, A. and Boritinger, A. *Environmental Science and Technology*, **8**, 1017 (1974).
10 Faure, J., Viallet, P., and Picat, P. *La Tribune de Cebedeau*, **28**, 385 (1975).
11 McKay, J.F. and Lathan, D.R. *Analytical Chemistry*, **45**, 1050 (1973).
12 Done, J.M. and Reid, W.K. *Separation Science*, **5**, 825 (1970).
13 Collins, S.W. and Webb, A.A. *Tappi*, **55**, 1335 (1972).
14 Sagfors, P.E. and Starck, B. *Water Science Technology*, **20**, 49 (1988).
15 Rona, M. and Schmuckler, G. *Talanta*, **20**, 237 (1973).

Chapter 12

Ion exchange chromatography

12.1 Organic compounds

12.1.1 Non saline waters

12.1.1.1 Non ionic surfactants

Nickless and Jones [1] and Musty and Nickless [2,3] evaluated Amberlite XAD-4 resin as an extractant for down to milligram per litre levels of polyethylated secondary alcohol ethoxylates, $(R(OCH_2CH_2)_nOH)$ surfactants and their degradation products from water samples. This resin was found to be an effective adsorbent for extraction of polyethoxylated compounds from water except for polyethylene glycols of molecular weight less than 300. Flow rates of 100ml min^{-1} were possible using 5g of resin, and interfering compounds can be removed by a rigorous purification procedure. Adsorption efficiencies of 80–100% at 10ppb were possible for non ionic detergents using distilled water solutions. The main purpose of the work of Nickless and Jones [1] was the investigation of secondary alcohol ethoxylate as it proceeds through the water system. Associated with this is polyethylene glycol, a likely biodegradation product. Alkylphenol ethoxylate was also considered but only as an interferent that should be differentiated in order to allow fuller characterisation of secondary alcohol ethoxylate residues.

Jones and Nickless [4] continued their study of Amberlite XAD–4 resins by examining polyethoxylated materials before and after passage through a sewage works. Samples from the inlet and outlet of the sewage works and from the adjacent river were subjected to a three-stage isolation procedure and the final extracts were separated into a non ionic detergent and a polyethylene glycol. The non ionic detergent concentration was 100 times lower (8ng mL^{-1}) in the river than in the sewage effluent. Thin layer chromatography and ultraviolet, infrared and NMR spectroscopy were used to identify, in the non ionic detergent compound, alkylphenol ethoxylates (the most persistent), secondary alcohol ethoxylate and primary alcohol ethoxylate.

12.1.2 Seawater

12.1.2.1 Carbohydrates

Josefsson [5] determined soluble carbohydrates in seawater by partition chromatography after desalting by ion exchange membranes. The electrodialysis cell used had a sample volume of 430mL and an effective membrane surface area of 52cm². Perinaplex A–20 and C–20 ion exchange membranes were used. The water-cooled carbon electrodes were operated at up to 250mA at 500V. The desalting procedure normally took less than 30h. After the desalting, the samples were evaporated nearly to dryness at 40°C *in vacuo*, then taken up in 2mL of 85% ethanol and the solution was subjected to chromatography on anion exchange resins (sulphate form) with 85% ethanol as mobile phase. By this procedure, it was possible to determine eight monosaccharides in the range 0.15–46.5µg L $^{-1}$ with errors of less than 10% and to detect traces of sorbose, fucose, sucrose, diethylene glycol and glycerol in seawater.

12.1.3 Sewage effluents

12.1.3.1 Miscellaneous organics

Among the limited applications of ion exchange chromatography to the separation of organics in water are the separation of organics in sewage effluents [6–8] and the determination of chlorinated biphenyls [9]. Pill *et al.* [8], using high resolution ion exchange chromatography, identified and quantified 56 and 13 organics, respectively, in primary and secondary sewage effluents. Many more unidentified compounds were present.

12.2 Organometallic compounds

12.2.1 Non saline waters

12.2.1.1 Organoarsenic compounds

Ion exchange chromatography has been used to achieve separations of monomethyl–arsenate, dimethylarsinic and tri- and pentavalet arsenic. Dietz and Perez [10] have described methods which separated the inorganic arsenic from each of the organic species using ion exchange chromatography. Here, further inorganic speciation relies on redox-based colorimetry [11]. Both the accuracy and precision suffer from the low As(III)/As(IV) and As(total)/P ratios normally encountered in the environment.

Grabinski [12] has described an ion exchange method for the complete separation of the above four arsenic species, on a single column containing both cation and anion exchange resins. Flameless atomic absorption spectrometry with a deuterium arc background correction is used as a detection system for this procedure. This detection system was chosen because of its linear response and lack of specificity for these compounds combined with its resistance to matrix bias in this type of analysis.

The elution sequence was as follows: 0.006M trichloroacetic acid (pH2.5), yielding first As(III) and then monomethylarsonate; 0.2M trichloroacetic acid yielding As(V): 1.5M NH_4OH followed by 0.2M trichloroacetic acid yielding dimethylarsinite.

Grabinski [12] spiked 0.500µg of each of the four arsenic species into filtered (0.45µM) lake water and distilled and deionised water. Arsenic recoveries for the entire procedure averaged 104%, 100% and 97% and 99% for As(III), monomethylarsonate, As(IV) and dimethylarsinite, respectively. Relative standard deviations for replicate determinations ranged from 0.7% for AS(III) and dimethylarsinite to 1.3% for As(V).

Aggett and Kadwani [13] report the development and application of a relatively simple anion exchange method for the speciation of arsenate, arsenite, monomethylarsonic acid and dimethylarsinic acid. As these four arsenic species are weak acids the dissociation constants of which are quite different it seemed that separation by anion exchange chromatography was both logical and possible.

Aggett and Kadwani [13] employed a two stage single column anion exchange method using sodium hydrogen carbonate and chloride as eluate anions. These species appear to have no adverse effects in subsequent analytical procedures. Its successful application is dependent on careful control of pH. Analyses were performed by hydride generation atomic absorption spectroscopy.

Separation of arsenic(III) and dimethylarsinic acid by elution with sodium hydrogen carbonate was satisfactory in the pH range 5.2–6.0. Elution of monomethylarsonic acid was accelerated and satisfactory separation from arsenic(V) achieved using saturated aqueous carbon dioxide (pH4.0–4.2) containing 10g L^{-1} of ammonium chloride. In order to avoid oxidation of trivalent to pentavalent arsenic on the column, it was necessary to pretreat the resin with 1M nitric acid and 0.1M EDTA before use.

Henry and Thorpe [14] separated monomethylarsonic acid, dimethylarsenic acid, As(III) and As(IV) on an ion exchange column from samples of pond water receiving fly ash from a coal-fired power station. They then determined these substances by differential pulse polarography. The above four arsenic species were present in non saline water systems. Moreover, a dynamic relationship exists whereby oxidation–reduction and biological methylation–dimethylation reactions provide the pathways for the intercoversions of the arsenicals.

12.3 Cations

12.3.1 Non saline waters

12.3.1.1 Cobalt

Boyle *et al.* [15] have described a method for determining cobalt in natural waters using cation exchange liquid chromatography. Cobalt was determined directly in 500µL fresh water samples with a detection limit of 20pM per kg.

12.3.1.2 Vanadium

An ion exchange chromatographic method has been described [16] for the determination of the various forms of vanadium in fresh water. These include tetravalent cationic, pentavalent anionic and neutral complexed forms of vanadium. Separation is achieved on two columns in series involving the absorption of the sample on Chelex 100 and Dowex 1×8 columns followed by the selective elution of the different vanadium species and their assay by neutron activation analysis. Experiments were carried out using vanadium–48 radiotracer. Recoveries of total, complexed tetravalent and pentavalent vanadium were respectively 31, 36 and 34% and total vanadium 100%.

12.3.1.3 Miscellaneous cations

Ion exchange chromatography has been employed to separate rare earth metals from more common metals [17], uranium, cobalt and cadmium [18] and anionic from cationic forms of metals [19].

12.3.2 Surface waters

12.3.2.1 Miscellaneous cations

Small *et al.* [20] have discussed a novel ion exchange chromatographic method for the determination of a wide range of elements in surface waters.

Ion exchange resins have a well known ability to provide excellent separation of ions, but the automated analysis of the eluted species is often frustrated by the presence of the background electrolyte used for elution. By using a novel combination of resins, these workers have succeeded in neutralising or suppressing this background without significantly affecting the species being analysed which in turn permits the use of a conductivity cell as a universal and very sensitive monitor of all ionic species either cationic or anionic. Using this technique, automated analytical schemes have been devised for Li^+, Na^+, K^+, Rb^+, Cs^+,

NH_4^+, Ca^{2+}, Mg^{2+}, F^-, Cl^-, Br^-, I^-, NO_3^-, NO_2^-, SO_4^{2-}, SO_3^{2-}, PO_4^{3-} and many amines, quaternary ammonium compounds, and organic acids. Elution time can take as little as 1.0 min/ion and is typically 3 min/ion.

12.3.3 Mineral, spa and spring waters

12.3.3.1 Lithium, sodium, potassium and caesium

Araki *et al.* [21] have described an ion exchange chromatographic technique for determining these elements. Detection was by hydrogen flame ionisation. Cerium and phosphate interfere in this method.

12.3.3.2 Nickel

Nickel has been determined in spring water in amounts down to 0.05µg by ion exchange chromatography on Dowex A–1 through Dowex 1–10 [22].

12.3.3.3 Copper

Toshio [23] determined copper in hot spring waters by cation exchange chromatography in ammoniacal pyrophosphate solution. A procedure may be included to overcome iron interference.

12.3.3.4 Boron

Boron has been determined in mineral waters [24] by ion exchange chromatography on Dowex 50WX–8 cation exchange resin and Dowex 3 basic anion exchange resin to remove interfering strong electrolytes. Boron is estimated in the percolate by mass spectrometry.

12.3.4 High purity waters

Jones *et al.* [25] have described ion exchange chromatographic separation systems using either a low capacity silica based cation exchange material with a lactate eluent, or a high capacity resin based cation exchanger with a tartrate eluent. Photometric detectors made use of post-column reactors for incorporating a reagent, the most successful of which was eriochrome black–T, which produced changes in absorption or fluorescence when mixed with metal species as they eluted from the column. These methods were applied to on-line monitoring of trace metals, notably manganese, iron, cobalt and nickel, in the primary coolant of a pressurised water reactor.

12.4 Anions

12.4.1 Non saline waters

12.4.1.1 Arsenate and arsenite

Henry et al. [26] reported a method for the determination of As(III), As(V) and total inorganic arsenic by differential pulse polarography. As(III) was measured directly in 1M perchloric acid or 1M hydrochloric acid. Total inorganic arsenic was determined in either of these supporting electrolytes after the reduction of electroinactive As(V) with aqueous sulphur dioxide. As(V) was evaluated by difference. Sulphur dioxide was selected because it reduced As(V) rapidly and quantitatively, and excess reagent was readily removed from the reaction mixture.

12.4.1.2 Chloride, bromide, iodide, chlorate, bromate and iodate

Salov et al. [27] have described a procedure for determining bromide (and chloride, iodide, chlorate, bromate and iodate) in water employing high performance liquid chromatography with an indirectly coupled argon plasma mass spectrometric detector.

Akaiwa et al. [28] have used ion exchange chromatography on hydrous zirconium oxide, combined with detection based on direct potentiometry with an ion selective electrode, for the simultaneous determination of chloride and bromide in non saline waters.

12.4.1.3 Fluoride

Okabayashi et al. [29] collected fluoride ion selectively on an anion exchange resin loaded with alizarin fluorine blue sulphonate.

12.4.1.4 Nitrate and nitrite

Davenport and Johnson [30] used ion exchange chromatography on Amberlite IRA–900 strongly basic resin to determine nitrate and nitrite in water. 0.01M perchloric acid was used as eluent and an electrochemical cadmium electrode detector was used.

12.4.1.5 Sulphate and chloride

Stainton [31] has described an automated method for the determination of sulphate and chloride in non saline waters. An ion exchange resin is used to convert the sulphates and chlorides to their free acids. Detection is achieved by electrical conductance. The use of silver-saturated cation exchange resin to precipitate chloride permits distinction between chloride and sulphate. High levels of nitrate, orthophosphate and fluoride give positive interference for sulphate; bromide and iodide similarly interfere with chloride estimates.

12.4.1.6 Miscellaneous

Shintani and Dasgupta [32] have reported that post-suppression membrane-based ion exchange chromatography with fluorescence detection permits detection limits superior to those obtained by conductivity detection in hydroxide eluent suppressed anion chromatography.

12.4.2 Aqueous precipitation

12.4.2.1 Chloride and bromide

Akaiwa et al. [28] have used ion exchange chromatography on hydrous zirconium oxide combined with a detection based on direct potentiometry with an ion selective electrode for the simultaneous determination of chloride and bromide in rain.

12.4.3 Potable water

12.4.3.1 Nitrate and nitrite

Sherwood and Johnson [33] have described an ion exchange chromatographic determination of nitrate and amperometric detection at a copperised cadmium electrode. The chromatograms obtained in this procedure resolve nitrate from dissolved oxygen.

Gerritse [34] separated nitrate and nitrite on a cellulose anion exchanger column and detected them in a spectrophotometric flow-through cell at 210nm. The detection limit was 1–5µg L^{-1} as nitrogen with a sample volume of 100–200µL and the linear range extends up to 20mg L^{-1} as nitrogen.

References

1 Nickless, G. and Jones, P.J. *Journal of Chromatography*, **156**, 87 (1978).
2 Musty, P.R. and Nickless, G. *Journal of Chromatography*, **89**, 185 (1974).
3 Musty, P.R. and Nickless, G. *Journal of Chromatography*, **120**, 369 (1976).
4 Jones, P. and Nickless, G. *Journal of Chromatography*, **156**, 99 (1978).
5 Josefsson, B.O. *Analytica Chimica Acta*, **52**, 65 (1970).
6 Lis, A.W., McLaughlin, R.K., Tran, J., Davies, G.D. and Anderson, W.R. *Tappi*, **59**, 127 (1976).
7 Katz, S., Pitt, W.W., Scott, C.D. and Rosen, A.A. *Water Research*, **6**, 1029 (1972).
8 Pill, W.W., Jolley, R.L. and Scott, C.D. *Environmental Science and Technology*, **9**, 1068 (1975).

9 Hanai, T. and Walton, H.F. *Analytical Chemistry*, **49**, 764 (1977).
10 Dietz, E.A. and Perez, M.T. *Analytical Chemistry*, **48**, 1088 (1976).
11 Johnson, D.L. and Pilson, M.E.O. *Analytica Chimica Acta*, **58**, 289 (1972).
12 Grabinski, A.A. *Analytical Chemistry*, **53**, 966 (1981).
13 Aggett, J. and Kadwani, R. *Analyst (London)*, **108**, 1495 (1983).
14 Henry, F.T. and Thorpe, T.M. *Analytical Chemistry*, **52**, 80 (1980).
15 Boyle, E.A., Handy, B. and Van Green, A. *Analytical Chemistry*, **59**, 1499 (1987).
16 Orvini, E., Ladola, L., Sabbioni, E., Pietra, R. and Goetz, L. *Science of the Total Environment*, **13**, 195 (1979).
17 Mieura, O. *Journal of Analytical Chemistry*, **62**, 1424 (1990).
18 Korkisch, J. and Godl, L. *Talanta*, **12**, 1035 (1974).
19 Welte, B. and Montiel, A. *Techn. Sci. Municip.*, **75**, 100 (1980).
20 Small, H., Stevens, T.S. and Bauman, W.C. *Analytical Chemistry*, **47**, 1801 (1975).
21 Araki, S., Suzuki, S., Hobo, T. *et al. Japan Analyst*, **17**, 847 (1968).
22 Nevoral, V. and Okae, A. *Cslka. Farm.*, **17**, 478 (1968).
23 Toshio, N. *Bulletin of the Chemical Society of Japan*, **42**, 3017 (1969).
24 Gorene, B., Marcel, J. and Tramsek, G. *Mikrochimica Acta*, **1**, 24 (1970).
25 Jones, P., Barron, K. and Ebdon, L. *Analytical Proceedings (London)*, **22**, 373 (1985).
26 Henry, F.T., Kirch, T.D. and Thorpe, T.M. *Analytical Chemistry*, **51**, 215 (1979).
27 Salov, V.V., Yoshimaga, J., Shibata, Y. and Maritana, M. *Analytical Chemistry*, **64**, 2425 (1992).
28 Akaiwa, H., Kawamoto, H. and Osumi, M. *Talanta*, **29**, 689 (1982).
29 Okabayashi, Y., Oh, R., Nakagawa, T., Tanaka, H. and Chicuma, M. *Analyst (London)*, **113**, 829 (1988).
30 Davenport, R.J. and Johnson, D.C. *Analytical Chemistry*, **46**, 1971 (1984).
31 Stainton, M.P. *Limnology and Oceanography*, **19**, 707 (1974).
32 Shintani, H. and Dasgupta, P.K. *Analytical Chemistry*, **59**, 1963 (1987).
33 Sherwood, G.A. and Johnson, P.C. *Analytica Chimica Acta*, **129**, 101 (1981).
34 Gerritse, R.G. *Journal of Chromatography*, **171**, 527 (1979).

Chapter 13

Column coupling capillary isotachoelectrophoresis and isotachelectrophoresis

At present, ion chromatography has a dominant position among the separation methods used for the analysis of inorganic anions [1–4]. However, when the separation efficiencies typically achieved by ion chromatography are compared to characterising capillary zone electrophoresis for analysis of a group of analytes, it is apparent that the latter technique has advantages. As the effective mobilities of inorganic anions are sufficiently due to differences in their ionic mobilities they can be influenced in a desired way via appropriately selecting (differences in pK_a values) and/or via the use of suitable additives [5], the selectivity factors [6] are also favourable leading to rapid resolutions of the analytes by capillary zone electrophoresis.

This technique offers very similar advantages to ion chromatography in the determination of anions in water, namely multipleion analysis, little or no sample pretreatment, speed, sensitivity and automation.

13.1 Organic compounds

13.1.1 Non saline waters

13.1.1.1 Carboxylates

Barth [7] used capillary isotachoelectrophoresis for the quantitative determination of volatile organic acids in waters associated with oil-bearing geological formations. The detection limit was 0.02M.

13.1.1.2 Chlorinated carboxylic acids

Onodera *et al.* [8] examined the applicability of isotachophoresis to the identification and determination of chlorinated mono- and dicarboxylic acids in chlorinated effluents. Four electrolyte systems for the separation of acids were evaluated. The potential unit values in each system were determined for the chlorinated acids. A mechanism for the reaction of phenol with hypochlorite in dilute aqueous solutions is suggested, based on results from the isotachophoretic analysis of diethyl ether extracts taken from phenol treated with hypochlorite.

13.1.1.3 Herbicides

Purkayastha [9] examined the applicability of paper electrophoresis to nine ionisable chlorinated phenoxyacetic acid type herbicides including 2,4–D, 2,4,5–T, also MCPA, Fenoprop, Dicamba (2-methoxy-3,6-dichloro-benzic acid), and Picloram (4-amino-3,5,6-trichloropicolinic acid). Solutions were applied to paper moistened with pyridine–acetic acid buffer solution, pH3.7, 4.4 or 6.5, and a voltage of 2–4kV was applied. After 30min the paper was air-dried, sprayed with ammoniacal silver nitrate solution, and exposed to ultraviolet radiation. Experimental variation that increased the mobility of the spots included the applied voltage from 2 to 4kV (potential gradients of 50–100V cm $^{-1}$), and adding a foreign electrolyte (eg potassium nitrate) to the buffer. Addition of methanol to the buffer resulted in decreased mobility as well as a variation in the relative mobilities of the compounds.

13.1.1.4 Aminoacetates

Yu and Dovichi [10] used capillary zone electrophoresis with thermo-optical absorbance detection to determine sub μg L $^{-1}$ concentrations of 18 amino acids.

13.2 Cations

13.2.1 Non saline waters

13.2.1.1 Chromium

Zelensky *et al.* [11] have studied the capability of capillary isotacho-phoresis in the trace determination of chromium(VI) in water samples at low (μg L $^{-1}$) concentrations. A column coupling configuration of the separation unit in the analyser was employed together with photometric detection at 405nm wavelength (Fig. 13.1). Losses of chromium(VI) due to adsorption on the walls of the glassware were prevented by the addition of sulphate (0.0001M) to the sample solutions. At lower pH addition of naphthalene–1,3,6–trisulphonate prevented adsorption on to the walls of the separation unit. With these suitable precautions detection limits were in range of 4–5μg L $^{-1}$ for a 30L sample.

Although in some instances it is possible to determine low concentrations of inorganic anions by capillary zone electrophoresis [12], problems arise when the concentrations of the sample constituents vary considerably. This is due to the fact that the determination of

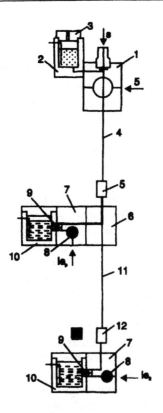

Fig. 13.1 Separation unit in a column coupling configuration as used for the analysis of anions in river water: 1, sampling block with a 30μL sampling valve; 2, terminating electrolyte compartment with a cap (3); 4, 0.85mm id capillary tube (pre-separation column); 5 and 12, conductivity detectors; 6, bifurcation block; 7, refilling block with needle valve (8); 9, mechanically supported membranes; 10, leading electrolyte compartments; 11, 0.30mm id capillary tube; s, positions for the sample introduction (valve or microsyringe); lep and le, positions for refilling of the columns used for the first and second stages respectively

Source: Reproduced with permission from Elsevier Science, UK [11]

microconstituents may require the sample load to have an impact on both the migration velocities and the resolution of analytes [13]. The use of indirect detection, as preferred in the case of inorganic anions [12,14] is also less capable of achieving adequate load capacities.

Very similar advantages are offered by capillary isotachophoresis [15,16]. However, with the exception of the work by Bocek *et al.* [17] in which the determinations of chloride and sulphate in mineral water were demonstrated, little attention has been paid to this subject.

13.3 Anions

13.3.1 Non saline waters

13.3.1.1 Chloride, fluoride, nitrite, nitrate, sulphate and phosphate

Zelensky *et al.* [11] applied the technique to the determination of $0.02–0.1mg$ L^{-1} quantities of chloride, nitrate, sulphate, nitrite, fluoride and phosphate in river water. Approximately 25min was required for a full analysis. These workers used a column coupling technique which, by dividing the analysis into two stages, enables a high load capacity and a low detection limit to be achieved simultaneously without an appreciable increase in the analysis time.

The separation unit of the capillary isotachophoresis instrument used is shown in Fig. 13.1. A 0.85mm id capillary tube made of fluorinated ethylene propylene copolymer was used in the pre-separation (first) stage and a capillary tube of 0.30mm id made of the same material served for the separation in the second stage. Both tubes were provided with conductivity detection cells [18] and an ac conductivity mode of detection [15] was used for making the separations visible.

The retardation of chloride and sulphate through complex formation with cadmium enabling the separation of anions of interest [16] to be carried out, was found to be unsuitable, as the high concentration of cadmium ions necessary to achieve the desired effect led to the loss of fluoride and phosphate (probably owing to precipitation).

Similarly, the use of calcium and magnesium as complexing co-counter ions [19] to decrease the effective mobility of sulphate was found to be ineffective as a very strong retardation of fluoride occurred.

Better results were achieved when a divalent organic cation was used as a co-counter ion in the leading electrolyte [20,21] employed in the first separation stage when, simultaneously, the pH of the leading electrolyte was 4 or less, and the steady state configuration of the constituents to be separated was chloride, nitrate, sulphate, nitrite, fluoride and phosphate.

Kaniansky *et al.* [21] have also reported on a method for the determination of nitrate, sulphate, nitrite, fluoride and phosphate by capillary zone electrophoresis coupled with capillary isotacho-electrophoresis in the column coupling configuration. Such distributions of these anions are typical for many environmental matrices, and it is shown that capillary isotachophoresis–capillary zone electrophoresis tandem enables the capillary isotachoelectrophoresis determination of the macroconstituents, while capillary isotachophoresis–preconcentrated microconstituents cleaned up from the macroconstituents can be determined by capillary zone electrophoresis with a conductivity detector. This approach was effective when the concentration ratio of macromicroconstituents was less than $(2–3) \times 10^4$. The limits of detection

Table 13.1 Electrophoretic methods for the determination of anions in various types of water

Anion	Electrophoretic method	Comments	Detection	LD	Ref.
		Non saline waters			
Chloride, sulphate	Column coupling capillary isotachoelectrophoresis			0.02–0.1mg L^{-1}	[22]
Chloride, fluoride, nitrate, sulphate, nitrite and phosphate	Column coupling capillary isotachoelectrophoresis	–	–	sub µg L^{-1}	[23]
		Sea water			
Sulphide	Capillary isotachoelectrophoresis	Sulphide liberated from sample with acid and collected in sodium hydroxide prior to electrophoresis	–	0.1mg L^{-1}	[24]
		Rain water			
Bromide	Capillary isotachoelectrophoresis	–	–	–	[25]
		Potable water			
Nitrate, sulphate	Capillary isotachoelectrophoresis	Interferences by carboxylates and sulphate overcome	–	–	[26]
Multi anions	Electrophoresis	Comparison with results obtained by ion chromatography	–	mg L^{-1}	[27]

Source: Own files

achieved for the microconstituents (200, 300 and 600ppt for nitrite, fluoride and phosphate, respectively) enabled their determinations at low parts per billion concentrations in instances when the concentrations of sulphate or chloride in the samples were higher than 1000ppm.
See also Table 13.1.

13.3.1.2 Other applications

The application of electrophoresis techniques to the determination of anions in various types of water is reviewed in Table 13.1.

References

1 MacCarthy, P., Klusman, R.W. and Rice, J.A. *Analytical Chemistry*, **61**, 269R (1989).
2 Frankenberger, W.T. Jr., Mehra, H.C. and Gjerde, D.T. *Journal of Chromatography*, **504**, 211 (1990).
3 MacCarthy, P., Klusman, R.W., Cowling, S.W. and Rice, J.A. *Analytical Chemistry*, **63**, 310R (1991).
4 Tarter, J.D. ed. *Ion Chromatography*. M. Dekker, New York and Basel (1987).
5 Fukushi, K. and Hiro, K. *Journal of Chromatography*, **518**, 189 (1990).
6 Foret, F. and Bocek, P. *Advances in Electrophoresis*, **3**, 271 (1989).
7 Barth, T. *Analytical Chemistry*, **59**, 2232 (1987).
8 Onodera, S., Udagawa, T., Tabata, M., Ishkura, S. and Suzuki, S. *Journal of Chromatography*, **287**, 176 (1984).
9 Purkayastha, R. *Bulletin of Environmental Contamination and Toxicology*, **4**, 246 (1969).
10 Yu, M. and Dovichi, N. *Analytical Chemistry*, **61**, 37 (1989).
11 Zelensky, I., Zelenska, V. and Kaniansky, D. *Journal of Chromatography*, **390**, 111 (1987).
12 Jackson, P.E. and Haddad, P.R. *Journal of Chromatography*, **640**, 481 (1993). ˙
13 Beckers, J.L. and Everaerts, F.M. *Journal of Chromatography*, **508**, 3 (1990).
14 Jones, W.R. and Jandik, P. *Journal of Chromatography*, **546**, 455 (1991).
15 Everaerts, D.M., Beckers, J.L. and Verheggen Th. P.E.M. In *Isotachophoresis – Theory Instrumentation and Applications*. Elsevier, Amsterdam, Oxford, New York (1976).
16 Hjalmarsson, S.G. and Baldesten, A. *Critical Reviews of Analytical Chemistry*, **11**, 261 (1981).
17 Bocek, P., Miedziak, I., Demi, M. and Janak, J. *Journal of Chromatography*, **137**, 83 (1987).
18 Kanianski, D., Koval, M. and Stankoviansky, S. *Journal of Chromatography*, **267**, 67 (1983).
19 Kaniansky, D. and Evereerts, F.M. *Journal of Chromatography*, **148**, 441 (1978).
20 *The Testing of Water*. E. Merck, Darmstdt, Germany (1980).
21 Kaniansky, D., Zelinsky, I.M., Hyhenova, A. and Onuska, F.I. *Analytical Chemistry*, **66**, 4258 (1994).
22 Bocek, P., Miedziak, I., Demi, M. and Janak, J. *Journal of Chromatography*, **137**, 83 (1987).
23 Zelinski, I., Kanmiensky, D., Havassi, P. and Lednarov, U. *Journal of Chromatography*, **294**, 317 (9184).

24 Fukushi, K. and Hiro, K. *Journal of Chromatography*, **393**, 433 (1987).
25 Fukushi, K. and Hiro, K. *Bunseki Kagaku*, **36**, 712 (1987).
26 Kaniansky, D., Zelinsky, I. and Cerovsky, M. *Journal of Chromatography*, **367**, 274 (1986).
27 Choi, C.K. and Cho, J.S. *Analytical Science and Technology*, **8**, 839 (1995).

Chapter 14

Thin layer chromatography

This technique has been applied exclusively to organic and organometallic compounds with no applications to the determination of anions and cations.

14.1 Organic compounds

14.1.1 Non saline waters

14.1.1.1 Petroleum spills

Thin layer chromatography has replaced paper chromatography in many applications because of its greater versatility and reproducibility. It has been applied as a rapid method of classifying petroleum and natural and synthetic oils by chromatographing 0.1mg quantities of sample oils [1]. These were spotted on to a silica gel plate and developed with an ascending solvent composed of 70:30 vol/vol chloroform–benzene. After 1h the solvent was evaporated, and the resulting chromatogram evaluated in ultraviolet light (366nm) and the fluorescent areas noted. This was followed by spraying the chromatogram with concentrated sulphuric acid and baking it at 120°C for 15min. Coloured zones formed under visible and ultraviolet light were noted. With this procedure, oils were classified into four groups:

1 hydrocarbon oils (eg petroleum products);
2 synthetic ester oils;
3 naturally occurring oils;
4 oils of different composition (eg silicone oil and low–molecular–weight polyethylene oxides).

The relatively polar solvent, compared to, say, hexane, provides little separation of the hydrocarbon components of petroleum products, and therefore while they can be distinguished from the more polar oils investigated, insufficient information is obtained for differentiation between petroleum products.

Krieger [2] employed horizontal silica gel plates and hexane as developing solvent. This resulted in an approximately radial chromatogram which was visualised by spraying with 0.03% fluorescein solution and viewing under ultraviolet light. Crude oil and heavy fuel oil, light petroleum, various tar oils, road tar and olive oil were examined by the technique and some distinction made between them. Natural oil, due to its more polar nature, was not developed, while the remaining products gave generally similar patterns, although in some cases it was possible to match chromatograms of sample and suspect oils.

Normal hexane, a relatively non polar solvent, was adopted by Lambert [3] to produce more separation between petroleum components after ascending development on a silica gel plate. As many as 8–10 samples were chromatographed on a single plate during about 30min, and identification was achieved by comparison of pollutants with standard petroleum materials. Motor oil, light and heavy fuel oil, various greases and olive oil were among many oils investigated, initially under ultraviolet light and then after spraying with concentrated sulphuric acid–formaldehyde solution. Well-defined coloured zones were obtained with these products with reasonable differentiation between most of the oils examined, although light and heavy fuel oil could be confused. An extension of Lambert's technique has been achieved [4] in which some commercially available petroleum fractions and greases were examined.

Gas oil, motor oil, light fuel oil, motor spirit, paraffin, hydrant grease, submersible pump grease and bitumastic grease were chromatographed on silica gel plates using petroleum ether (bp 60–80°C) as developing solvent. Chromatograms were evaluated initially under ultraviolet light and characteristic colours and patterns noted. Further detail was collected by spraying the plates with 1% formalin in concentrated sulphuric acid and baking them at 60°C for 16h. Examination under visible and ultraviolet light revealed differently coloured areas.

Matthews stated that thin layer chromatography is highly suited to the rapid identification of the heavier petroleum and coal tar oils, and their residues [5], and has developed a scheme of systematic quantitative analysis [6]. The scheme is based on the varying absorption of oils on different solid supports, for example, silica gel, alumina and kieselguhr, and on the use of solvents of varying polarity, such as petroleum ether (bp 40–60°C), ethanol, toluene and chloroform. Ascending thin layer chromatography was adopted for the scheme, the initial step being separation on alumina and development with acetone.

Channel thin layer chromatography eliminates most of the horizontal spreading of a vertically moving component on the plate, and thereby increases the sensitivity. It has been applied to analysis of oil pollutants [7]. In this modification, two parallel scratches are made 2mm apart on a silica gel coated plate. The sample is applied so that it develops between these scratches with carbon tetrachloride. This permits the thin layer chromatography of smaller quantities and the resulting rectangular spots are easily measured. The procedure is limited to oils that boil above 150°C. One litre samples are extracted with only 2mL carbon tetrachloride, and loss of any solvent into the water is claimed not to affect the subsequent determination. Therefore exact separation is unnecessary, and the separated extract is chromatographed. On exposure to iodine crystals, brown rectangular areas appear which are easily measured. The developing solvent used, again carbon tetrachloride, probably does not separate saturated and lower aromatic hydrocarbons significantly, and therefore the composition of the oil is less important. A non volatile gas oil is proposed as a standard hydrocarbon material, and various concentrations are chromatographed to calibrate the plate in terms of area spot (mm) produced by 0.1mg of the gas oil. Determination of petrolemn oils in the range 15mg L^{-1} are possible.

Semenov et al. [8] determined small amounts of petroleum products in chloroform extracts of non saline water by extracting the sample (200–500mL) followed by thin layer chromatography on alumina. He developed the chromatogram with light petroleum–carbon tetrachloride–acetic acid (35:155:1), and examined the plate in ultraviolet radiation; the petroleum products exhibit three zones (pale blue, yellow and brown). Each zone is then extracted with chloroform, the fluorescence of the extracts measured and the results referred to a calibration graph. The sensitivity is 0.1mg L^{-1}. The infrared and fluorescence spectra of the zone obtained with various petroleum products are discussed.

Sauer and Fitzgerald [9] have described thin layer chromatographic technique for the identification of water-borne petroleum oils. Aromatic and polar compounds are removed from the sample by liquid–liquid extraction with acidified methanol, the extract is chromatographed on a silica gel thin layer plate, and the separated components are detected by their fluorescence under long- and short-wave ultraviolet light. Unsaturated non-fluorescing compounds are detected by iodine staining.

Silica gel in conjunction with cyclohexane as solvent [10] has been found to be the most suitable absorbent for the separation of saturated hydrocarbons from aromatic hydrocarbons in the gas oil to lubricating oil range. Better separation was claimed if the cyclohexane is saturated with less than 0.5% dimethyl sulphoxide. Under the conditions used, paraffins travelled with the solvent front, followed respectively by the mono-aromatic, diaromatics, polynuclear aromatics and polar substances, which generally remained on the starting point. Chromatograms can be revealed by ultraviolet light in the normal manner to show aromatic and heterocyclic compounds, and then spraying with dichromate–sulphuric acid reagent to reveal all components, including saturated hydrocarbons [10]. Rhodamine B in fluoreceine gives an absorbing spot on a fluorescing background under ultraviolet light at 350 or 254nm. A solution of phosphorus pentachloride in carbon tetrachloride gave some classification of hydrocarbon type in lubricating oils in the following manner [11]:

1 paraffins – light brown;
2 monoaromatics – wine red;
3 diaromatics – dark blue–green;
4 polar aromatics – dark brown.

Triems [12,13] classified oils by fractionating the top residue of the oil, freed from compounds boiling at more than 200°C, on silica gel with 2,2,4-trimethylpentane, benzene and acetone as eluants. Three classes of oil were differentiated, viz. paraffinic–naphthenic, naphthenic–aromatic and aromatic–naphthenic. Each class is subdivided into three groups according to their sulphur content.

Farrington et al. [14] used column chromatography and thin layer chromatography to isolate hydrocarbons, arising from marine contamination, from fish lipids. The hydrocarbon extracts were then examined to select those that can be determined by gas chromatography–mass spectrometry, by combinations of spectrophotometric methods, or by wet chemistry. As a screening method gas chromatography was shown to be fairly accurate and precise for hydrocarbons boiling in the range 287–450°C and of suitable polarity.

See also Table 14.1.

14.1.1.2 Carboxylates

Khomenkô et al. [15,16] used thin layer chromatography for determining non volatile organic acids dissolved in non saline water. The organic acids are extracted from the water and concentrated, then separated on a silica gel column into four groups which are concentrated to 0.1–0.2mL and thin layer chromatography is carried out on layers of silica gel KSK previously air-dried for 20min and activated for 30min at 105°C. The acids in the first, second and third fractions are developed in butanol–benzene–acetic acid (10:20:3, 15:85.2 and 15:35:8 respectively) and in the fourth fraction in ethyl acetate–water–formic acid (9:1:1). After drying the chromatograms for 1.5h at 120°C, the organic acids are detected by spraying with 0.4% solution of bromocresol green in 20% ethanolic alkali and the spot areas are measured for a semi-quantitative determination of the acids.

14.1.1.3 Phenolic acids

Rump [17] has described a cellulose thin layer method for the detection of phenolic acids such as m-hydroxybenzoic acid, m-hydroxyphenylacetic acid and m-hydroxyphenylpropionic acid, in water samples suspected to be contaminated with liquid manure. The phenolic acid is extracted with ethyl acetate from a volume of acidified sample equalling 1mg of oxygen consumed (measured with potassium permanganate). The ethyl acetate is evaporated and the residue dissolved in ethanol. After spotting of a 1μm aliquot on a cellulose plate the chromatogram is developed by capillary ascent with the solvent n-propanol-n-butanol–25% NH_3–water (4:4:1:1 by vol). The solvent front is allowed to advance 10cm. The air-dried plate is sprayed with a diazotised p-nitroaniline reagent to make the phenolic acids visible.

14.1.1.4 Chlorinated hydrocarbons

Hollies et al. [18] found that choro-n-paraffins could be chromatographed on a silica gel plate from which an image of the chromatogram could be 'printed' on an aluminium oxide plate by heating the two face to face so that the high sensitivity of detection on aluminium oxide could be utilised.

In these methods the samples are cleaned up by liquid–solid adsorption chromatography, and thin layer chromatography but those rich in lipids require preliminary solvent extraction. The methods distinguish between chloro-n-paraffins based on long carbon chains (C_{20}–C_{30}) and those based on shorter chains (C_{13}–C_{17}). The methods cover the ranges 500ng L^{-1} to 8μg L^{-1} for water (ie from about the solubility limit upwards) and 50μg kg^{-1} to 16mg kg^{-1} for sediments and biota. The precision of the methods ranges from ±50% relative at the lowest concentrations to ±12% relative at the highest. Recoveries are about 90% for water, 80% for sediments, and between 80 and 90% for biota according to sample type.

14.1.1.5 Phenoxy acetic acid herbicides

Bogacka and Taylor [19] determined 2,4–D and MCPA herbicides in water using thin layer chromatography. In this method a 1L sample of filtered water is treated with 50g of sodium chloride and 5mL of hydrochloric acid and the herbicides are extracted into ethyl ether (200, 100 and 100mL). The extract is dried with anhydrous sodium sulphate, concentrated to a few millilitres and passed through a column (180mm × 15mm) of silica acid with 90% methanol–acetic acid (9:1) as stationary phase and the herbicides are eluted with 150mL of light petroleum saturated with the methanol–acetic acid mixture. The first 30mL of eluate is rejected. The remaining eluate is evaporated to dryness, and the residue is dissolved in ether and concentrated to about 0.1mL before thin layer chromatography on silica gel G–kieselguhr G (2:3) (activated for 30min at 120°C) with light petroleum–acetic acid–liquid paraffin (10:1:2) as solvent. The developed plates are air dried, sprayed with 5% silver nitrate solution and dried, then sprayed with 2m potassium hydroxide–formaldehyde (1:1), dried at 130–135°C for 30min, sprayed with nitric acid, and observed in ultraviolet illumination. For the determination the spots are compared with standards. Evaluation by the method of standard addition gave recoveries of 95.1% and 88.8% respectively with standard deviations of 14.2% and 14.3% for 2,4–D and MCPA respectively.

These workers [20] also examined thin layer chromatography of 2,4–DP (Dichloroprop) and MCPP (mixture of Mecoprop and 2-(2-chloro-4-methylphenoxy)propionic acid). In this method the ethyl ether extract of the sample is purified on a column of silicic acid and the herbicides are separated by thin layer chromatography on silica gel–kieselguhr (2:3) with light petroleum–acetic acid–kerosene (10:1:2) as solvent. The sensitivity is 3μg of either compound per litre, the average recoveries of Dichlorprop and MCPP are 85.7% and 87.4%, respectively, and the corresponding standard deviations were 13.9% and 15.5%.

Phenoxyacetic and herbicides are also discussed in Table 14.1.

14.1.1.6 Carbamate herbicides

N–methylcarbamate and N,N'–dimethylcarbamates were determined in water samples by hydrolyses with sodium bicarbonate and the resulting amines reacted with 4-chloro-7-nitrobenzo-2,1,3-oxadiazole in isobutyl methyl ketone solution to produce fluorescent derivatives [21]. These derivatives were separated to thin layer chromatography on silica gel G or alumina with tetrahydrofuran–chloroform (1:49) as solvent. The fluorescence is then measured in situ (excitation at 436nm, emission at 528 and 537nm for the derivatives of methylamine and dimethylamine, respectively). The method was applied to non saline water samples containing parts per 10^9 levels of carbamate. The disadvantage of the method is its inability to differentiate between carbamates of any one class.

Thin layer chromatographic methods have been described [22] for determining N–arylcarbamate and urea herbicides in non saline water (Barban, Chloroprophan, Diuron, Fenuron, Fenuron TCA, Linuron, Monouron, Monouron TCA, Neburan, Propham, Siduron and Swep) and for determining O–arylcarbamate pesticides in water (Aminocarb, Carbaryl, Methiocarb, Mexacarbate and Propoxur). To determine N–arylcarbamates a measured volume of water sample is extracted with methylene chloride and the concentrated extract is cleaned up with a Florasil column. Appropriate fractions from the column are concentrated and portions are separated by thin layer chromatography. The herbicides are hydrolysed to primary amines, which in turn are chemically converted to diazonium salts. The layer is sprayed with 1-naphthol and the products appear as coloured spots. Quantitative measurement is achieved by visually comparing the response of sample extracts to the responses of standards on the same thin layer plate. Direct interferences may be encountered from aromatic amines that may be present in the sample.

To determine O–arylcarbamate insecticides a measure volume of water is extracted with methylene chloride. The concentrated extract is cleaned up with a Florasil column. Appropriate fractions from the column are concentrated and portions are separated by thin layer chromatography. The carbamates are hydrolysed on the layer and the hydrolysis products are reacted with 2,6-dibromoquinone chlorimide to yield specific coloured products. Quantitative measurement is achieved by visually comparing the responses of sample extracts with the responses of standards on the same thin layer. Identification is confirmed by changing the pH of the layer and observing colour changes of the reaction products. Direct interferences may be encountered from phenols that may be present in the sample. These materials react with chromogenic reagent and yield reaction products similar to those of the carbamates.

Carbofuran and its degradation products carbofuranphenol, 3-keto-carbofuran, 3–hydroxycarbofuranphenol, N–hydroxymethylcarbofuran and 3-hydroxycarbofuran, have been determined in water samples by thin layer chromatography of ether extracts of the sample [23].

Carbamate herbicides are also discussed in Table 14.1.

14.1.1.7 Miscellaneous herbicides

Smith and Fitzpatrick [24] have also described a thin layer method for the detection in water and soil of herbicides residues, including Atrazine, Barban, Diuron, Linuron, Monouron, Simazine, Trifluralin, Bromoxynil, Dalapon, Dicamba, MCPB, Mecoprop, Dicloram, 2,4–D, 2,4–DB, Dichlor-prop, 2,4,5–T and 2,3,6–trichlorobenzoic acid.

Neutral and basic herbicides were extracted from water made alkaline with sodium hydroxide or from soil, with chloroform; extracts of soil were cleaned up on a basic alumina containing 15% of water. Acidic herbicides were extracted with ethyl ether from water acidified with hydrochloric acid or from an aqueous extract of soil prepared by treatment with 10% aqueous potassium chloride that was 0.05m in sodium hydroxide and filtration into 4m hydrochloric acid. The concentrated chloroform solution of neutral and basic herbicides was applied to a pre-coated silica gel plate containing a fluorescent indicator and a chromatogram was developed two-dimensionally with hexane–acetone (10:3) followed after drying by chloroform–nitromethane (1:1). The spots were detected in ultraviolet radiation. Atrazine, Barban, Diuron, Linuron, Monouron, Simazine and Trifluralin were successfully separated and were located as purple spots on a green fluorescent background. The ether extracts were dried over sodium sulphate, concentrated, and applied to a similar plate, which was developed two-dimensionally with chloroform–anhydrous acetic acid (19:1) followed after by drying by benzene–hexane–anhydrous acetic acid (5:10:2). The spots were detected by spraying with bromocresol green. Bromoxynil and (as the acids) Dalapon, Dicamba, MCPA, MCPB, Mecoprop, Dicloram, 2,4–D, 2,4–DB, Dichlorprop, 2,4,5–T and 2,3,6–trichlorobenzoic acid were seen as yellow spots on a blue background. The limits of detection were 1ppm in soil and 0.1ppm in natural water.

In a further thin layer chromatographic method for determining carbamate and urea herbicides in water at the parts per 10^9 level Frei *et al.* [25] extracted a 500mL sample with dichloromethane (2 × 50mL) and evaporated the combined extract to 1mL at room temperature in a rotary evaporator and then to dryness at 40°C. The residue was dissolved in acetone (1 or 2 drops) and 0.5mL of sodium hydroxide and heated to 80°C for 30–40min, cooled and shaken with 0.2mL of hexane. 10μL of the hexane layer is applied to a 0.25mm layer of silica gel G–CaSO$_4$, 0.2% dansyl chloride in acetone is applied to the sample spot, and the chromatogram developed by the ascending technique with benzene–triethylamine–acetone (75:24:1). The plate is sprayed with 20% triethanolamine in isopropyl alcohol or 20% liquid paraffin in toluene, then dried. The fluorescence of the spots of the dansyl derivatives of the aniline moieties is measured *in situ*. Results are reported for carbamate pesticides, eg Propham, Chloropropham and Barban, and the urea pesticides, Linuron, Diuron, Chlorbromuron and Fluometuron. Detection limits are about 1ng. Two-dimensional chromatography was used to eliminate interference.

See also Table 14.1.

14.1.1.8 Urea herbicides

Deleu *et al.* [26] have used two-dimensional thin layer chromatography and gas chromatography to separate and identify 10 urea herbicides in concentrations down to 4μg L^{-1} in river waters (Fig. 14.1). Four different adsorbents were compared. The eluting solutions were:

1 diethyl ether–toluene (1:3);
2 diethyl ether–toluene (2:1);
3 equal volumes of 1 and 2;
4 chloroform–nitromethane (3:1).

See also Table 14.1.

14.1.1.9 Chlorophylls

Several chromatographic methods have been published, and most of these utilise thin layer chromatography to separate the chlorophylls. Thin

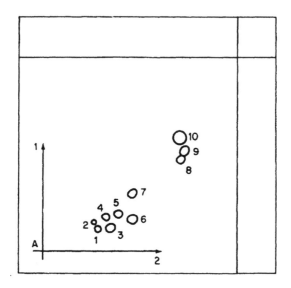

Fig. 14.1 Two-dimensional chromatogram of: 1 = Metoxuron; 2 = Fenuron; 3 = Monouron; 4 = Isoproturon; 5 = Chlorotoluron; 6 = Diuron; 7 = Metabenzthiazuron; 8 = Neburon; 9 = Linuron, Chlorobromuron; 10 = Buturon A.
Source: Reproduced by permission from Elsevier Science, UK [26]

layer materials employed have included the following: layer of kieselguhr G impregnated with triolein, caster oil, or paraffin oil [27,28] layers prepared from glucose shaken with ether [29], plates coated with kieselguhr G impregnated with peanut oil dissolved in iso–octane [30] and layers of powdered confectioner's icing sugar containing 5% cornflour suspended in light petroleum [31].

14.1.1.10 Other applications

The thin layer chromatographic determination of various other types of organic compounds in non saline waters is reviewed in Table 14.1.

14.1.2 Seawater

Abayachi and Riley [100] used high performance liquid chromatography to determine chlorophylls, and their degradation products, and carotenoids in phytoplankton and marine particulate matter.

Pigment extraction is carried out with acetone and methanol. After evaporation of the combined extracts under reduced pressure, the pigments are separated on a Partisil 10 stationary phase with a mobile

249

Table 14.1 Thin layer chromatography of organics in non saline waters

Organic	Stationary phase	Eluent	Comments	Detection	LD	Ref.
Hydrocarbons	Silica gel	–	Separates alkanes, unsaturates, aromatics	–	–	[32]
Petroleum distillates	Silica gel	n-hexane	–	Aq fluorescein spray	–	[8,33]
Polyaromatic hydrocarbons	–	–	2-dimensional TLC	–	ng	[34–39]
Polyaromatic hydrocarbons	Silica gel	Cyclohexane	–	Fluorescence	1–10ng	[40]
Polyaromatic hydrocarbons	–	–	–	–	–	[41]
Non ionic detergents	Silica gel	Acetone or chloroform	–	Dragendorff reagent	0.5–5µg	[42]
Phenols	Activated carbon	–	–	–	–	[43]
Hexane 1,6 diamine	–	–	–	–	0.5µg	[44]
Amines, nitrosamines	Silica gel	–	Derivitivisation of amines and nitrosamines with 1-dimethylamino-5-napthalene sulphonyl chloride and 7-chloro-4-nitro-benzo-oxy 1,3 diazole	Fluorescence after UV irradiation at 365nm	–	[45]
Cyclohexylamine	–	–	Extraction from water on C_{18} microcolumn then TLC	Ninhydrin and exposure to UV	–	[46]
Alkylchloro and nitrophenols	–	–	Comparison of results by TLC and gas chromatography	–	–	[47]

Table 14.1 continued

Organic	Stationary phase	Eluent	Comments	Detection	LD	Ref.
Pentachloro phenol and other chloro-phenols	–	–	–	–	–	[48–50]
Chlorophenols	Kieselgel G	Benzene	–	Diatolised sulphon-itic acid spray or 1:1 15% ferric chloride and 1% $K_3Fe(CN)_6$	–	[51]
Dinitro-o-cresol	Cellulose MN	Ammonia saturated butanol	Cresols extracted from water with diethyl ether then TLC	Bright yellow spots produced with ammonia, spots extracted from plate; with 50% acetic acid-methanol and extinction measured at 370nm	–	[52.53]
Glutamic aldehyde (aldehydes and ketones)	–	–	Chloroform extract of water sample converted to hydrazone	–	–	[54]
Caprolactam	–	–	–	Absorbtive voltammetry	–	[55]

Table 14.1 continued

Organic	Stationary phase	Eluent	Comments	Detection	LD	Ref.
Fluorescent eluting agents	–	–	–	–	–	[56]
Chemical warfare agents, eg quinnuclidinyl benzylate, thiocholine, methylethoxy phosphorate, dibenzoxazepine	–	–	–	–	–	[57]
Fungicides, Bayor2, Bayleton2, Baystan3, Cultar2, Impact	Litufol	Chloroform dimethyl ketone	Extracted from water with chloroform then TLC	Bromocresol blue spray	0.1–0.5 mg L^{-1}	[58]
Polychlorobiphenyls	Kieselguhr G	Acetonitrile: acetone: methanol:water (20:9:20:1) saturated with liquid paraffin	–	Spray with 0.85% ethanolic silver nitrate and exposed to UV black spots formed	–	[59]
Polychlorobiphenyls	Silica gel impregnated with 8% paraffin oil	Acetonitrile: acetone:methanol water (20:9:20:1)	–	Mixture of silver nitrate and 2-phenoxyethanol	0.5–1 μg	[60]
Organochlorine insecticides	Silica gel G or alumina	Hexane or hexane–acetone (4:9:1)	Preliminary solvent extraction of water and Florasil clean-up	–	–	[61]
Chlorophos	–	–	–	–	–	[62]
Organochlorine insecticides	–	–	Scheme for separation based on column chromatography and TLC	–	–	[63,64]

Table 14.1 continued

Organic	Stationary phase	Eluent	Comments	Detection	LD	Ref.
p,p'DDT, p,p'DDD, p,p'DDE, dichloro-biphenyl	Alumina coated sheets	Aqueous soln. of sodium sulphate with polyamide	–	–	–	[65]
Chlorinated insecticides	Silica gel	n-heptane or n-heptane containing 0.3% ethanol	–	Plate dried at 65°C and sprayed with silver nitrate plus 2-phenoxyethanol, dried and exposed to UV light	–	[66]
Chlorinated insecticides	–	–	–	–	–	–
Methoxychlor	–	Hexane:acetone (10:10)	–	Spray with ammoniacal silver nitrate and UV irradiation	–	[67.68]
o,p cresols, phenol 2-napthol, 2,5,3,4 and 3,5 xylenols	Kieselgel G impregnated with potassium carbonate	–	First couple phenols with fast red salt AL (CI azoic diazo component 36)	Spray with diazolised sulphonilic acid	–	[69]
Polyhydric phenols	Kieselgel G coated with potassium carbonate	Dioxan–benzene anhydrous acetic (25:9:4) or benzene:acetone (9:10)	–	Spray with diazolised sulphanilic acid or molybdo-phosphoric acid	–	[70]

Table 14.1 continued

Organic	Stationary phase	Eluent	Comments	Detection	LD	Ref.
Phenols, cresols, xylenols, naphthols	Kieselgel G coated with potassium carbonate	–	Performed on active carbon extracts of water, phenols coupled with Fast blue salt BB (Cl azoic diazo component 20)	–	–	[71]
Phenoxyacetic acid herbicides, 2,4'D, 2,4,5 T and MCPA	–	–	–	Detected as violet spots by spraying with aqueous chromotropic acid containing sulphuric acid and heating to 160°C or spraying with silver nitrate then UV	0.05–2µg	[72]
S-triazines Simazino, atrazine, prometryn herbicides	– Silica gel G impregnated with fluorescein	– Chloroform: acetone (9:1)	– Performed on dichloro-methanol or diethyl ether extract of water sample, then eluted with diethyl ether (0.5% water on basic alumina column	– Spray with acetone soln. of 5% brilliant green (Cl basic green) and expose to bromine vapour	– –	[73,74] [75,76]
[¹⁴C]Ametryne	Silica gel	–	–	Located under UV then liquid scintill-ation counting	5ng	[77]
Triazine and chlorophenoxy acetic acid herbicides	–	–	–	–	10µg L^{-1}	[46]

Table 14.1 continued

Organic	Stationary phase	Eluent	Comments	Detection	LD	Ref.
Atrazine, simazine, 24D, silvex, 2,4,5 T	—	—	—	—	10ppb	[46]
Thiocarbamic acid	—	—	—	—	—	[78,79]
Diuron (3-(3,4 dichlorophenyl 1:1: dimethyl urea	—	—	—	—	—	[80]
Malathion, dimethoate, ethion	Silica gel	Hexane–diethyl ether (9:2) (Malathion) or benzene–diethyl ether (9:2) (Malathion) or benzene–ethyl acetate (9:2) (Malathion), hexane–diethyl ether (9:2) (Dimethoate) or benzene– diethyl ether (5:1) (Dimethoate) benzene (Ethion) or hexane (Ethion)	Extracted from water sample with chloroform or dichloroethane Malathion Rf0.4–0.35 Dimethoate Rf0.3–0.35 Ethion Rf0.8–0.85	Bromine vapour then 3,5 dibromo -p-quinone in dimethyl/formamide bromophenol blue in acetone	0.1–0.2µg	[81]

255

Table 14.1 continued

Organic	Stationary phase	Eluent	Comments	Detection	LD	Ref.
Malathion, trichlorophen, Dichlorvos	Silica gel G containing silver nitrate or in the case of Malathion, fluorescein	Chloroform: ethyl acetate (9:1)	Extracted from water with chloroform. R_f. Trichlorophon 0.1–0.12 Dichlorvos 0.55–0.6 Malathion 0.5–0.6	Spray with butter yellow then expose to UV Malathion exposure to bromine vapour	1–2µg	[82]
Malathion, Parathion, Parathion-methyl, Bromophos	Silica gel F	Hexane: dichloromethane (1:1)	Extracted from water with dichloromethane	Yellow spots produced on spraying with palladium(II) chloride	–	[83]
23 organophosphorus insecticides	Alumina G (Merck type E)	Acetone:hexane (7:93)	–	Enzymic choline esterase inhibition method or tetrabromomethane in acetone or citric acid in acetone	–	[84–86]
16 phosphorothioate insecticides	Silica gel	Hexane:acetone or hexane: diethyl ether	–	Fluorescent spots with palladium(II)– calcein or palladium –calcein blue complex	5–15ng dimethoate	[87]
Parathion, parathion methyl phosphamidon, Azinphos methyl	Silica gel G containing 10% copper sulphate and 2% aq ammonia (1:5)	Hexane diethyl ether (7:3)	–	Heating for 10min at 100°C	4–25µg	[88]

Table 14.1 continued

Organic	Stationary phase	Eluent	Comments	Detection	LD	Ref.
Abate and abate residues	Silical gel G	Hexane:acetone (10:1)	Water extracted with chloroform R_f Abate 0.10–0.12 (red spot on yellow background	Red spots with 10% N,N,dimethyl-p-phenylazo) aniline spray (CI solvent yellow 2) in 95% iso propyl alcohol	9µg	[89]
Abate	Reverse phase TLC plate	–	–	Densitometry	–	[90]
12 acidic herbicides, 19 nitrogenous herbicides (carbamates, ureas and triazines)	–	–	–	–	–	[91]
Polidim	–	–	–	–	–	[92]
Di fenzoquat (Avenge)	–	–	–	–	–	[93]
Misc. herbicides	–	–	–	Hill reaction inhibition with mixture of bean leaves (phaseolus vulgaris) and redox indicator 2,6 dichlorophenol	0.01mg L^{-1}	[94]

Table 14.1 continued

Organic	Stationary phase	Eluent	Comments	Detection	LD	Ref.
Goltix insecticides	Silica gel	Chloroform: methanol (85:15)	Extraction from water with chloroform in presence of sodium chloride	–	60μg L^{-1}	[95]
Pyrazon (5-amino-4-chloro-2-phenyl-3-pyridizone)	–	–	–	–	–	[96–99]

Source: Own files

phase consisting of light petroleum (bp 60–80°C), acetone, dimethyl sulphoxide and diethylamine (75:23.25:1.5:0.25 by volume). When chlorophyll c is present, a further development is performed with a similar, but more polar, solvent mixture. Detection is carried out spectro-photometrically at 440nm. The method has a sensitivity for the chlorophylls of ca. 80ng and for the carotene of ca. 5ng. The coefficient of variation of the chromatographic stage of the procedure lies in the range of 0.6–1.8%.

Abayachi and Riley [100] compared results obtained by the high performance liquid chromatographic method with those obtained by a reflectometric thin layer chromatographic method and the SCOR/UNESCO polychromatic procedure. The results obtained from the latter were evaluated by the SCOR/UNESCO equations. The carotenoids were determined collectively from the absorption of the 90% acetone extract at 480nm by means of the equations of Strickland and Parsons [101]. The results of these comparative studies (Table 14.2) show that there is satisfactory agreement for all pigments between the two chromatographic methods. However, although the results for chlorophyll a by the polychromatic method were in reasonable accord with those derived chromatographically, many of those for the other chlorophylls were highly discrepant. Obviously, the polychromatic method, in particular, is unsatisfactory with respect to the interference of chlorophyll degradation products, as these are nearly always present in environmental samples.

Garside and Riley [102] have used thin layer chromatography to achieve a preliminary separation of chlorophylls on solvent extracts of water and algae prior to a final determination of spectrophotometry of fluorometry. Garside and Riley [102] filtered seawater samples (0.5–5L) through Whatman GF/C glass fibre coated with a layer, 1–2mm thick, of light magnesium carbonate.

Shoaf and Lium [103] used thin layer chromatography to separate algal chlorophylls from their degradation products. Chlorophyll is extracted from the algae with dimethyl sulphide and chromatographed on commercially available thin layer cellulose sheets, using 2% methanol and 98% petroleum ether as solvents, before determination by either spectrophotometry or fluorometry.

Typical values for chlorophylls and degradation products obtained by Shoaf and Lium [103] are shown in Table 14.3. Recoveries of pure chlorophylls a and b were 98% and 96% respectively.

Boto and Bunt [104] also used thin layer chromatography for the preliminary separations of chlorophylls and phaeophytins and combined this with selective excitation fluorometry for the determination of the separated chlorophylls a, b and c and their corresponding phaeophytin components. An advantage of the latter technique is that appropriate selection of excitation and emission wavelengths reduces the overlap

Table 14.2 Comparison between algal pigment analyses carried out by the high performance liquid chromatographic method, by reflectometric thin layer chromatography, and by the polychromatic method

Pigment	HPLC μg dL⁻¹	HPLC % of total carotenoids	TLC μg dL⁻¹	TLC % of total carotenoids	Polychromatic SCOR/UNESCO μg dL⁻¹	Ref [102] μg dL⁻¹
Phaeodactylum tricornutum						
Chlorophyll a	2.68		2.80		2.9	3.0
Chlorophyll b	0.00		0.00		0.005	−0.34
Chlorophyll c	0.44		0.47		0.59	0.43
β—Carotene	0.44	10.5	0.46	10.5		
Fucoxanthin	2.84	68.0	2.88	69.9	4.6	
(total carotenoids)	4.18		4.37			
Diadinoxanthin	0.90	21.5	1.03	23.6		
Chl, a Chl, c	1:0.17	1:0.50	1:0.16		1:0.2	1:0.14
Dunaliella tertiolecta						
Chlorophyll a	4.53		4.40		4.3	4.5
Chlorophyll b	2.18		2.20		0.04	0.17
β—Carotene	1.10	33.3	1.01	30.2		
Lutein	0.53	16.0	0.55	16.6	2.9	
(total carotenoids)	3.32		3.32			
Violaxanthin	1.27	38.5	1.33	40.2		
Neoxanthin	0.41	12.2	0.43	13.0		
Chl. a; Chl. b	1:0.48	1:0.50			1:0.52	1:0.42
Oscillatoria spp.						
Chlorophyll a	17.07		17.75		15.6	16.11
Chlorophyll b	0.00		0.00		−0.54	−2.27
Chlorophyll c	0.00		0.00		1.89	6.3
β—Carotene	1.97	5.3	2.08	46.8		
Echineone	1.65	37.9	1.66	37.4	0.8	
(total carotenoids)	4.34		4.44			
Myxoxanthophyll	0.73	16.8	0.70	15.8		

Source: Reproduced by permission from Elsevier Science, UK [100]

Table 14.3 R_f values for chlorophylls and degradation products (relative to solvent)

Compound	R_f	Colour
Phaeophytin *a*	0.89	Grey
Chlorophyll *a*	0.76	Blue green
Phaeophytin *b*	0.61	Greenish yellow
Chlorophyll *b*	0.34	Yellowish green
Phaeophytin *c*	0	Yellowish green
Chlorophyll *c*	0	Yellowish green

Note: Highly purified chlorophylls *a* and *b* were purchased from a commercial source. Chlorophyll *c* was extracted from a fresh water species of *Cyclotella*. The phaeophytins were formed by acidification of the chlorophylls with hydrochloric acid.

Source: Reproduced by permission from J. Research US Geological Survey [103]

between the emission spectra of each pigment to a greater extent than is possible with broad-band excitation and the use of relatively broad-band filters for emission.

14.1.3 Potable waters

14.1.3.1 Phenols

Thielemann [105] has used thin layer chromatography to study the effect of chlorine dioxide on 1- and 2-naphthols in potable water. The coloured products obtained are thought to be condensation products of chloroderivatives of 1,2– or 2,6–naphthaquinone. Thielemann applied paper chromatography to a study of the reaction products of polyhydric phenols with chlorine dioxide.

14.1.3.2 Chlorinated insecticides

Mosinska [106] has described a semi-quantitative thin layer chromatographic method for the determination of trichlorphon in potable water. The sample is extracted with redistilled chloroform. The extract is dried with sodium sulphate, reduced in volume to 5mL *in vacuo* and then evaporated to dryness in a stream of air. The residue is dissolved in acetone and chromatographed on chloride-free silica gel G plates (activated at 100°C for 1h) with benzene–methanol (17:3) as solvent. The spots, revealed with ammoniacal silver nitrate in acetone, are compared with those of standards for semi–quantitative determination. The detection limit is 0.02mg L $^{-1}$ and the efficiency of extraction is 70%.

Sackmauerva *et al.* [66] used thin layer chromatography on silica plates to confirm the identity of chlorinated insecticides previously identified by gas chromatography. The compounds can be separated by single or repeated one-dimensional development in *n*-heptane or in *n*-heptane containing 0.3% ethanol. The plate was dried at 65°C for 10min and detected by spraying with a solution of silver nitrate plus 2-phenoxy-ethanol. Thereafter, the plate was dried at 65°C for 10min and illuminated with an ultraviolet light ($\lambda = 254$nm) until spots representing the smallest amounts of standards were visible (10–15min). The pesticide residues may be evaluated semi-quantitatively by simple visual evaluation of the size and of the intensity of spot coloration and by comparing extracts with standard solutions.

Sherma and Slobodien [107] determined chloropyrofos insecticide and its metabolite 3,5,6-trichloro-2-2-pyridinol in potable water at 5µg L^{-1} by thin layer chromatography. Pre-adsorbent silica gel layers were used for resolution and silver nitrate for detection prior to reflectance scanning. Recovery of 3,5,6-trichloro-2,2-pyridinol from water was 84%. Recovery of chloropyrifos from potable water was 87.5%.

14.1.3.3 Polyaromatic hydrocarbons

Grimmer *et al.* [108] described a semiquantitative test for the detection of PAHs in potable water. The chromatogram is developed by acetonitrile–methylene chloride–water (9:1:1) and assessed by visual observation under an ultraviolet lamp. The method was capable of distinguishing samples containing more than 50ng PAH compounds per litre.

Kunte [109] investigated the interference effects of 44 PAHs on a thin layer chromatographic method for determining fluoranthene, benzo(b) fluoranthene, benzo(k)fluoranthene, benzo(a)pyrene, benzo(ghi)perylene and indeno(1,2,3-cd)pyrene according to the West German potable water regulations. Overlapping occurred but these overlaps did not interfere with the determination of these six compounds, nor was quantitative determination of the six compounds influenced by the addition of a soil extract or a mixture of 18 PAHs occurring in the environment.

Harrison *et al.* [110] examined the effects of water chlorination upon levels of some PAHs in water, including a study of the effects of pH, temperature, contact time and chlorinating agent.

14.1.4 Sewage effluents

14.1.4.1 Abate insecticides

Thin layer chromatography, has also been applied to the determination of Abate residues in sewage effluents [111]. The sample of surface water or sewage, acidified with sulphuric acid [111], was extracted with chloroform. The extract was evaporated under nitrogen at 60–70°C and the residue was dissolved in acetone. This solution was applied to a layer of activated silica gel G for 1h at 90°C together with Abate standards. The plate was developed for 1h in hexane–acetone (10:1), air dried for 1min, exposed to bromine vapour for 1min, exposed to air for 3min, and sprayed with 1% N,N–dimethyl-p-(phenylazo)aniline (C1 solvent yellow 2) solution in 95% isopropyl alcohol. Abate gave distinct red spots (R_f 0.11 ± 0.01) on a bright yellow background. Spot areas were related to concentration by a calibration graph. Down to 9µg of Abate could be readily determined.

14.1.5 Waste waters

14.1.5.1 Petroleums

Many of the thin layer chromatographic procedures for qualitative analysis of petroleum products can be used for semiquantitative analysis. However, a more useful approach is that of Geobgen [112] who employed the common ascending thin layer technique on silica gel plates, using n-hexane as the developing solvent. One litre waste water samples were extracted directly with carbon tetrachloride, and a portion of the extract chromographed during about 15min. A non polar developing solvent such as n-hexane will affect a separation between saturated hydrocarbons, which tend to travel near the solvent front, and the more strongly absorbed aromatic hydrocarbons. The saturated hydrocarbon areas were evaluated as quickly fading yellow areas upon a blue background after spraying with bromothymol blue solution. Aromatic hydrocarbons were visualised by examination of their fluorescence under ultraviolet light. However, saturated hydrocarbon areas were used for quantitative determinations, being compared to a range of known concentrations of a standard solution containing 1g eicosan, 0.1g phenanthrene and 1g paraffin oil. Geobgen [112] claimed that such a standard solution is satisfactory, since experience has shown that virtually all waste water samples contain less than 5–10% aromatic hydrocarbons. With this procedure concentrations of <5mg L^{-1} can be determined. Overloading of the plates can cause the polar material contaminating the oil sample, especially in sewage samples, to tail into the saturated paraffin area and cause interference.

Goretti et al. [113] discussed the thin layer chromatographic determination of the constituents of ether extracts of industrial waste water. A light petroleum extract of the sample is evaporated and the residue is weighed, dissolved in ethyl ether (5–50µg) and applied to silica gel in a lane formed by scratching two lines 3mm apart along the plate and widening them beyond the width of the applied spot at the start. The chromatogram is developed with chloroform–benzene (1:3) for 13cm at 18°C in 75min.

14.1.5.2 Phenols

Ragazzi and Giovanni [114] determined phenols in waste water, from processing of olives, by a thin layer chromatographic method. The phenols were separated in amounts from 10 to 100μg on silica gel or cellulose with various solvent systems. The compounds were detected by spraying the plate with 20% sodium carbonate solution and then with Folin–Ciocalteu reagent. The material from each spot was removed from the plate and suspended in water or methanol, then Folin–Ciocalteu reagent and 20% sodium carbonate solution was added. The solution was diluted to approximate volume and centrifuged. The extinctions of the blue solution were measured at 725nm against blank solution obtained by similarly treating blank areas of the plate.

14.1.5.3 Quinones

Thielemann [115] discussed the determination of hydroquinone and its oxidation product 1,4-benzoquinone, both of which are toxic constituents of coal industry waste water. He reviews methods for the quantitative detection of benzoquinone and describes a semiquantitative method for its determination by thin layer chromatography on Kieselgel G using 2% ethanolic solution of 4-aminoantipyrine (1-phenyl-2,3-dimethyl-4-aminopyrazol-5-one) as spray reagent.

14.1.6 Trade effluents

14.1.6.1 Quinones

Thielemann [116] used 4-aminopyrine as a spray reagent for the thin layer chromatographic identification of *p*-benzoquinone and anthraquinone in extracts of coal processing plant effluents. The separation is achieved on Kieselgel G with benzene–acetone (9:1) as solvent, and the spots are revealed by spraying with an aqueous solution of the reagent. Red–brown and violet colours are obtained with benzoquinone and anthraquinone respectively. The intensity of the colour may be increased by spraying the plates with 0.1M hydrochloric acid before development of the chromatogram.

14.1.6.2 Dibutylethylene carboxylate, stilbenzene, azobenzene and azoxybenzene

A thin layer chromatographic method has been developed [117] for the determination of organic compounds capable of *cis–trans* isomerisation, using dibutyl ethylenedicarboxylate, stilbenzene, azobenzene and azoxybenzene as examples. The *cis*–isomers were formed from the *trans* by irradiating solutions with ultraviolet light. For chromatography, plates with a layer of silica gel fixed with gypsum and plates with a similar layer fixed with starch were used. The presence of isomerising compounds was indicated by a characteristic arrangement of spots at the corners of a square on the prolongation of one of the diagonals passing through the deposition point and the point of intersection of solvent fronts, or the 'magic square' technique. This method can be used without a standard, and may be suitable for analysis of complex effluent samples containing compounds with close R_f values and identical coloration.

14.1.7 High purity waters

14.1.7.1 Octadecylamine

Octadecylamine, a corrosion-inhibiting boiler water additive, has been isolated from water by solvent extraction with ethylene dichloride or by trapping on a micro Chromosorb column [118]. The levels of octadecylamine were determined using high performance thin layer chromatography and reflectance densitometry. Recovery at 3mg L^{-1} averaged 81.3% using solvent extraction and 94.0% with the column. Recovery at 0.3mg L^{-1} was 94.2% using the column procedure. Other permitted boiler water additives did not interfere with the analysis.

14.1.7.2 Cyclohexylamine

Sherma and Pallasta [46] determined cyclohexylamine in water by solid phase extraction and high performance thin layer chromatography .

14.2 Organometallic compounds

14.2.1 Non saline waters

14.2.1.1 Organolead compounds

Potter *et al.* [119] have applied gas chromatography and thin layer chromatography to the detection and determination of alkyllead compounds and alkyllead salts in non saline waters. Samples suspected of containing alkyllead salts were applied from an appropriate solution to plates coated with 0.25mm layer of MN Aluminoxid G (Camlab, Cambridge) equilibrated with the atmosphere. Chloroform was found to be a suitable solvent for R_3PbX and water or acetone for R_2PbX_2. Plates were eluted with acetic acid–toluene (1:19, v/v) and the spots were developed by spraying with a solution of dithizone in chloroform (0.1%

w/v). R_3PbX gave a yellow spot and R_2PbX_2 gave a salmon red spot; inorganic lead gave a crimson spot on the baseline and tetraalkyllead was not detected. Under these conditions the R_f values for triethyllead chloride and trimethyllead chloride were 0.5 and 0.2, respectively and for diethyllead dichloride and dimethyllead dichloride were 0.3 and 0.1 respectively. The mixed methyllead salts gave distinct spots with intermediate R_f values. The limit of detection of this method was 0.5–1.0µg alkyllead salt.

14.2.1.2 Organotin compounds

The presence of Bu_3Sn and Bu_2Sn species in chloroform or benzene extracts from hydrobromic acid acidified non saline waters can be demonstrated qualitatively by thin layer chromatography on Eastman chromatogram sheets using hexane–acetone–acetic acid 40:4:1 as an eluent [120]. After spraying with a 0.1% dithizone solution in chloroform, Bu_3Sn species are visualised as a yellow spot (R_f 0.75) and Bu_2Sn species as a red spot (R_f 0.50). The detection limit is improved by keeping the thin layer strip for 10s in bromine vapour after elution. Bromine breaks down carbon–tin bonds, and both Bu_3Sn and Bu_2Sn species are now detected as red spots after spraying. The detection limit after bromination is about 0.5µg tin per spot.

Waggon and Jehle [121,122] have reported on the quantitative detection of triphenyl and tri-, di-, and monobutyltin species in aqueous solution by a combination of liquid–liquid extraction, thin layer chromatography and anodic stripping voltammetry.

The presence of Bu_3Sn and Bu_2Sn species in chloroform or benzene extracts from hydrobromic acid acidified non saline waters can be demonstrated qualitatively by thin layer chromatography on Eastman chromatogram sheets using hexane–acetone–acetic acid 40:4:1 as an eluent [123]. After spraying with a 0.1% dithizone solution in chloroform, Bu_3Sn species are visualised as a yellow spot (R_f value 0.75) and Bu_2Sn species as a red spot (R_f value of 0.50). The detection limit is improved by keeping the thin layer strip for 10s in bromine vapour after elution. Bromine breaks carbon–tin bonds, and both Bu_3Sn and Bu_2Sn species are now detected as red spots after spraying. The detection limit after bromination is about 0.5µg tin per spot.

14.2.2 Sewage effluents

14.2.2.1 Organomercury compounds

Takeshita [124] has used thin layer chromatography to detect alkyl-mercury compounds and inorganic mercury in sewage. The dithizonates were prepared by mixing a benzene solution of the alkylmercury compounds and a 0.4% solution of dithizone. When a green coloration was obtained the solution was shaken with M sulphuric acid followed by aqueous ammonia and washed with water. The benzene solution was evaporated under reduced pressure, and the dithizonates, dissolved in benzene, were separated by reversed phase chromatography on layers of corn starch and Avicel SF containing various proportions of liquid paraffin. Solutions of ethanol and 2-methoxyethanol were used as developing solvents. The spots were observed in daylight. The detection limits were from 5 to 57ng (calculated as organomercury chloride) per spot.

14.2.3 Trade effluents

14.2.3.1 Organomercury compounds

Thin layer chromatography has been used to evaluate [125] organomercury compounds in industrial effluents. $C_1–C_6$ n-alkylmercury chlorides were separated on layers prepared with silica gel plus sodium chloride using as development solvent cyclohexane–acetone–28% aqueous ammonia (60:40:1). The R_f values decrease with increasing carbon chain length and the phenylmercury acetate migrated between the C_1 and C_2 compounds. The spots are detected by spraying with dithizone solution in chloroform.

References

1 Crump, G.B. Nature (London), **193**, 674 (1963).
2 Krieger, H. Gas– u. Wassfach, **104**, 695 (1963).
3 Lambert, G. Bull. Cent. Belge Stud. Docm. Eaux, **20**, 271 (1966).
4 Cross, B.T. and Hart, M. Water Research Association, Medmanham, UK, International Laboratory Report 1RL62 (1962).
5 Matthews, P.J. Water Pollution Control, **67**, 588 (1968).
6 Matthews. P.J. Appl. Chemistry (London), **20**, 87 (1970).
7 Benyon, L.R. Methods for the Analysis of Oil and Water in Soil. The Hague, Striching CONCAWE, 40pp (1968).
8 Semenov, A.D., Stradomanskaya, A.G. and Zurbina, L.F. Gidrokhim Mater., 53, 51 (1971) Ref. Zhur. Khim 19GD (1971).
9 Sauer, W.A. and Fitzgerald, G.E. Environmental Science and Technology, 10, 893 (1976).
10 Killer, F.C.A. and Amos, R. Journal of Institute of Petroleum (London), **52**, 315 (1966).
11 Coats, J.P. Journal of Institute of Petroleum (London), **57**, 209 (1971).
12 Triems, K. Chem. Techn. Berl., **20**, 596 (1968).
13 Triems, K. and Heinze, O. Analytical Abstracts, **16**, 2556 (1969).
14 Farrington, J.W., Teal, J.M., Quinn, J.G., Wade, T. and Burnes, K. Bulletin of Environmental Contamination and Toxicology, **10**, 129 (1973).

15 Khomenkô, A.N., Goncharova, I.A. and Stradomskaya, A.N. *Analytical Abstracts*, **14**, 7929 (1967).
16 Khomenko, A.N. *Analytical Abstracts*, **14**, 7927 (1967).
17 Rump, O. *Water Research*, **8**, 889 (1974).
18 Hollies, J.J., Pinnington, P.F., Handley, A.J., Baldwin, M.K. and Bennett, D. *Analytica Chimica Acta*, **111**, 201 (1979).
19 Bogacka, T. and Taylor, R. *Chemia Analyt.*, **15**, 143 (1970).
20 Bogacka, T. and Taylor, R. *Chemia Analyt.*, **16**, 215 (1971).
21 Lawrence, J.F. and Frei, R.W. *Analytical Chemistry*, **44**, 2046 (1972).
22 Supplement to the 15th edn. of *Standard Methods for the Examination of Water and Waste Water*. Selected Analytical Methods Approved and Cited by the US Environmental Protection agency. American Public Health Association, American Waterworks Association, Water Pollution Control Federation, Sept. (1978). Methods S60 and S63. Methods for benzidine, chlorinated organic compounds, pentachlorophenol and pesticides in water and waste water (Interim, Pending issuance of methods for organic analysis of water and wastes, Sept. 1978), Environmental Protection Agency, Environmental Monitoring and Support Laboratory (EMSL).
23 Ya, C.C., Booth, G.H., Hanson, D.J. and Larsen, J.P. *Journal of Agriculture and Food Chemistry*, **22**, 431 (1974).
24 Smith, A.E. and Fitzpatrick, A. *Journal of Chromatography*, **57**, 303 (1971).
25 Frei, R.W., Lawrence, J.F. and Le Gay, D.S. *Analyst (London)*, **98**, 9 (1973).
26 Deleu, R., Barthelemy, J.P. and Copin, A. *Journal of Chromatography*, **134**, 483 (1977).
27 Daley, R.J., Gray, O.B.J. and Brown, S.R. *Journal of Chromatography*, **76**, 175 (1973).
28 Riley, J.P. and Wilson, T.R. *Journal of Marine Biological Association*, **45**, 583 (1965).
29 Madgwick, J.C. *Deep Sea Research*, **12**, 233 (1965).
30 Jones, I.D., Butler, L.S., Gibbs, E. and White, R.C. *Journal of Chromatography*, **70**, 87 (1972).
31 Weber, J.H. and Wilson, S.A. *Water Research*, **9**, 1079 (1975).
32 Hunter, D. *Environmental Science and Technology*, **9**, 241 (1975).
33 Ruebelt, C.Z. *Analytical Chemistry*, **221**, 299 (1966).
34 Borneff, J. and Kunte, H. *Archives Hygiene Bakt.*, **153**, 202 (1968).
35 Gilchrist, C.A., Lynes, A., Steel, G. and Whitham, B.T. *Analyst (London)*, **97**, 880 (1972).
36 Kay, J.F. and Latham, D.R. *Analytical Chemistry*, **45**, 1050 (1973).
37 Stepanova, M.I., Ii'ina, R.I. and Shaposhnikov, *Zh. Analyt. Khim.*, **27**, 1201 (1972).
38 Nowacka-Barezyk, K., Adamiak-Ziemba, J. and Janina, A.Z. Chemia *Analyt.*, **18**, 223 (1973).
39 Crane, R.I., Crathorne, B. and Fielding, M. *Determination of Polycyclic Aromatic Hydrocarbons in Water*, Water Research Centre, UK, LR 4407 (1980).
40 Schossner, H., Falkenberg, W. and Althaus, H. *Zeitschrift für Wasser und Abwasser Forshung*, **16**, 132 (1983).
41 Poole, S.K., Dean, T.A. and Poole, C.F. *Journal of Chromatography*, **400**, 323 (1987).
42 Patterson, S.J., Hunt, E.C. and Tucker, K.B.E. *Journal of Proceedings of the Institute of Sewage Purification*, **1**, 90 (1966).
43 Edeline, F., Deswaef, F. and Lambert, G. *Tribune de Cebedeau*, **31**, 137 (1978).
44 Dyat-Lovitskaya, F.G. and Botvinova, L.E. Khim. Volakna, 1, 64 (1970). Ref. Z. Khim, 19GD (13) Abstract No. 13G308 (1970).

45 Kostynkovskii, Y.L. and Melamed, R. *Zhur. Anal. Chem.*, **42**, 924 (1987).
46 Sherma, J. and Pallasta, L. *Journal of Liquid Chromatography*, **9**, 3439 (1986).
47 Ch'mil, V.D. *Zhur Anaticheskoi Khimu*, **36**, 729 (1981).
48 Henshaw, B.G., Morgan, J.W.W. and Williams, N. *Journal of Chromatography*, **110**, 37 (1975).
49 Zigler, M.G. and Phillips, W.F. *Environmental Science and Technology*, **1**, 65 (1967).
50 Wolkoff, A.W. and Larose, R.J. *Journal of Chromatography*, **99**, 731 (1974).
51 Thielemann, H. and Luther, M. *Pharmazie*, **25**, 367 (1970).
52 Chambon, P. and Chambon, R. *Journal of Chromatography*, **87**, 787 (1973).
53 Makarova, C.U. Cig. Sanit., 5, 61 (1971) Ref. Zhur Khim., 19GD (20) Abstract No. 20G, 191 (1971).
54 Maktaz, E.D., Botvinova, L.E. and Kruchinina, A.A. *Soviet Journal of Water Chemistry*, **6**, 59 (1984).
55 Tachsteinova, A. and Kopanica, M. *Analytica Chimica Acta*, **199**, 77 (1987).
56 Abe, A. and Yoshima, H. *Water Research*, **13**, 1111 (1979).
57 Bodega, P., Marutolu, C., Bresug, A., Sarbu, C. and Gocan, S. Rome RO 89223, B1 3pp. (1986).
58 Girenko, D.B., Kilsenko, M.A. and Moraru, L.E. *Gig. Sanit.*, (**9**), 41 (1988).
59 de Vos, R.H. and Peet, E.W. *Bulletin of Environmental Contamination and Toxicology*, **6**, 164 (1971).
60 Sackmauereva, O.M., Pal'usova, O. and Szokolay, A. *Water Research*, **11**, 551 (1977).
61 Arias, C., Vidal, A., Vidal, C. and Daria, J. *An Bromat (Spain)*, **22**, 273 (1970).
62 Novikov, V.V., Boldina, Z.N. and Kudrin, L.V. *Gig. Sanit.*, **12**, 76 (1987).
63 Suzuki, T., Nagayoshi, H. and Kashiwa, T. *Agricultural Biological Chemistry*, **38**, 279 (1974).
64 Suzuki, T. *Analytical Abstracts*, **26**, 2374 (1974).
65 Armstrong, D.W. and Terrill, R.W. *Analytical Chemistry*, **51**, 2160 (1979).
66 Sackmauereva, M., Pal'usova, O. and Szokolay, A. *Water Research*, **11**, 551 (1977).
67 Weil, L. *Analytical Abstracts (London)*, **24**, 1259 (1973).
68 Fehringer, N.V. and Westfall, J.E. *Journal of Chromatography*, **57**, 397 (1971).
69 Chen, Jo-Yun T. and Dority, R.W. *J. Am. Oil Color Chem. Ass.*, **55**, 15 (1972). ·
70 Hutzinger, O., Jamieson, W.D. and Safe, S. *J. Am. Oil Color Chem. Ass.*, **54**, 178 (1971).
71 Kawataski, J.A. and Frasch, D.L. *J. Ass. Off. Anal. Chem.*, **52**, 1108 (1960).
72 Meinard, C. *Journal of Chromatography*, **61**, 173 (1971).
73 Ch'mil, V.D. *Zhur. Anal. Khim.*, **42**, 2048 (1987).
74 Fishbein, L. *Chromat. Rev.*, **12**, 167 (1970).
75 Zawadzka, H., Adamezewska, M. and Elbanowska, H. *Chemia Analit.*, **18**, 327 (1973).
76 Abbott, O. *Anal. Abstr.*, **13**, 5917 (1966).
77 Frei, R.W. and Duffy, J.R. *Mikrochim. Acta.*, **3**, 480 (1969).
78 Grushevskaya, N.Y. and Zazarinova, N.F. *Zhur Anal. Khim.*, **42**, 164 (1987).
79 Ch'mil, V.D. and Vasyargina, R.D. *Zhur. Anal. Khim.*, **42**, 1691 (1987).
80 Guthrie, R.K., Cheery, D.S. and Ferebee, R.W. *Water Research Bulletin*, **10**, 304 (1974).
81 Rodica, T. *Revta Chem.*, **20**, 259 (1969).
82 Zycinski, D. *Roczn. Panst. Zakl. Hig.*, **22**, 189 (1971).
83 Jovanovic, D.A. and Prosie, Z. *Acta Pharm.*, **22**, 91 (1972).
84 Leoni, V. and Pucceti, G. *Farmaco. Ed. Prat.*, **26**, 383 (1971).

85 Schutzmann, O. and Barthel, R. *Analytical Abstracts*, **18**, 3473 (1966).
86 Kovacs, O. *Analytical Abstracts*, **13**, 2016 (1966).
87 Bidleman, T.F., Nowlan, B. and Frei, R.W. *Analytica Chimica Acta*, **60**, 13 (1972).
88 Guven, K.C. and Aktugla, A. *Eczacilik Bult.*, **14**, 44 (1972).
89 Howe, L.H. and Petty, C.F. *Journal of Agricultural and Food Chemistry*, **17**, 401 (1969).
90 Sherma, J. and Boymel, J.L. *Journal of Chromatography*, **247**, 201 (1982).
91 Abbott, D.C. and Wagstaff, P.J. *Journal of Chromatography*, **43**, 361 (1969).
92 Uskova, L.A., Panasyuk, T.D. and Prakhorchuk, E.A. *Khim. Sel'sk Khoz.*, **8**, 69 (1987).
93 Marutolu, C., Viassa, M., Sarbu, C. and Nagy, S. HRCCC. *Journal of High Resolution Chromatography Communication*, **10**, 465 (1987).
94 Kovac, J., Tekel, J. and Kurucova, M.Z. *Lebensm-Centers, Forsch.*, **184**, 96 (1987).
95 Primak, A.P., Ushakova, T.V. and Biryukova, N.F. *Khim. Sel'sk Khoz.*, (**1**), 73 (1987).
96 Drescher, N. *Methodensamminiung zur Ruckstandardsanalytik von Pflanzenschutmittein*, Verlag Chimie, Weinheim (1972).
97 Drescher, N. Bestimmung der Ruckstande von Pyramin in Pflanze und Boden, 8–9 January 1964, BASF, Ludwigshafen am Rhein, p. 78 (1964).
98 Zaborowska, W., Witkowska, I. and Kozak, H. *Rocz. Panstu. Zakl. Hig.*, **24**, 735 (1973).
99 Palusova, O., Sackmauereva, M. and Madarix, A. *Journal of Chromatography*, **106**, 405 (1975).
100 Abayachi, J.K. and Riley, J.P. *Analytica Chimica Acta*, **107**, 1 (1979).
101 Strickland, J.D.H. and Parsons, T.F. *A Practical Handbook of Seawater Analysis*. Fisheries Research Board of Canada, Ottawa (1968).
102 Garside, C. and Riley, J.P. *Analytica Chimica Acta*, **46**, 179 (1969).
103 Shoaf, W.T. and Lium, B.I.W. *Journal of Research of US Geological Survey*, **5**, 263 (1977).
104 Boto, K.G. and Bunt, J.S. *Analytical Chemistry*, **50**, 392 (1978).
105 Thielemann, H. *Mikrochimica Acta*, **5**, 669 (1972).
106 Mosinska, K. *Pr. Inst. Przen. Org*, **1971**, 253 (1972).
107 Sherma, J. and Slabodien, R. *Journal of Liquid Chromatography*, **7**, 2735 (1984).
108 Grimmer, G., Hellman, H., de Jong, B. et al. *Zeitschrift für Wasser und Abwasser Forshung*, **13**, 108 (1980).
109 Kunte, H. *Fresenius Zeitschrift für Analytische Chemie*, **301**, 287 (1980).
110 Harrison, R.M., Perry, R. and Wellings, R.A. *Environmental Science and Technology*, **10**, 1151 (1976).
111 Howe, L.J. and Petty, C.F. *Journal of Agricultural and Food Chemistry*, **17**, 401 (1969).
112 Geobgen, H.G. *Hanstech. Essen. Vortrag*, **231**, 55 (1970).
113 Goretti, G., Morsella, J. and Petronio, B.U. *Rio Ital. Sostanze Grasse*, **51**, 66 (1974).
114 Ragazzi, E. and Giovanni, V. *Journal of Chromatography*, **77**, 369 (1973).
115 Thielemann, H. *Zeitschrift für Wasser und Abwasser Forshung*, **7**, 91 (1974).
116 Thielemann, H. *Z. Anal. Chemie*, **253**, 38 (1971).
117 Kerzhner, B.K., Urubel, T.L. and Kofanov, U.I. *Soviet Journal of Water Chemistry and Technology*, **4**, 73 (1982).
118 Sherma, J., Chandler, K. and Ahringer, J. *Journal of Liquid Chromatography*, **7**, 2743 (1984).
119 Potter, M.R., Jarview, A.W.P. and Markell, R.N. *Water Pollution Control*, **76**, 123 (1977).

120 Andreae, M.O., Asmond, J.F., Foster, P., Van't dack, L. *Analytical Chemistry*, **53**, 1766 (1981).
121 Waggon, H., Jehle, D. *Die Nahrung*, **17**, 739 (1973).
122 Waggon, H., Jehle, D. *Die Nahrung*, **19**, 271 (1975).
123 Maguire, R.J. and Huneault, H. *Journal of Chromatography*, **209**, 458 (1981).
124 Takeshita, R., Agaki, H., Fujita, M. and Sakegami, O. *Journal of Chromatography*, **51**, 283 (1970).
125 Murakiami, T. and Yoshinaga, T. *Japan Analyst*, **20**, 1145 (1971).

Chapter 15

Gas chromatography

Gas chromatography has been applied mainly to the determination of organic and organometallic compounds with limited applications to the determination of anions and cations.

15.1 Organic compounds

15.1.1 Non saline waters

15.1.1.1 Aliphatic hydrocarbons, inland oil pollution incidents

A general classification of oil eg crude oil, petroleum, gas oil, is often satisfactorily achieved by gas chromatographic techniques possibly coupled with mass spectrometry or infrared spectroscopy applied to a sample of the oil pollutant. The true identification invariably requires samples from potential sources for comparison with the pollutant.

Capillary columns provide greater resolution and therefore more detail for comparison between a polluting oil and suspect sample. The enormous separation power available has been demonstrated in their application to petroleum analysis. Gouw [1] describes a versatile 10m × 0.25mm capillary column coated with CV–101, and its application to the separation of hydrocarbon mixtures in the C_4–C_{58} n–alkane range. Columns of 152m × 0.25mm coated with 1-octadecene and operated at 30°C have been recommended for identification of crude oils [2].

A polar liquid phase was found more suitable for studying the major components of petrol, gas oil and diesel oil [3], forming true solutions in water. With such a phase, saturated hydrocarbons tended to elute before aromatic hydrocarbons, which were found to be the principal components in true solution, and therefore their investigation was facilitated, in the case of gas oil and diesel oil, forming true solutions in water. In the case of gas oil and diesel oil, no saturated hydrocarbons could be detected in solution. These authors reached the important conclusion that the determination of the origin of oil components in true aqueous solution could be more difficult because of selective solution of certain components. This effect was more likely to apply to the lower distillates, which tended to be relatively more water soluble, rather than the non volatile petroleum products. Distinction between petrol and gas oil or diesel oil seemed possible, but appeared difficult between similar products such as gas oil and diesel oil.

Invariably, dual packed columns have been employed, and one of the earliest articles devoted to the identification of petroleum products is that of Lively [4], who used dual 1.2m × 6mm columns packed with 20% SE–30 as the liquid phase and a Chromosorb solid support. Most subsequent workers in this field have employed the same or a similar liquid phase, usually at a lower loading of 4–5% and somewhat different column dimensions 1.2–3.6m × 3–6mm [5–9]. Chromosorb, or occasionally a similar solid support [7] was usually the solid support utilised after first being acid washed and treated with hexamethyl disilazine (HMDS) or dimethyl dichlorosilane (DMCS). Liquid phases of similar properties that have been employed are 5% and 10% OV–1, 20% SE–52, 5% E301 [6], and 2.2% [10], 10% [13] and 20% [8] Apiezon L. These liquid phases are essentially non polar substances, but more polar phases, 5% and 10% polyethylene glycol 1500, have been used for investigations of the water-soluble components in diesel and gas oil and in petrol.

In many publications, flame ionisation detectors and temperature programming were employed, the latter especially in the identification of less volatile oils. For routine investigations, isothermal conditions have been adopted in order to allow quicker analysis, but at the sacrifice of some fine detail in the chromatogram [8].

Specific n-alkane peaks are usually identified by comparison with standard solutions of some known compounds and a 1% w/v solution has been recommended [11].

Detailed salient points involved in classifying and identifying oils have been compiled [6,12]. Most distillate products, eg petrol, paraffin, fuel oils, diesel oil, etc, and even very similar products such as turbojet fuel and kerosene, can be classified in the absence of excessive weathering effects. Various chromatograms have been published: kerosene, turbojet fuel, steam-cracked naphtha; lubricating oil–gas oil-weathered paraffin [12]; white spirit, turpentine substitute, paraffin, 30s fuel oil [8]; standard gasoline; petrol; diesel fuel and gas oil, various gasolines, diesel fuels and aviation fuels [13]. Lubricating, cutting transformer oils, etc, heavy fuel oils, asphaltic and bituminous materials are difficult or impossible to classify with any certainty by gas chromatography.

A very practical description of the application of gas chromatography to quantitative analysis of petroleum products in aqueous and soil samples is contained in the report by CONCAWE [11]. Well-detailed procedures are given for three hydrocarbon mixtures with the different boiling point ranges.

Jeltes [14] has described a gas chromatographic method for the determination of mineral oil–water in which a true solution of diesel fuel and gas oil in water were analysed by isothermal gas chromatography at 110°C with polyoxyethylene glycol 1550 as stationary phase and Chromosorb W as support. For two-phase oil–water systems, the oil was extracted with carbon tetrachloride and the extracts were analysed by temperature programmed gas chromatography with SE–30 as stationary phase by a method similar to that of Benyon [11]. Components of mineral oils, ie water–soluble types boiling below 300°C, could be identified and determined, and it is possible sometimes to establish the origin of the oil.

An interesting gas chromatographic technique of identifying petroleum products, including lubricating oils, is that of Dewitt Johnson and Fuller [9]. A gas chromatograph column effluent was split in order that it could be simultaneously sensed by a double-headed flame photometric detector, specific for both sulphur and phosphorus compounds, and by dual-flame ionisation detectors for carbon compounds in general. Since most petroleum products are claimed to contain sulphur and phosphorus compounds, three chromatographic traces were obtained for the products examined. In the cases of lubricating oils, which normally give poorly resolved peaks on chromatograms with flame ionisation detection, sulphur and phosphorus detection gave considerably more detail for identifying purposes. The apparatus is, however, complex and expensive for routine application.

McAuliff [15] reported that the sampling errors caused in the injection of water–hydrocarbon mixture on to a gas chromatographic column can be overcome by the addition of acetone for solubilisation only when the hydrocarbon chain length is less than 12. The emulsified sample is injected directly on to a column 1.2m × 6mm of 10% of SE–30 on Chromosorb W HMDS (80–100 mesh) operated at 170–210°C, with hydrogen flame ionisation detection. Results are given for the hydrocarbons from C_{11} to C_{18}.

Adlard and Matthews [16] applied the flame photometric sulphur detector to pollution identification. A sample of the oil pollutant was submitted to gas chromatography on a stainless steel column 1m × 3mm packed with 3% of OV–1 on AW–DMCS Chromosorb G (85–100 mesh). Helium was used as carrier gas (35mL min^{-1}) and the column temperature was programmed from 60 to 295°C and 5°C per minute. The column effluent was split between a flame ionisation and a flame photometric detector. Adlard and Matthews [16] claim that the origin of oil pollutants can be deduced from the two chromatograms. The method can also be used to measure the degree of weathering of oil samples.

Lur'e et al. [17] have described a method which involves extraction of the sample (about 500mL, containing less than 200mg of hydrocarbons) with hexane, concentration of the dried extract to 1ML and gas chromatography of an aliquot (less than 2μL) of this solution. For lower levels of hydrocarbons, 2L of sample is allowed to percolate through a column (13cm × 1cm) on activated carbon, the carbon is dried in air, then the hydrocarbons are extracted with chloroform and the extract is concentrated for gas chromatography. The gas chromatography is carried out on a column (3–4m × 4–5mm) packed with 10% of SE–30 or 20% of 3, 3' oxydipropionitrile supported on Chromosorb with helium as carrier and a katharometer detector.

Cole [18] found that in the identification of kerosine and aviation fuels, the usual packed columns provided sufficient detail only up to C_{13} n-alkane, a range readily altered by evaporation effects. A suitable capillary column was developed which revealed extra detail of minor components above C_{13} n-alkane, and consisted of a 45m × 0.25mm stainless steel column coated with OV–101. This gave satisfactory resolution and stability in the 50–310°C temperature range.

Jeltes and Van Tonkelaar [19] compared gas chromatographic and infra-red methods for the determination of dissolved mineral oil in water. They saturated various petroleum fractions with water by shaking for 4min, then after 2 days, the clear aqueous phases were extracted with carbon tetrachloride. The polar compounds were removed from the carbon tetrachloride extracts by shaking with Florasil and decanting after the Florasil had settled. These extracts were analysed by infrared and gas chromatographic methods, both before and after treatment with Florasil, and also after the two-phase systems had been exposed for eight weeks to light and air.

Jeltes and Van Tonkelaar [20] investigated problems of oil pollution, the nature of the contaminants and the chemical methods used for their detection. In particular, the use of gas chromatography to obtain 'fingerprint' chromatograms of oil pollutants in water, and of infrared spectrophotometry to determine the oil contents of soils and sediments, is discussed.

Kawahara [21] has discussed the characterisation and identification of spilled residual fuel oils on surface water using gas chromatography and infrared spectrophotometry. The oily material was collected by surface skimming and extraction with dichloromethane, and the extract was evaporated. Preliminary distinction between samples was made by dissolving portions of the residue in hexane or chloroform. If the residue was soluble in chloroform but not in hexane it was assumed to be crude oil, a grease, a heavy residual fuel oil or an asphalt; if it was soluble in both solvents it was assumed to be very light or heavy naphtha, kerosine, gas oil, white oil, diesel oil, jet fuel, cutting oil, motor oil or cutter stock. The residue was also examined by infrared spectrophotometry; wavenumber values of use for identification purposes are tabulated.

Lysyj and Newton [22] have described a multi-component pattern recognition and differentiation method for the analysis of oils in non saline waters. The method is based on that described earlier by Lysyj and

Newton [23] which depends on the thermal fragmentation of organic molecules followed by gas chromatography. Dried algae and outboard motor oil were analysed and a specific pattern or numerical 'fingerprint' was obtained for by each polygraphic means. The algal pattern comprised three specific peaks and seven peaks common to those of the oil, whereas the oil pattern comprised two specific peaks and seven peaks common to those of the algae.

Garra and Muth [24] characterised crude, semi-refined and refined oils by gas chromatography. Separation followed by dual-response detection (flame ionisation for hydrocarbons and flame photometric detection for sulphur-containing compounds) was used as a basis for identifying oil samples. By examination of chromatograms, it was shown that refinery oils can be artificially weathered so that the source of oil spills can be determined.

McAucliffe [25] found various impurity peaks in the direct injection of aqueous solutions of hydrocarbons, which limited sensitivity to about 1m L^{-1} of individual hydrocarbons. He employed a 3.66m × 6mm column containing 25% SE–30 on firebrick and a precolumn containing ascarite to absorb the water and improve the sensitivity. Other workers have used similar precolumns to remove water.

Desbaumes and Imhoff [26] have described a method for the determination of volatile hydrocarbons and their halogenated derivatives in water.

Bridie et al. [27] have studied the solvent extraction of hydrocarbons and their oxidative products from oxidised and non-oxidised kerosine–water mixtures, using pentane, chloroform and carbon tetrachloride. Extracts are treated with Florasil to remove non-hydrocarbons before analysis by temperature programmed gas chromatography. From the results reported it is concluded that, although each of the solvents extracts the same amount of hydrocarbons, pentane extracts the smallest amount of non hydrocarbons. Florasil effectively removes non hydrocarbons from pentane extracts, but also removes 10–25% of aromatic hydrocarbons. However, as the other solvents are less susceptible than pentane to treatment with Florasil, pentane is considered by those workers to be the most suitable solvent for use in determining oil in water.

Other applications of gas chromatography to the determination of aliphatic hydrocarbons in non saline waters are discussed in Table 15.9.

15.1.1.2 Aromatic hydrocarbons

Wasik and Tsang [28,29] have described a method for the determination of traces of arene contaminants using isotope dilution gas chromatography. They used perdeuterated benzene as the isotope source and analysed solutions containing 10–20mg L^{-1} of benzene and toluene. This method is most effective when the isotope can be completely separated from the parent compound and other contaminant, otherwise the isotope ratio must be determined by mass spectrometry.

Gas chromatography has found some applications in the determination of simple aromatics in water. Mel'kanovitskaya [31] has described a method for determining C_6–C_8 aromatics in subterranean waters. In this method the sample (25–50mL) is adjusted to pH8–9 and extracted for 3min with 0.5 or 1.0mL of nitrobenzene; the extract is washed with 0.3mL of 5% hydrochloric acid or 5% sodium hydroxide solution and with 0.3mL of water adjusted to pH7. The purified extract is subjected to gas chromatography at 85°C on a column (1m × 4mm) packed with 15% of polyoxyethylene glycol 2000 on Celite 545 (60–80 mesh) and operated with nitrogen (10mL min⁻¹) as carrier gas, decane as internal standard and flame ionisation detection.

Other applications of gas chromatography to the determination of aromatic hydrocarbons in non saline waters are discussed in Table 15.9.

15.1.1.3 Polyaromatic hydrocarbons

Chatot *et al.* [32] coupled thin layer chromatography with gas chromatography to determine PAHs. The hydrocarbons were separated by two-dimensional thin layer chromatography on prewashed alumina–cellulose acetate (2:1), with pentane as solvent in the first direction and ethanol–toluene–water (17:4:4) as solvent in the second direction. Zones fluorescent under ultraviolet radiation were extracted with benzene, the extracts were evaporated, and a solution of each residue in benzene (20μL) was injected on to a stainless steel column (2m × 3mm) packed with 4.5% of SE–52 on Chromosorb G (DMCS). The carrier gas was nitrogen (60mL min⁻¹), the column was temperature programmed from 100 to 290°C at 3°C per minute, and detection was by flame ionisation.

Harrison *et al.* [33] studied the factors governing the extraction and gas chromatographic analysis of PAHs in water. Factors such as initial concentration, presence of suspended solids and prolonged storage of the samples affected considerably extraction efficiencies. It is recommended that water samples should be collected directly into the extraction vessel and that analysis should be carried out as soon as possible after extraction.

Caddy and Meek [34] combined gas chromatography with a fluorcence detection technique for the determination down to 10mg L⁻¹ of PAHs in non saline water. The water sample was adjusted to pH2.5 then stirred with Celite. The Celite was then removed and extracted with cyclohexane. The cyclohexane was evaporated to a small bulk and this injected on to the gas chromatographic column (5% OV–1 or 5% Dexsil on 80–100 mesh DCMS Chromosorb W) and temperature programmed from 70 to 350°C.

Table 15.1 Determination of polyaromatic hydrocarbons in river water

Compound	PAH measured in undosed sample μg L^{-1}	PAH added μg L^{-1}
Thin layer chromatography		
Fluoroanthene	0.11	0.05
Benzo(k)fluoroanthene	0.06	0.06
Benzo(a)pyrene	1.10	0.07
Perylene	0.03	1.10
Indeno(1,2,3-cd)pyrene	0.04	0.04
Benzo(ghi)perylene	0.04	0.08
Gas chromatography		
Fluoranthene	0.18	0.05
Pyrene	0.26	0.13
Benzo(a)anthracene chrysene	0.14	0.16
Benzo(k)fluoroanthene	–	–
Benzo(j)fluoroanthene	0.24	0.06
Benzo(b)fluoroanthene	–	–
Benzo(a)pyrene	0.21	0.011
Perylene	0.04	0.10
Indeno(1,2,3-cd)pyrene	1.10	0.04
Benzo(ghi)perylene	0.04	0.08

Source: Reproduced by permission from Elsevier Science, UK [35]

Acheson et al. [35] compared the gas chromatographic and the thin layer chromatographic methods on river water and motorway drainage samples. Samples were extracted with dichloroethane. Seven compounds were identified on a thin layer plate (Table 15.1). Gas–liquid chromatographic analysis was also performed on the dichloromethane extracts. Results (Table 15.1) show a less accurate measurement of added PAHs and indicated a generally lower precision thin layer chromatographic analysis, probably as a result of the difficulty of assessment of gas chromatographic peak areas in the presence of a variable background. In a small proportion of cases measurement of peak areas was rendered impossible by a high background, and no estimate of PAH concentration was made.

Basu and Saxena [36] investigated the isolation of benzo(a)pyrene, fluoranthene, benzo(j)fluoranthene, benzo(k)fluoranthene, indeno(1,2,3–cd)pyrene and benzo(ghi)perylene from non saline water. A Florasil column clean-up procedure is described which is capable of removing impurities introduced from water and polyurethane-foam plugs to the extent necessary for their interference-free detection. This isolation step is followed by gas chromatography using a flame ionisation detector operated in a differential mode. The matched columns consisted of 1.8m

× 3mm stainless steel packed with 3% Dexsil–300 on Chromosorb W (AW, DMCS treated, 100–120 mesh).

Other applications of gas chromatography to the determination of poly-aromatic hydrocarbons in non saline waters are reviewed in Table 15.9.

15.1.1.4 Non ionic surfactants

Favretto *et al.* [37] applied gas–liquid chromatography to an evaluation of the polydispersity of polyoxthylene non ionic surfactants. A 500mL volume of water was extracted three times with 50mL of 1,2-dichloro-ethane. The combined organic layers were extracted with 5.0mL of 0.10M sulphuric acid and then with 5.0mL of 1.10M sodium hydroxide solution. The purified extract was concentrated under vacuum, transferred into a conical glass mini-vial and evaporated to a small volume (0.1mL) by means of a stream of nitrogen. An aliquot of this solution was injected into the gas chromatographic column.

Chromatographic peaks attributable to the polydisperse surfactant $C_m \pm E_n$ were first recognised by using as internal standard a similar $C_m \pm E_n$ surfactant (the comparison of the elution temperature of the peaks is a doubtful criterion as elution temperatures vary with the age of the column) mixed (approximately 1:1 by mass) with a portion of residue solution. The values of n of the peaks pertaining to the distribution were then determined by introducing a mono-disperse $C_m \pm E_n$ as an internal standard in another aliquot of the residue solution (1:10).

The gas–liquid chromatographic evaluation of n can obviously be performed on a single distribution $C_m \pm E_n$ but not on mixtures of dis-tributions. Commercial surfactants always consist of mixed distribution. Only when the procedure is applicable, $n = \Sigma x_n n$ is calculated from the observed distribution of the molecular fraction (x_n) for various values of n. It should be checked that $\Sigma x_n = 1$.

Stephanou [38] identified non ionic surfactants in non saline water by gas chromatography coupled with chemical ionisation mass spectrometry. Tertiary octylphenol and lauryl alcohol ethoxylates were qualitatively detected using their chemical ionisation mass spectra and the results used to compare the chemical ionisation and electron impact mass spectrometry techniques.

15.1.1.5 Phenols

Semenchenko and Kaplin [39] acidified the sample with hydrochloric acid, then saturated it with sodium chloride. The phenols were extracted with diethyl ether. The extract was shaken with aqueous sodium bicarbonate (5%) to remove any organic acids, then the phenols extracted into aqueous sodium hydroxide and methylated with dimethyl sulphate.

The methyl derivatives were analysed in a glass column (1.5m × 3.5mm id) packed with 10% of tritolyphosphate–Apiezon L (1:) supported on Chromosorb W; the column was operated at 120°C, with nitrogen (40mL min^{-1}) as carrier gas and a flame ionisation detector. Determination was by means of peak–area measurement. Down to 1μg of phenols per litre could be determined (in 200mL water samples) with a relative error of less than ±15%.

Eichelberger et al. [40] collected the phenols from water samples on columns of activated carbon. The phenols were stripped from the carbon with chloroform prior to gas chromatography of the cleaned-up extract. Clean-up was achieved by a treble extraction of the chloroform phase with aqueous sodium hydroxide. The aqueous extracts were combined, acidified with concentrated hydrochloric acid to pH2 and extracted three times with ethyl ether, and the combined ethyl ether and extracts were passed through a 10cm Florasil column topped with anhydrous sodium sulphate (2cm). The phenols were eluted with ether, and the eluate was concentrated by evaporation and analysed by gas chromatography. An aluminium column (3m × 3mm) packed with 10% of Carbowax 20M terephthalic acid on HMDS–treated Chromosorb W (60–80 mesh) was used, operated at 210°C for the majority of phenols with nitrogen (50mL min^{-1}) as carrier gas and flame ionisation detection. Recoveries were, on average, 85% for alkylphenols and 94% for chlorophenols.

Murray [41] determined low concentrations of phenols, cresols and xylenols in chloroform extracts of water. The trimethylsilyl derivatives of the phenols were formed and analysis completed by gas chromatography. The method was rapid and required a minimum of sample manipulation. The lower limit of detection was 0.100mg L^{-1} for phenol, 0.025mg L^{-1} for cresols and 0.050mg L^{-1} for xylenols.

The method used by Yorkshire Water Authority [42] for determining phenols in potable waters has a criterion of detection of 3–70μg depending on the phenol. It is capable of determining phenols and cresols, xylenols, dihydric phenols, monochlorophenols, dichlorophenols and trichlorophenols. The phenols are extracted from the water sample with ethyl acetate using a liquid–liquid extractor. After drying, the extract is concentrated to a small volume then treated with bis(trimethylsily)trifluoroacetamide to produce the phenol trimethylsilyl ethers. Retention times of various phenols on these gas chromatographic columns are tabulated in Table 15.2. Unfortunately this method is rather lengthy, requiring about 2 days per sample.

Voznakova and Popl [43,44] departed from the usual approach to phenol analysis by gas chromatography made by earlier workers, ie direct injection gas chromatography or solvent extraction followed by gas chromatography. These workers described a method for determining phenol, o-cresol and 2,6–xylenol in water in which the phenols are sorbed

Table 15.2 Retention times of phenol trimethylsilyl (TMS) ethers on 7% Carbowax 20M at (a) 130°C, (b) 180°C

Phenol TMS ether of	Retention time (a) (min from injection)	Relative retention time (compared to phenol TMS ether)
Phenol	3.22	1.00
o-Cresol	4.07	1.26
p-Cresol	4.37	1.36
p-Cresol	4.65	1.45
2-Ethylphenol	4.95	1.54
2,4-Xylenol	5.97	1.85
3,5-Xylenol	5.97	1.85
4-Ethylphenol	6.50	2.02
2,6-Xylenol	6.75	2.10
Catechol	6.93	2.16
2-Chlorophenol	7.77	2.41
4-Chlorophenol	9.22	2.87
Resorcinol	10.37	3.22
Quinol	11.80	3.67
4-Chloro-2-methylphenol	12.42	3.86
5-Indanol	18.37	5.71
2,4-Dichlorphenol	>30	–
2,6-Dichlorophenol	>30	–
2,4,6-Trichlorophenol	>30	–
	Retention time (b)	
Pyrogallol	3.27	–
-octylphenol	6.03	–
-cyclohexylphenol	6.83	–
-phenylphenol	11.30	–

Source: Reproduced by permission from the Yorkshire Water Authority [42]

on a macroporous polymer sorbent (Separon SE) and then undergo thermal desorption for gas chromatographic analysis. This method is claimed to be quick and reproducible and have a detection limit of between 1 and 10µg L $^{-1}$ depending on the phenol.

Goldberg et al. [45] showed that in the solvent extraction of phenols from water samples it is not unusual to lose 50% or more of the extracted solute during evaporation from 500mL to around 5mL.

Goldberg and Weiner [46] used solvent-heavier-than-water, two-cycle liquid extractors to concentrate phenols at the µg L $^{-1}$ level from water into dichloromethane. The non-aqueous solution containing the extract was concentrated by Kuderna–Danish concentrators, and this was followed by gas chromatography. Overall concentration factors were around 1000 with efficiencies 23.1–87.1%. Determinations could be made with accuracy of only 15–20% because of solute losses during concentration. The range of concentration used was of the order of 1µg L $^{-1}$.

The Department of the Environment (UK) [47] has published details of five procedures for the determination of monohydric and some other phenols in water, the first two for media in which the phenol concentration exceeds 100µg L $^{-1}$ and the remaining three for lower concentrations. The procedures are based on the three gas chromatographic methods with the reaction conditions modified to suit the nature of the sample in the case of the method employing 4-aminoantipyrene (also known as 4-aminophenazone). A table showing the relative responses and possibilities of the aminoantipyrene and 3-methyl-2-benzothlazolinone hydrazone methods in respect of a range of phenolic compounds is included. None of the above methods is capable of achieving the very low levels of detection required in connection with the implementation of EEC Directives.

Janda and Krijt [48] have described a method for the isolation of phenols from water by distillation–extraction. The water sample (150ml) is acidified and strongly salted, then continuously steam-distilled with 3mL diethyl ether for 1.5h. The extract is analysed by splitless gas chromatography. The recovery, for a concentration range of about 0.1–30mg L $^{-1}$ approaches 100%. The small amount of ether used avoids the need for preconcentration and the dual isolation technique reduces the probability of simultaneous isolation of potentially interfering compounds. The phenols could be isolated into diethyl ether from suspensions and from water with high suspended solids content. The detection limit, using splitless injection and glass capillary columns, is approximately 10µg L $^{-1}$.

Other applications of gas chromatography to the determination of phenols in non saline waters are discussed in Table 15.9.

15.1.1.6 Carboxylic acids

Bethge and Lindstroem [49] first removed metal cations from a 10mL sample of water by elution with water from a Dowex 50W–X8 ion exchange column and the eluate was titrated to pH8 with standard tetra-butylammonium hydroxide. A calculated amount (as determined from the titration) of hexanoic acid was added as internal standard, the solution was concentrated to a syrup, the syrup was dissolved in acetone, and α-bromotoluene was added in slight excess. After 2h to ensure complete reaction, 1µL of the acetone solution was injected into a stainless steel column (2m × 2mm) packed with 3% of butane–1,4–diol succinate with nitrogen (30ml min $^{-1}$), as carrier gas and flame ionisation detection was used. The column was kept at 120°C for 17min, then temperature programmed to 150°C at 2.5°C min $^{-1}$. Down to 50μm concentration of the acids could be determined by this procedure.

Other applications of gas chromatography to the determination of carboxylic acids in non saline waters are discussed in Table 15.9.

15.1.1.7 Carbohydrates

Ochiai [50] used gas chromatography to determine dissolved carbohydrates in non saline water. The carbohydrates are first hydrolysed to alditol acetates of monosaccharides by refluxing for 7h with 1M hydrochloric acid under nitrogen. The hydrolysate is then reduced with sodium hydroxide at 60°C. The gas chromatograph used was equipped with a flame ionisation detector. A glass column (2m × 3mm id) packed with 5% OV–275 on Chromosorb W was employed at a nitrogen flow rate of 40mL min $^{-1}$. A temperature-programmed analysis from 160°C to 240°C min $^{-1}$ required 40min to elute the acetyl derivatives of the monosaccharides and the internal standard inositol.

15.1.1.8 Haloforms and aliphatic halogen compounds

Murray and Riley [51,52] described gas chromatographic methods for the determination of trichloroethylene, tetrachloroethylene, chloroform and carbon tetrachloride in non saline waters. These substances were separated and determined on a glass column (4m × 4mm) packed with 3% of SE–52 on Chromosorb W (AW DMCS) (80–100 mesh) and operated at 35°C, with argon (30ml min $^{-1}$) as carrier gas. An electron capture detector was used, with argon–methane (9:1) as quench gas. Chlorinated hydrocarbons were stripped from water samples by passage of nitrogen and removed from solid samples by heating in a stream of nitrogen. In each case the compounds were transferred from the nitrogen to the carrier gas by trapping on a copper column (30cm × 6mm) packed with Chromosorb W (AW DMCS) (80–100 mesh) coated with 3% of SE–52 and cooled at –78°C, and subsequently sweeping on to the gas chromato-graphic column with the stream of argon. A limitation of this procedure is that compounds which boil considerably above 100°C could not be determined [53].

A different approach was to pass the water through a bed of activated carbon which was subsequently extracted exhaustively in a Soxhlet unit, and the extract was evaporated and analysed; this measured perchloroethylene and hexachloroethane, but the results are uncertain quantitatively [54]. A method has been published by which the watersample was codistilled with cyclohexane and the organic phase was then injected into an electron capture detector gas chromatograph [55]. Extraction with n-pentane followed by gas chromatography has also been used [56] although the extraction was easy and effective, the chromatographic conditions described were time-consuming and unsuitable for compounds heavier than perchloroethylene.

Chlorinated normal paraffins up to C_{30} carbon number range are of low volatility and are thermally unstable, producing hydrogen chloride on decomposition; hence direct gas chromatography is not attractive.

Deetman *et al.* [57] have devised an electron capture gas chromatographic technique, applicable to water, for the determination of down to 1ng L^{-1} of 1,1,1-trichloroethane, trichloroethylene, perchloroethylene, 1,1,1,2-tetrachloroethane, 1,1,2,2-tetrachloroethane, pentachloroethane, hexachloroethane, pentachlorobutadiene, hexachlorobutadiene, chloroform and carbon tetrachloride. These workers used extraction of the water samples with *n*-pentane as a means of isolating the chlorinated compounds from the sample. Recoveries of 95% were obtained in a single extraction. To dry the extract anhydrous sodium sulphate was found to be effective. Furthermore this drying agent could be freed from electron capturing contaminants by heating [58] and did not absorb the chlorinated compounds. Fig. 15.1 shows a typical chromatogram obtained by this method.

Hagenmaier *et al.* [59] have described a method for the quantitative gas chromatographic determination of volatile halogenated hydrocarbons in lake water samples. Sample enrichment is effected by liquid–liquid extraction with pentane, followed by separation on a capillary gas–liquid chromatographic column, with electron capture detection. A 1:25 pentane–water ratio was employed in conjunction with a standard solution of a reference compound (1-bromobutane) for estimating extraction efficiency. The detection limit using split injection was about 0.005µg L^{-1} and could be increased to less than 1ng L^{-1} by on-column injection. The method was applied to samples of water from Constance lake. The water contained appreciable amounts of trichloroethylene (8–20µg L^{-1}) and tetrachloroethylene (2–5µg L^{-1}).

Other applications of gas chromatography to the determination of haloforms and aliphatic halogen compounds in non saline waters are discussed in Table 15.9.

15.1.1.9 Unsaturated aliphatic halogen compounds

Alberti and Jonke [60] describe a gas chromatographic method for the determination of vinyl chloride in surface waters using a flame ionisation detector and a Poropak Q or Chromosorb 101 column. The detection limit is 0.3mg L^{-1} and samples of waste waters from vinyl chloride or PVC factories can be injected direct into the gas chromatograph, while water samples with lower concentrations require preliminary enrichment for which a gradient–tube method is described.

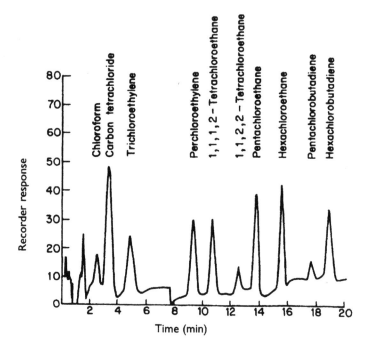

Fig. 15.1 Typical chromatogram of a water sample
Source: Reproduced by permission from Elsevier Science, UK [57]

Burgasser and Calaruotolo [61] have described a gas chromatographic method for determining semi- or non volatile chlorinated organics such as hexachlorobutadiene, hexachlorocyclopentadiene, octachlorocyclopentene and hexachlorobenzene in amounts down to 0.1µg L^{-1} in non saline waters. These compounds fall into the category of those which are preferentially soluble in non aqueous solvents. These workers used a Brinkmann Polytron homogeniser to perform the extraction and a Sorval refrigerated centrifuge to speed up the phase separation process. The extraction of chlorinated organic compounds from water can be carried out in a single vessel in one step, taking only 10min to complete.

In the direct gas chromatographic method [62] vinyl chloride, arenes and other volatile halogen compounds are separated from the water sample by stripping them in a closed system. The stripped compounds were absorbed on Poropak N in a glass tube within the closed system and eluted with methanol. They were then separated by gas chromatography on a 3m Chromosorb 102 column. The solvent is compatible with either electron capture or photoionisation detectors. Using photoionisation detectors both arenes and vinyl chloride were determined with detection limits of 1µg L $^{-1}$. The procedure was made semi-automatic by the use of autosamplers on the gas chromatograph, enabling 25 samples a day to be analysed.

Other applications of gas chromatography to the determination of unsaturated aliphatic halogen compounds in non saline waters are discussed in Table 15.9.

15.1.1.10 Chlorophenols

Rudling [63] determined down to 100µg L $^{-1}$ pentachlorophenol (PCP) in water by an electron capture gas chromatographic method. Hexane (0.5mL) and fresh acetylation reagent (pyridine (2mL) plus acetic anhydride (0.8mL) stored in the cold) (40µL) are added to the combined aqueous extracts and shaken for 1min. The hexane phase is analysed by gas chromatography on a glass column (1m × 1.5mm) packed with 5% of QF–1 on Varaport 30 (100–120 mesh) operated at 150°C with nitrogen as carrier gas (25mL min $^{-1}$).

Renberg [64] has used an ion exchange technique for the determination of chlorophenols and phenoxyacetic acid herbicides in water. The water sample is mixed with Sephadex QAE A–25 anion exchanger and the adsorbed materials are then eluted with a suitable solvent. The chlorinated phenols are converted into their methyl ethers and the chlorinated phenoxy acids into their methyl or 2-chloroethyl esters for gas chromatography.

Renberg [64] recommends the use of γ-BHC (lindane), DDE or DDT as a gas chromatographic internal standard. The relative retention times of the derivatives corresponding to these internal standards are shown in Table 15.3. The detection limits for the different substances for 1L of a water sample are 0.0001–0.1ppb.

Chriswell and Cheng [65] have shown that chlorophenols and alkyl-phenols in the pbb to ppm range in non saline waters and treated drinking water can be determined by sorption on macroporous anion exchange resin, elution with acetone and measurement by gas chromatography.

Other applications of gas chromatography to the determination of chlorophenols in non saline waters are discussed in Table 15.9.

15.1.1.11 Polychlorobiphenyls

Bauer [66] has described a gas chromatographic procedure for the determination of PCBs in water. The PCBs were extracted from water with hexane, and the extract was dried and concentrated before gas chromatographic analysis. Extraction and concentration of PCBs were

Table 15.3 Retention times relative to γ-BHC, p,p,-DDE and p,p-DDT

	OV–17	QF + SF 96
γ-BHC	7 min at 160°C	9min at 150°C
Methyl ethers of		
2,4,6-trichlorophenol	0.096	0.14
2,3,4,6-tetrachlorophenol	0.25	0.32
pentachlorophenol	0.64	0.70
γ-BHC	1 min at 200°C	3min at 180°C
p,p-DDE	2.93	3.02
Methyl esters of		
2,4-dichlorophenoxyacetic acid	1.24	0.94
2,4,5-trichlorophenoxyacetic acid	1.83	1.29
2-Chloroethyl esters of		
2,4-dichlorophenoxyacetic acid	1.70	2.65
2,4,5-trichlorophenoxyacetic acid	2.95	4.00
Methyl ether of		
2,hydroxy-2',4,4'-trichlorodiphenyl ether	2.93	3.27
	SF96	
p,p-DDT	2min at 200°C	
Dimethyl ether of hexachlorophene	3.42	

Source: Reproduced by permission from the American Chemical Society [64]

also effected by the use of slow filtration through sand or algae columns. The sand was extracted with 60% acetone–hexane, acetone was removed by washing the extract with water, and the washed hexane was passed through a column filled with layers of sodium sulphate, alumina and Florasil. The eluate was analysed by gas chromatography. Algal material was dried, ground with sand, and treated in a manner similar to that used for sand alone. The gas chromatography was carried out on glass columns (3m × 2.8mm) packed with 2.5% QF–1, 2.5% silicone rubber and 0.5% Epikote 1001 on Chromosorb W AW–DCMS (80–100 mesh) and operated at 200°C with nitrogen (90ml min $^{-1}$) as carrier gas and an electron capture detector.

Berg et al. [67] separated PCBs from chlorinated insecticides on an activated carbon column prior to derivatisation and gas chromatographic separation on a capillary column. Separation is based on the observation that PCBs adsorbed on activated charcoal cannot be removed quantitatively with hot chloroform but can be with cold benzene. Insecticides of the DDT group and a variety of others (eg g-BHC, Aldrin, Dieldrin, Endrin and Heptachlor and its epoxide) can be eluted from the charcoal with acetone–ethylether (1:3). Typical recoveries from a mixture of *p,p'*-DDE (1,1-dichloro-2,2-*bis*–(4,chlorophenyl)ethylene), *p,p'*-TDE, and *o,p'*-DDT and a PCB (Arochlor 1254) by successive elution with 90mL of 1:3 acetone–ether and 60mL of benzene were 91, 92, 92, 94 and 90% respectively. Identification and determination of the PCBs was effected by catalytic dechlorination to bicyclohexyl or perchlorination to decachloro-biphenyl, followed by gas chromatography. For bicyclohexyl, gas chromatography was carried out on a column (2.4m × 6mm) of 10% DC–710 on Chromosorb W, operated at 90°C for 2.5min then temperature programmed at 10°C per min, with flame ionisation detection. For deca-chlorobiphenyl, the column (0.6m × 3mm) is 5% SE–30 on Chromosorb W, operated at 215°C, with nitrogen as carrier gas and a tritium detector.

Beezhold and Stout [68] studied the effect of using mixed standards on the determination of PCBs. Mixtures of Arochlors 1254 and 1260 were used as comparison standards and gas chromatograms of these mixtures were compared with those obtained from a hexane extract of the sample after clean-up on a Florasil column. Polychlorinated biphenyls were separated from DDT and its analogues on a silica gel column activated for 17h and with 2% (w/w) of water added. The extracts were analysed on a silanised glass column packed with 5% DC–200 and 7.5% QF–1 on Gas Chrom Q (80–100 mesh) operated at 195°C with nitrogen as carrier gas (50–60mL min $^{-1}$) and a tritium detector.

Glass–wall coated open tubular capillary columns have been used for the gas chromatography of PCBs [69,70]. Fig. 15.2 shows a chromatogram obtained by this technique [69] for a mixture of Arochlors in hexane. The column comprised 25mL × 0.25mm of WCOT glass CP^{tm} Sil 7. It can be seen that 0.1ppm of these substances is easily detected. Schulte and Acker [70] gas chromatographed PCBs with a glass capillary column at temper-atures up to 320°C. They used a 60m column impregnated with SE–30–SC and helium as carrier gas with a flame ionisation detector. The column was maintained briefly at 80°C after injection of the sample, heated to 180°C and then temperature programmed to 260°C at 20°C per minute.

Other applications of gas chromatography to the determination of PCBs in non saline waters are discussed in Table 15.9.

15.1.1.12 Aliphatic amines

Hermanson *et al.* [71] used an aluminium column (276cm × 4mm) packed with 80–100 mesh Chromosorb W supporting 8.9% of amine 220 at 95°C with nitrogen as carrier gas and flame ionisation detection. A rectilinear response was obtained between peak area and amount of propylamine, dipropylamine, and propanol; between 0.2 and 2.0µg.

Gas flame ionisation chromatography has been used to determine dimethylamine [72,73], dimethylformamide [72], propylamine [73] and diisopropylamine [71] in river water and industrial effluents. To separate

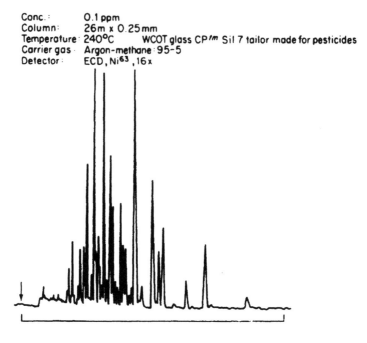

Conc.:   0.1 ppm
Column: 26m x 0.25mm
Temperature: 240°C   WCOT glass CP /m Sil 7 tailor made for pesticides
Carrier gas Argon-methane: 95–5
Detector: ECD, Ni^{63}, 16x

Fig. 15.2 Chromatogram of Arochlor 1260
Source: Reproduced by permission from Springer Verlag, Heidelberg [70]

C_1–C_4 mono-, di- and trialkylamines, Onuska [73] adjusted the pH of the sample to between 5 and 8. A 1µL aliquot of the filtrate was injected on to a stainless steel column (185cm × 2mm id) packed with 28% of Pennwalt 223 and 4% of potassium hydroxide on Gas–Chrom R (80–100 mesh) and maintained at 134°C. A dual-flame ionisation detector was used and the carrier gas was helium (flow rate 52.2ml min $^{-1}$). The detector response was rectilinear between 10ng and at least 100µg of dimethylamine, and the reproducibility was good. The column could be regenerated by increasing the column temperature to greater than 180°C.

Kimoto *et al.* [74] used gas chromatography to determine simple aliphatic amines in dichloromethane extracts of water samples. They used a gas chromatograph equipped with a nitrogen–phosphorus detector and a 1.8m × 2mm id glass column packed with Carbopack B–Carbowax 20–potassium hydroxide. The flow rates were helium 20mL min $^{-1}$ and air 80mL min $^{-1}$. The injector and detector temperatures were 200°C and 250°C respectively. The column was programmed from 70 to 150°C at 4°C min $^{-1}$ and held at the upper temperature for an additional 4min. Then 2µL samples were injected.

Other applications of gas chromatography to the determination of aliphatic amines in non saline waters are discussed in Table 15.9.

15.1.1.13 Nitrosamines

Nikaido *et al.* [75] give details of procedures for the recovery of low levels of dialkylnitrosamines from non saline waters, including lake water and sewage effluents before subsequent detection by gas chromatography. The recovery technique involves the addition of potassium carbonate to the sample and concentration of the nitroso compounds on Amberlite XAD–2 resin. Greater than 90% recoveries were obtained for dimethylnitrosamine and diethylnitrosamine. The recoveries obtained by this procedure were 97% or higher and were particularly good for nitrosamine levels of about 10 parts per 10^9.

Tompkins and Griest [76,77] have described a gas chromatographic method for the determination of N-nitrosodimethylamine in contaminated ground waters. This procedure utilises a new solid–phase extraction procedure which extracts N-nitrosodimethylamine at the nanogram per litre level from aqueous samples using a C_{18} (reversed phase) membrane extraction disc layered over a carbon-based extraction disc. The reversed phase disc removes non polar water insoluble neutrals and is set aside; the carbon-based disc is extracted with a small volume of dichloromethane. N-nitrosodimethylamine is quantified in the organic extract using a gas chromatograph equipped with both a short-path thermal desorber and a chemiluminescence nitrogen detector. The detection limit for the procedure is 3ng L^{-1} N-nitrosodimethylamine with a recovery of about 57%.

Mills and Alexander [78] have discussed the factors affecting the formation of dimethylnitrosamine in samples of water and soil.

Other applications of gas chromatography to the determination of nitrosamines in non saline waters are reviewed in Table 15.9.

15.1.1.14 Nitriloacetic acid

Chau and Fox [79] concentrated nitriloacetic acid in lake water samples by passing them down a Dowex 1 column (formate form) and elution with 2.5M to 8M formic acid. The nitriloacetic acid is then esterified, with heptadecanoic acid added as internal standard, by heating for 1h at 100°C in a sealed ampoule with propanol saturated with hydrogen chloride. The propyl esters are analysed on a stainless steel column (1.8m × 6mm) packed with 3% of OV–1 on Chromosorb WHP (80–100 mesh), temperature programmed from 180 to 225°C min^{-1}, and operated with nitrogen as carrier gas (65mL min^{-1}) and flame ionisation detection. The calibration graph is rectilinear for up to 20µg of nitriloacetic acid as ester.

The limit of detection is 0.01µg and at the level of 20µg L $^{-1}$ the standard deviation was 1.33µg L $^{-1}$ and the coefficient of variation was ±6.3%.

Warren and Malec [80] determined nitriloacetic acid and related amino-polycarboxylic acids (iminodiacetic acid, glycine and sarcosine in inland waters and sewage effluents by converting to the butyl or N-trifluoroacetyl esters followed by chromatography on dual glass U-shaped columns (1.9m × 2mm) packed with 0.65% of ethanediol adipate on acid-washed Chromosorb W (80–100 mesh), temperature programmed from 80 to 220°C and operated in the differential mode with flame ionisation detectors. The signal was fed to a digital integrator and then to both channels of a dual-pen recorder.

Williams et al. [81] applied a nitrogen-specific detector to a survey of the levels of nitrolacetic acid to its tri-n-butyl ester in US tap water supplies at concentrations approaching 1ppb (US). These workers based their method on that described by Aue [82], described earlier. Williams et al. [81] used a Perkin–Elmer Model 910 gas chromatograph, equipped with a single column, a two-way effluent splitter, a flame ionisation detector and a nitrogen–phosphorus detector operating in the nitrogen mode. The column was 1.8m × 6mm od glass, packed with either 5% OV–101 or 3% OV–210 on 80–100 mesh Chromosorb WHP. The carrier gas was helium at a flow rate of 60mL min $^{-1}$ and the effluent splitter diverted 60% to the flame ionisation detector and 40% to the nitrogen detector. Hydrogen and air flows were optimised for each detector. The injector and detector temperatures were 240 to 280°C respectively and the column and interface temperatures 200 and 250°C.

Aue et al. [82] claimed a limit of detection of 1ppb nitriloacetic acid for a 50mL water sample, but preliminary investigations of Williams et al. [81] with standard solutions of the tri-n-butyl ester showed that quantitation at this level was difficult due to interference from the solvent peak when using acetone as the injection solvent as specified by Aue et al. [82]. The use of alternative injection solvents gave some improvement but quantitation was still difficult. Analysis of standard solutions of the tri-n-butyl ester, equivalent to 1ppb nitrolacetic acid in a 50mL water sample, showed that the sensitivity of the nitrogen-specific detector was adequate, quantitation was straightforward, and there was minimal interference from the injection solvent, acetone. The nitrogen-selective detector gave a linear response over the range of 1–1000ng injected of the tri-n-butyl ester of nitriloacetic acid.

Reichert and Linckens [83] have reviewed gas chromatographic methods for the determination of nitriloacetic acid in potable waters. They point out that esterification of the nitriloacetic acid is required to enable it to be volatilised in the gas–liquid chromatographic column, and they compared a range of esterification reagents and conditions for simplicity and speed of operation. The method chosen involved treatment of a concentrated sample with a mixture of *n*-propanol–acetyl chloride (10:1) and the resulting nitroacetic acid–propyl ester injected into a column, which was fitted with a nitrogen-sensitive detector. The detection limit for nitriloacetic acid in potable water is about 1μg L $^{-1}$.

15.1.1.15 Ethylene diamine tetraacetic acid

Gardiner [84,85] described a gas chromatographic method for the determination of EDTA in aqueous environmental samples. The separation of the major peaks is increased by preparing the ethyl derivatives of the sample compounds, 1,6-hexanediamine tetraacetic acid (HDTA) being used as internal standard. The lower limit of detection of the method is approximately 15μg L $^{-1}$ with 25mL samples. In this method the ethyl derivatives of the sample components were prepared so that the major peaks would be well separated. The ethyl esters of fatty acids up to and including the C_{18} fatty acids eluted well before the EDTA derivatives and did not interfere.

Kari [86] and Schürch and Dubendorfer [87] both point out that gas chromatographic procedures need lengthy sample pretreatments.

15.1.1.16 Organosulphur compounds

Simo *et al.* [88] have discussed the determination of nanomolar concentrations of dimethyl sulphoxide, along with dimethyl sulphide and dimethyl sulphoniopropionate, at nanomolar levels in non saline waters. After removal of dimethyl sulphide by purge and cryotrapping, dimethyl sulphoniopropionate is removed by the same method after alkaline hydrolysis, and dimethyl sulphoxide is reduced to dimethyl sulphide using a combination of sodium borohydride and hydrochloric acid. The dimethyl sulphide produced is stripped, cryotrapped and analysed by gas chromatography. Detection of 3pmol of dimethyl sulphoxide was achieved, resulting in a detection limit of 005nM for a 50mL sample.

Andreae [89] has described a gas chromatographic method for the determination of nanogram quantities of dimethyl sulphoxide in non saline waters, sea water and phytoplankton culture waters. The method involves a chemical reduction to dimethyl sulphide, which is then determined gas chromatographically using a flame photometric detector.

Andreae [89] investigated two different apparatus configurations. One consisted of a reaction-trapping apparatus connected by a six-way valve to a gas chromatograph equipped with a flame ionisation detector; the other apparatus combined the trapping and separation functions in one column, which was attached to a flame photometric detector. The gas chromatographic flame ionisation detector system was identical to that described by Andreae [89] for analysis of methylarsenicals, with the

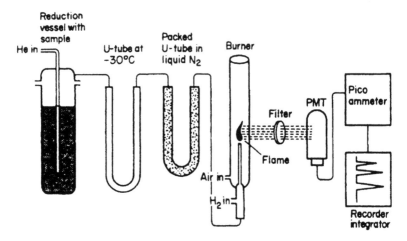

Fig. 15.3 Apparatus for the reduction of DMSO to DMS and the flame photometric detection of DMS (FPD system)
Source: Reproduced by permission from the American Chemical Society [89]

exception that a reaction vessel which allowed the injection of solid sodium borohydride pellets was used. The flame photometric system (Fig. 15.3) is modified after a design by Braman *et al.* [90].

Other applications to the determination of organic sulphur compounds in non saline waters are discussed in Table 15.9.

15.1.1.17 Alkyl and aryl phosphates

Murray [91] has described a gas chromatographic method for the determination in water of triarylphosphate esters (1mol S–140, tricresyl phosphate, cresol phosphate). These substances are used commercially as lubricant oil and plastic additives, hydraulic fluids and plasticisers. The method involves extraction from the samples, hydrolysis and measurement of the individual phenols by gas chromatography as the trimethylsilyl derivatives. The lower detection limit was about 3ppm.

In this procedure a weighed sample was placed in a 20ml ampoule and 10ml of 5% potassium hydroxide in 95% methanol added. The ampoule was sealed and autoclaved at 25psi (172kPa) for 90min. When cool, the ampoule was opened and the contents washed into 950ml of distilled water, acidified with about 5ml 6M hydrochloric acid to pH1–2 and made up to 1000ml. This was transferred to a separatory funnel, 5µL of *o*-xylene were added by syringe as an internal standard, and the mixture was extracted once with 50ml of chloroform. The solvent layer was evaporated to 1–2ml and treated with Tri–Sil concentrate to form tri-methylsilyl derivatives. This was allowed to react overnight and analysed by gas chromatography the following day. The concentrations of the individual phenols were calculated from calibration graphs and the composition of the ester was determined.

A gas chromatograph with flame ionisation detectors was used for the analysis and the column used was 2.4m × 3mm stainless steel packed with 5% Imol on Chromosorb W, AW and SCMS treated 80–100 mesh. A temperature programme from 80–120°C at 4°C min^{-1} was used.

15.1.1.18 Chlorinated insecticides and polychlorobiphenyls

Work on the determination of chlorinated insecticides has been almost exclusively in the area of gas chromatography using different types of detection systems, although a limited amount of work has been carried out using liquid chromatography and thin layer chromatography.

By its nature, gas chromatography is able to handle the analysis of complex mixtures of chlorinated insecticides. It is not surprising therefore, that much of the published work discussed in this section is concerned with the analyses of mixtures of different types of chlorinated insecticides as found in environmental samples. Work on the determination of individual insecticides is reported at the end of this section.

The previous pages dealt with methods for PCBs alone. In actual practice, environmental samples which are contaminated with PCBs are also highly likely to be contaminated with chlorinated insecticides. Many reports have appeared discussing co–interference effects of chlorinated insecticides in the determination of PCBs and vice versa and much of the published work takes account of this fact by dealing with the analysis of both types of compounds. This work is discussed below.

Various solvents have been used for the preliminary separation of chlorinated insecticides from non saline waters prior to gas chromato-graphic analysis. The most commonly used solvents include hexane, petroleum ether, mixtures of hexane and benzene or toluene, benzene, diethyl ether, methylene dichloride acetonitrile (Table 15.4).

Brodtmann [120] carried out a long-term study on the qualitative recovery efficiency of the carbon adsorption method versus that of a continuous liquid–liquid extraction method for several chlorinated insecticides. Comparative results obtained by electron capture gas chromatography indicate that the latter method may be more efficient.

In this method river water samples were passed through carbon for a predetermined period. The carbon extract was then obtained by chloroform extraction of the carbon in a modified Soxhlet apparatus for 36h using glass distilled, pesticide grade petroleum ether. The neutral fraction of the chloroform extract was then prepared for gas

Table 15.4 Extraction solvents for concentration of chlorinated insecticides prior to gas chromatographic analysis

Solvent	Insecticides mentioned	Gas chromatography	Ref.
n-hexane	α-, β-Endosulphan, Heptachlor	SE–30 on Chromosorb W	[91–95]
n-hexane	α-BHC, β-BHC, α-BHC, Aldrin, o,p'-DDE, α-Endosulphan, p,p'-DDE, Dieldrin, o,p'-DDD, Endrin, β-Endosulphan, p,p'-DDD, o,p'-DDT, p,p'-DDT	Capillary column	[96]
n-hexane	p,p'-DDD, p,p'-DDD, p,p'-DDE, p,p'-DDT, Aldrin, Dieldrin, Endrin, Heptachlor, Heptachlor epoxide, Lindane, Isodrin, Methoxychlor, Chlordane, chlordene, hexachlorobicyclo-heptadiene, hexachlorocyclopentadiene		[97,98]
Hexane, hexane, benzene and hexane–toluene	Dichlorvos		[99]
Hexane	α-BHC, γ-BHC, o,p'-DDD, p,p'-DDD, Dieldrin, Endrin, Heptachlor, Heptachlor epoxide	Electron capture detection	[100]
n-hexane or acetonitrile	α-BHC, β-BHC, γ-BHC, δ-BHC, Heptachlor, Heptachlor epoxide, Aldrin, Endrin, p,p'-DDE, p,p'-TDE, p,p'-DDT, o,p'-DDE, p,p'-TDE, o,p'-DDT, Mirex, Methoxychlor, Photodieldrin	Electron capture, capillary column	[101]
n-hexane	15 organochlorine pesticides	DCII on chromosorb WAWDMCS	[102]
Petroleum ether	DDT	DCII on chromosorb WAWDMCS	[102]
Petroleum ether	DDT, γ-BHC, Heptachlor epoxide, Dieldrin, Methoxychlor	DC–200 on chromosorb WHMDS	[103]
Petroleum ether	Dieldrin	flame ionisation and electron capture detectors	[104]
Petroleum ether	α-BHC, β-BHC, γ-BHC, γ-BHC, p,p'-DDE, o,p'-DDT, p,p'-DDD, p,p'-DDT	1.5% silicone OV–17 plus fluoroalchyl –siloxane on Chromosorb W, electron capture detection	[105]

Table 15.4 continued

Solvent	Insecticides mentioned	Gas chromatography	Ref.
Petroleum ether	8 organochlorine pesticides	Electron capture detection	[106]
Benzene	γ-BHC, Methychlor		[107,108]
Benzene	8 organochlorine pesticides		[109]
Benzene	15 organochlorine pesticides		[110]
Benzene	9 organochlorine pesticides		[107]
Dichloromethane and 15% dichloromethane in n-hexane	18 organochlorine pesticides		[111]
15% methylene chloride in n-hexane	15 organochloride pesticides		[112]
Methylene chloride	Aldrin, α-BHC, β-BHC, γ-BHC, Chlorobenside, Chlordane, (Kepone), p,p'-DDD, p,p'-DDE, o,p'-DDT, p,p'-DDT, Dichlone, Dieldrin, Endrin, Endosulphan, Heptachlor, hexatchlorobenzene, 1-hydroxchlordene, Methoxychlor, Mirex	5% OV–210 or 1–5% OV–17/1.95% OV–210 on Chromosorb, electron capture detector	[113]
Diethylether	Miscellaneous organochlorine insecticides	3% QF–1 or 2% OV–1, electron capture detection	[114]
15% diethyl ether in n-hexane	19 aminochlorine insecticides		[115]
Liquid–liquid extraction methods	General discussion, Lindane, hexachlorobenzene		[116,117]
Liquid–liquid extraction methods	Lindane, hexachlorocyclohexane, α,β,γ, hexachlorobenzene		[118,119]

Source: Own files

chromatography by the methods of Breidenbach [121]. Brodtmann [120] used a continuous liquid–liquid extraction apparatus as described by Kahn and Wayman [122] and Goldberg *et al.* [45] for the extraction of non polar solutes from river water. Pesticide grade petroleum ether, used in all cases, was recycled internally (initial solvent charge of 350mL), thereby continuously exposing essentially fresh solvent to the river water.

A Florasil clean-up step using sequential elutions with 6% and 15% ethyl ether–petroleum ether solutions was employed.

A dual-column gas chromatograph equipped with two tritium electron capture detectors was employed by Brodtmann [120]. Both columns were acrylic glass (1.83m × 0.32cm id). Column A was packed with 5% DC–260 on 80/100 DCMS Chromosorb W. Flow rate of carrier gas (5% methane–argon) through this column was 80mL min $^{-1}$. Column B was packed with 1.5% OV–17–1.95% QF–1 on 80/100 Chromosorb W DCMS support. Flow rate of carrier gas (5% methane–argon) through this column was 50mL min $^{-1}$ Both pairs of injectors, columns and detectors were maintained at, respectively, 212, 184 and 204°C.

The American Public Health Association [97,98] published an early gas chromatographic method for the solvent extraction and gas chromatographic determination of 11 chlorinated insecticides in water samples in amounts, down to 0.005mg L $^{-1}$, *p,p'*-DDE, *p,p'*-DDT, Aldrin, Dieldrin, Endrin, Heptachlor, Heptachlor epoxide, Lindane, Isodrin and Methoxychlor. The insecticides Carbophenothion, Chlordane, Dioxathion, Diazinon, Ethion, Malathion, Parathion methyl, methyl Trithion, Parathion, Toxaphene and VC–13 may be determined when present at higher levels. Also, the chemicals chlordane, hexachlorobicycloheptadiene and hexachlorocyclopentadiene, which are pesticide manufacturing precursors, may be analysed by this method.

The insecticides are extracted directly from the water sample with *n*-hexane. After drying and removing the bulk of the solvent, the insecticides are isolated from extraneous material by microcolumn adsorption chromatography. The insecticides are then analysed by gas chromatography. This method is a modification and extension of the procedures developed by Lamar *et al.* [119,123]. For the analysis of insecticides in waters that are grossly polluted by organic compounds other than pesticides, a high-capacity clean-up procedure is used, as detailed in the Federal Water Pollution Control Administration *Method for Chlorinated Hydrocarbon Pesticides in Water and Wastewater* [124]. Mean recoveries obtained by this procedure for 12 insecticides from surface water samples range between 87.2% (endrin) and 103% (Lindane).

Sackmauereva *et al.* [105] have described a method for the determination of chlorinated insecticides (BHC isomers, DDE, DDT and hexachlorobenzene) in water. In this method the water sample (1–3L) is extracted with petroleum ether. A vacuum is used to transfer the water sample (separation funnel No. 1) and petroleum ether (separation funnel No. 2) into a glass spiral filled with 8mm glass balls. The petroleum ether layer is then concentrated to a volume of about 0.5mL using a vacuum, and purified on an alumina column (Woelm, neutral, activated by heating at 300°C for 3h and deactivated by adding 11% water). Thereafter, insecticides were eluted with 15% dichloromethane in petroleum ether. The eluate was concentrated in a vacuum rotary evaporator to a volume of 1mL and then used for gas–liquid chromatography [125–128], on a column operated under the following conditions: temperature of the column 180–200°C, temperature of the injection port 210°C, temperature of the electron capture detector (^{63}Ni) 200–225°C, nitrogen flow rate 60–80mL min $^{-1}$, EC detector voltage 20–70V. A column filled with 1.5% silicone OV–17 plus silicone oil (fluoralchylsiloxane) on Chromosorb W (80–100 mesh) is used for separation of the o,p'-DDT, p,p'-DDE, p,p'-DDD and pp'-DDT. α-BHC and hexachlorobenzene (HCB) have a common peak. They can be separated on a column filled with 2.5% silicone oil XE–60 (β-cyanoethylmethylsilicone) on Chromosorb W (80–100 mesh).

Using the gas chromatography methods, Sackmauereva et al. [105] obtained from spiked samples the four BHC isomers at 93–103.5% recovery. Both DDT and DDE were yielded in 85.6–94% from water. Surface water from the Danube was examined for the content of BHC, DDE and DDT residues. The average content of γ-BHC was 0.117μg L $^{-1}$, and that of β-BHC 0.040μg L $^{-1}$, and that of the other BHC isomers plus HCB was 0.049μg L $^{-1}$. The average content of DDE was 0.050μg L $^{-1}$ and that of DDT 0.125μg L $^{-1}$.

Aspila et al. [129] reported the results of an interlaboratory quality control study involving five laboratories on the electron capture gas chromatographic determination of 10 chlorinated insecticides in standards and spiked and unspiked seawater samples (Lindane, Heptachlor, Aldrin, γ-Chlordane, α-Chlordane, Dieldrin, Endrin, p,p'-DDT, Methoxychlor and Mirex). The methods of analyses used by these workers were not discussed, although it is mentioned that the methods were quite similar to those described in the Water Quality Branch *Analytical Methods Manual* [130]. Both hexane and benzene were used for the initial extraction of the water samples.

Heptachlor revealed a poor recovery, which confirmed its degradation [131–135]. The degradation product [130–133] is known to be 1-hydroxychlordene.

Suzuki et al. [819] studied the determination of chlorinated insecticides in river and surface waters using high resolution electron capture gas chromatography with glass capillary columns. They compared resolution efficiencies of organochlorine insecticides and their related compounds with wall-coated open tubular (WCOT) and support-coated open tubular (SCOT) glass capillary columns with those of conventional

Table 15.5 Multi-residue methods

Ref.	Year	Author	Insecticides included	Extraction	Clean up	Recovery studies
[109]	1968	Pionke et al.	4 OGP and 8 OGC	Benzene	Act. silica gel, modified	Some data
[110]	1968	Kadoum	150 OGC compounds			
[137]	1968	Kadoum	4 OGP compounds		Aqueous acetonitrile partition	No data in abstract
			5 OGP compounds			
[107]	1969	Konrad et al.	9 OGP compounds	Benzene	None used	Data reported on all compounds
			5 OGP compounds			
[138]	1969	Askew et al.	40 OGP compounds	Used hexane, benzene and chloroform. The latter preferred.	Nuchar as needed but very rarely	Data reported on all compounds
[106]	1970	Johnson	8 OGC compounds	Pet. ether	Silica gel column	None reported
			15 OGC compounds	Hexane		
			3 OGP compounds			
[139]	1970	Herzel	Lindane, Aldrin, DDT, and metabolites	Hexane	None	None reported in abstract
[140]	1970	Ahling & Jensen	BHS, Lindane, DDE, DDD, DDT, PCB	Adsorption on undecan and Carbowax filter, then pet. ether	H_2SO_4	50 to 100%
[115]	1971	Ballinger	19 OGC compounds	15% ether in hexane	Florasil	Some data
[113]	1977	Thompson et al.	15 OCG compounds	15% methylene dichloride in hexane	Florasil	Some data

Source: Reproduced by permission from Springer Verlag, Heidelberg [113]

packed glass columns. These columns were coated with silicone V–101 as the liquid phase.

Thompson et al. [113] have described a gas chromatographic procedure for the multi-class, multi-residue analyses of organochlorine insecticides in water. It involves extraction with methylene chloride, separation into groups on a partially deactivated silica gel column, and sequential elution with different solvents. Final determinations of halogenated compounds and derivatised carbamates are made by gas chromatography with electron capture detection, and for organophosphorus compounds a flame photometric detector is used. This study included 42 organochlorine insecticides, 33 organophosphorus insecticides and seven carbamate herbicides. Table 15.5 illustrates the dearth of broadly applicable multi-residue methods in the literature. Of these references, only five are intended as multi-class, multi-residue procedures, outstanding among which is the work of Sherma and Shafik [136]. To correct this deficiency Thompson et al. [113] developed the multi-class multi-residue method which will provide the analyst with a means of simultaneously monitoring a water sample for a wide variety of pesticides.

The concentrated methylene chloride extract of the water sample is applied to a silica gel column, which is then eluted with n-hexane to provide fraction I, then with 60% v/v benzene–40% v/v n-hexane to provide fraction II, then with 5% v/v acetonitrile–95% v/v benzene to provide fraction III and, finally, with 25% v/v acetone–75% v/v methylene chloride to provide fraction IV. The fourth fraction only necessary if there is reason to suspect the presence of crufomate, dimethoate, mevinphos, phosphamidon or the oxygen analogues of diazinon or malathion. The majority of the organochlorine insecticides will be detected in fractions I and II with a few of the more polar compounds in fraction III (Table 15.6). Most of the organophosphorus insecticides will be found in fractions I and II, a very few in fraction IV and none in fraction III. Carbamate insecticides are found in fractions II and III. Of the organochlorine insecticides examined a high proportion of reproducible recoveries of 80% or better were obtained in the majority of cases. Thompson et al. [113] used a gas chromatograph with electron capture and flame photometric detectors, the latter operated in the phosphorus mode at 526nm. Columns were 1.8m × 4mm id, borosilicate glass packed with 1.5% OV–17–1.95% OV–210 or 5% OV–210, both coated on Gas–Chrom Q (80–100 mesh).

Other workers [97] have used electron capture gas chromatography coupled with the use of glass capillary columns for the separation of chlorinated insecticides.

Shevchuk et al. [99] investigated extraction efficiency by extractive chromatography of the extraction of Dichlorvos using hexane, and hexane with benzene and toluene. Hexane alone was not effective, the degree of extraction at pH2.5–7.5 reached a maximum 26%. The effect of

Table 15.6 Recoveries of 42 organochlorine compounds

Compound	Concn. (ppb)	Extraction only	I	II	III	IV	Total
		Recoveries (%) silica gel partitioning elution fraction					
Aldrin	0.20	89	88				88
α-BHC	0.09	91	76	4			80
β-BHC	0.47	99		88			88
γ-BHC (Lindane)	0.12	90	16	51			67
Chlorbenside	0.47	91	62				62
Chlordane	1.54	85	89	1			90
Chlordecone (Kepone)	3.64	72		18	8		26
p,p'-DDD	0.80	97	94				94
p,p'-DDE	0.45	96	101				101
o,p'-DDT	1.05	94	93				93
p,p'-DDT	1.58	104	98				98
Dichlone	13.4	85		79			79
Dieldrin	0.72	97		96			96
Endrin	1.11	105		98			98
Heptachlor	0.18	90	79				79
Heptachlor epoxide	0.31	91		89			89
Hexachlorobenzene (HCB)	0.20	74	96				96
1-Hydroxchlordene	0.34	81				82	82
Methoxychlor	5.70	97			104		104
Mirex	2.35	83	83				83

Source: Reproduced by permission from Springer Verlag, Heidelberg [113]

sodium, calcium, aluminium, lanthanum, gallium, yttrium and gadolinium and nitrate on the extraction with hexane showed that salting out significantly affected the distribution of Dichlorvos between the aqueous and the organic phase. Mixtures of solvents were more effective at extracting the Dichlorvos, and a synergistic effect was observed. The most effective mixture was hexane–benzene at a volumetric ratio of 2:3. Addition of 3.0m aluminium nitrate increased the extraction efficiency to 91%. Surface active agents and organic anions decreased the salting-out effect. Gas chromatography with flame ionisation detector was used to determine the concentration of Dichlorvos and compared with a thin layer chromatographic method. Preconcentration allowed an increase in the sensitivity of the determination by an order of magnitude.

Analysis of mixtures of chlorinated insecticides containing polychlorobiphenyl

The Department of the Environment (UK) [141] has used a tentative method for the determination of organochlorine insecticides and PCBs in non saline and drinking waters and sewage effluents. The first part of this method is concerned with the extraction and determination of the amounts of individual substances present; the second part is concerned with methods for verifying the identity of the various substances quantified in the first part. Representative data are presented for a number of typical insecticides using different chromatographic stationary phases (Table 15.7). Detection limits range from 8µg L $^{-1}$ (p,p'-DDT) to 106µg L $^{-1}$ (PCBs).

Polychlorobiphenyls have gas chromatographic retention times similar to the organochlorine insecticides and therefore complicate the analysis when both are present in a sample. Several techniques have been described for the separation of PCBs from organochlorine insecticides. A review of these methods has been presented by Zitko and Choi [142]. These techniques are time-consuming and, in general, semi-quantitative. In addition, differential adsorption or metabolism of the Arochlor isomers in marine biota prevent accurate analysis of the PCBs. The gas chromatographic determination of chlorinated insecticides together with PCBs is difficult. Chlorinated insecticides and PCBs are extracted together in routine residue analysis, and the gas chromatographic retention times of several PCB peaks are almost identical with those of a number of peaks of chlorinated insecticides, notably of the DDT group. The PCB interference may vary, because the PCB mixtures used have different chlorine contents, but it is common for PCBs to be very similar to many chlorinated insecticides and the complete separation of chlorinated insecticides from PCBs is not possible by gas chromatography alone [143–147]. Fig. 15.4 illustrates the possibility of the interference of DDT-type compounds in the presence of PCBS [138]. In an early paper on the determination of PCBS in water samples which also contain chlorinated insecticides, Ahling and Jensen [140] pass a sample through a filter containing a mixture of Carbowax 4000 monostearate on Chromosorb W. The adsorbed insecticides are eluted with light petroleum and then determined by gas chromatography on a glass column (160cm × 0.2cm) containing either 4% SF–96 or 8% QF–1 on Chromosorb W pretreated with hexamethyldisilazane, with nitrogen as carrier gas (30mL min $^{-1}$) and a column temperature of 190°C. When an electron capture detector is used the sensitivity was 10ng of Lindane per cubic metre, with a sample size of 200L. The recoveries of added insecticides range from 50 to 100%; for DDT the recovery is 80% and for PCBs 93–100%.

Dolan and Hall [148] have described a Coulson electrolytic conductivity detector of enhanced sensitivity for the gas chromatographic determination of chlorinated insecticides in the presence of PCBs. The detector was modified by the replacement of the silicone–rubber septum and stainless steel fitting at the exit of the pyrolysis furnace with PTFE fitting, by the reduction in diameter of the PTFE transfer tube, and by the

Table 15.7 Retention lines of organochlorine insecticides relative to dieldrin, on some GLC columns

Insecticide	Column packing								
	1% Apiezon Mor L	2.5% methyl silicone (eg OV-1)	2.5% phenyl-methyl silicone (eg OV-17)	2.5% cyano-silicone gum rubber XE-60	5% trifluoro propyl silicone oil QF-1 (FS-1265)	1% neo-pentyl-glycol succinate (NPGS)	1% FFAP	1.5% QF-1 +1% OV-1 (2)	2.0% OV-1 +3.0% QF-1 (1)
α-HCH	0.20	0.19	0.17	0.23	0.17	0.23	0.20	0.19	0.22
γ-HCH (lindane)	0.26	0.23	0.22	0.35	0.22	0.37	0.32	0.24	0.26
β-HCH	0.31	0.21	0.22	1.11	0.28	1.63	0.79	0.25	0.28
δ-HCH	0.35	0.26	0.43	1.09	0.31	1.34	0.27	0.29	0.31
Chlordane									0.34*
Heptachlor	0.36	0.41	0.36	0.23	0.23	0.22	0.22	0.35	0.36
Aldrin	0.49	0.53	0.46	0.26	0.28	0.22	0.25	0.43	0.44
Heptachlor epoxide	0.60	0.66	0.63	0.64	0.60	0.64	0.60	0.64	0.65
Endosulphan A	0.88	0.85	0.80	0.69	0.79	0.73	0.67	0.83	
Endosulphan B	1.45	1.16	1.35	2.33	1.66	2.69	2.39	1.32	
Dieldrin	1.00	1.00	1.00	1.00	1.00	1.00	1.00	1.00	1.00
p,p'-DDE	1.26	1.06	1.01	0.83	0.65	1.01	1.09	0.88	0.85
Endrin	1.49	1.28	1.55	0.71	1.22	1.12	1.07	1.26	1.15
o,p'-TDE	1.23	1.04	1.07	1.37	0.87	1.81	1.72	0.96	0.95
p,p'-TDE	1.83	1.30	1.40	2.44	1.21	3.39	3.03	1.30	1.24
o,p'-DDT	1.62	1.38	1.42	1.12	0.94	1.40	1.69	1.17	1.12
p,p'-DDT	2.40	1.72	1.85	2.07	1.32	2.84	2.99	1.60	1.51

Note: Chlordane is a multi-peak compound with other smaller peaks at RRT 0.37, 0.69, 0.76 and 1.23. Glass columns 1m × 3mm internal diameter 60–80 mesh acid-washed DMCS treated Chromosorb W support at 180°C and 20–40ml nitrogen min⁻¹

Source: Reproduced by permission of Department of the Environment, London [141]

Stage (a)

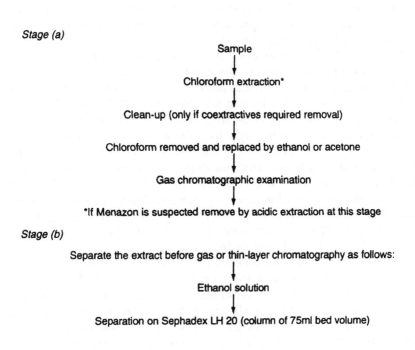

Sample

↓

Chloroform extraction*

↓

Clean-up (only if coextractives required removal)

↓

Chloroform removed and replaced by ethanol or acetone

↓

Gas chromatographic examination

↓

*If Menazon is suspected remove by acidic extraction at this stage

Stage (b)

Separate the extract before gas or thin-layer chromatography as follows:

↓

Ethanol solution

↓

Separation on Sephadex LH 20 (column of 75ml bed volume)

Fraction 1	*Fraction 2*	*Fraction 3*
(Elution volume 40 to 60ml, RV 50-80)	(Elution volume 60 to 70ml, RV 81-93)	(Elution volume 71 to 100ml RV 94-14
Chlorfenvinphos	Demeton-O-methyl	Azinphos-ethyl
Crufomate	Demeton-S-methyl	Azinphos-methyl
Demeton-S	Dibrom	Bromophos
Diazinon	Dichlofenthion	Carbophenthion
Dimefox	Dichlorvos	Coumaphos
Mevinphos	Disulfoton	Dimethoate
Oxydemeton-methyl	Ethion	Fenchlorphos
Phosphamidon	Ethoate-methyl	Fenitrothion
Schradan	Malathion	Haloxon
Sulfotep	Macarbam	Morphothion
TEPP	Phorate	Parathion
Vamidotion	Pyrimithate	Phenkapton
Also	Thionazon	Phosalone
trimethyl phosphate,	Trichlophon	
triethyl phosphate and		
tributyl phosphate		

Fig. 15.4 Screening technique described by Askew

Source: Reproduced by permission from the Royal Society of Chemistry, London [138]

replacement of the 4mm (id) reaction tube with one of 0.5mm. These modifications reduced hydrogen chloride adsorption and tailing, and improved sensitivity and reproducibility; sensitivity was also enhanced by increasing the cell voltage to 44V.

Musty and Nickless [149] used Amberlite XAD-4 for the extraction and recovery of chlorinated insecticides and PCBs from water. In this method a glass column (20cm × 1cm) was packed with 2g of XAD-4 (60–85 mesh) and 1L of tap water (containing one part per 10^9 of insecticides) was passed through the column at 8mL min $^{-1}$. The column was dried by drawing a stream of air through, then the insecticides were eluted with 100ml of ethyl ether–hexane (1:9). The eluate was concentrated to 5mL and was subjected to gas chromatography on a glass column (1.7m × 4mm) packed with 1.5% OV–17 and 1.95% QF–1 on Gas–Chrom Q (100–200 mesh). The column was operated at 200°C, with argon (10mL min $^{-1}$) as carrier gas and a ^{63}Ni electron capture detector (pulse mode). Recoveries of BHC isomers were 106–114%; of Aldrin, 61%; of DDT isomers, 102–144%; and of PCBs 76%.

Elder [150] evaluated mixtures of PCBs and DDE in terms of mixtures of commercial preparations from peak heights of packed column gas chromatograms using a programmable calculator. They proposed a method for evaluating gas chromatograms of multicomponent PCB mixtures and superimposed single components simultaneously. Apparent concentrations relative to calibration mixtures are assigned to a number of suitable peaks, and the apparent concentrations are related to the true concentrations by a set of linear equations which are solved by least squares approximation.

Södergren [151] investigated the simultaneous detection of PCBs, chlorinated insecticides and other compounds by electron capture and flame ionisation detectors combined in series using an open tube system.

Bacaloni et al. [152] used capillary column gas chromatography for the analysis of mixtures of chlorinated insecticides, PCBs and other pollutants. Graphitised capillary columns were used in this work. These columns have the interesting feature that according to the amount of stationary phase coated on the walls, they operate in gas–solid, gas–liquid and gas–liquid–solid chromatographic modes. Bacaloni et al. [152] used columns coated with PEG 20M and it is shown that selective columns can be obtained that give different performances for particular applications. Bacaloni et al. [152] gas chromatographed mixtures of 15 chlorinated pesticides on columns of different lengths and loadings of stationary phase under the same operation conditions at 160°C. Bacaloni et al. [152] concluded that a glass capillary column loaded with a large amount of PEG 20M is suitable for the analysis of volatile chlorinated compounds, phenols and amines, whereas for the analysis of chlorinated compounds, including the PCBs a column with a low loading of stationary phase is desirable.

Leoni [153] separated 50 organochlorine insecticides and PCBs into four groups by silica gel multicolumn chromatography. The separation was undertaken to simplify the chromatograms obtained by gas chromatography from contaminated samples. The sample (10–15L of surface water or 20–25L of potable water) was acidified with hydrochloric acid to pH3, and subjected to continuous extraction, twice with light petroleum (boiling range 40–60°C) and once with benzene. The benzene extract was evaporated and the residue was dissolved in light petroleum and added to the other extracts. The combined extracts were concentrated to 15mL and were partially purified by extraction with acetonitrile saturated with light petroleum (4 × 30mL). The extract was diluted with 2% sodium chloride solution (700mL) and the solution was again extracted with light petroleum, (2 × 100mL). This extract was evaporated and the residue was dissolved in 1mL of hexane. This solution followed by 1mL of hexane used for rinsing the container, was applied to a column (100mm × 4.2mm) of silica gel (Grace 950, 60–200 mesh, dried at 30°C for 2h then deactivated with 5% of water). The column was then percolated (at 1mL min $^{-1}$) in turn with hexane (20mL), benzene–hexane (3:2) (8mL), benzene (8mL) and ethyl acetate–benzene (1:1) (14mL). Each eluate was evaporated and the residue dissolved in 1mL of hexane and subjected to gas chromatography on OV–17 as stationary phase with electron capture detection. Quantitative recovery was achieved for all insecticides except Malathion, Disulfoton, Dimethoate and Phorate.

Other applications of gas chromatography to the determination of chlorinated insecticides and polychlorobiphenyls in non saline waters are discussed in Table 15.9.

15.1.1.19 Triazine herbicides

Some triazine herbicides are listed below:

- Atrazine (2-chloro-4-ethylamino-6-isopropylamino-1,3,5-triazine)
- Propazine
- Simazine (2-chloro-4,6-*bis*-ethylamino-1,3,5-triazine)
- Prometon
- Prometryne
- Atraton (2-ethylamino-4-isopropylamino-6-methoxy-1,3,5-triazine)
- Ametryne (2-ethylamino-4-isopropylamino-6-methylthio-1,3,5-triazine)
- Terbutryne
- Terbutylazine (4-tert-butylamino-2-chloro-5-ethylamino-1,3,5-triazine)
- GS26571 (2-amino-4-tert–butylamino-5-methoxy-1,3,5-triazine)
- GS30033 (2-amino-4-chloro-5-ethylamino-1,3,5-triazine)
- Terbumeton
- Secbumeton

Not unexpectedly, gas chromatography is the method of choice for the analysis of herbicides. McKone *et al.* [154] compared gas chromatographic methods for the determination of Atrazine (2-chloro-4-ethylamino-6-iso-propylamino-1,3,5-triazine), Ametryne (2-ethylamino-4-isopropylamino-6-methylthio-1,3,5-triazine) and Terbutryne in water. The herbicides were extracted from water with dichloromethane and the dried extracts were evaporated to dryness at a temperature below 35°C. By gas chromatography on a glass column (1m × 4mm) of 2% neopentyl glycol succinate on Chromosorb W (80–100 mesh) operated at 195°C and with a RbBr–tipped flame ionisation detector the three herbicides could be separated and 0.001ppm of each detected. This method was found to be superior to spectrophotometric and polarographic methods.

Purkayastha and Cochrane [155] compared electron capture and electrolytic conductivity detectors in the gas chromatographic determination of Prometon, Atraton (2-ethylamino-4-isopropylamino-6-methoxy-1,3,5-triazine), Propazine, Atrazine (2-chloro-4-ethylamino-6-isopropyl-amino-1,3,5-triazine), Prometryne, Simazine (2-chloro-4,6-*bis*-ethylamino-1,3,5-triazine) and Ametryne (2-ethylamino-4-isopropylamino-6-methyl-thio-1,3,5-triazine) in inland water samples. They found that the electrolytic conductivity detector seemed to have a wider application than a ⁶³Ni electron capture detector; use of the latter detector necessitated a clean-up stage for all the samples studied. The conductivity detector could be used in analysis of water without sample clean-up. Good recoveries of Atrazine added to water were obtained by extraction with dichloromethane.

Ramsteiner *et al.* [156] compared alkali flame ionisation, microcoulometric, flame photometric and electrolytic conductivity detectors for the determination of triazine herbicides in water. Methanol extracts were cleaned up on an alumina column and 12 herbicides were determined by gas chromatography with use of conventional columns containing 3% Carbowax 20m on 80–100 mesh Chromosorb G.

Hormann *et al.* [157] monitored various European rivers for levels of Atrazine, Terbumeton and dealkylated metabolites GS26571 (2-amino-4-tertbutylamino-5-methoxy-1,3,5-triazine) and GS30033 (2-amino-4-chloro-5-ethylamino-1,3,5-triazine). The compounds were extracted into dichloromethane and quantitated by gas chromatography with nitrogen-specific detection. Selected results were verified by gas chromatography with mass fragmentographic detection. The limit of detection was usually 0.4μg L⁻¹.

A gas chromatographic method has been issued by the US Environmental Protection Agency [158] for the determining of the microgram per litre level the following herbicides in water and waste water: Ametryne, Atraton, Atrazine, Prometon, Prometryne, Propazine, Secbumeton, Simazine and Tertbutylazine. The method describes an efficient sample extraction procedure and provides, through use of column chromatography, a method for the elimination of non-pesticide interferences and the preparation of pesticide mixtures. Identification is made by nitrogen-specific gas chromatographic separation, and measurement is accomplished by the use of an electrolytic conductivity detector or a nitrogen-specific detector.

Steinheimer and Brooks [159] developed a multi-residue method for the simultaneous determination of seven triazine herbicides in surface and ground water at a nominal detection limit of micrograms per litre. The technique uses solvent extraction, gas chromatographic separation and nitrogen-selective detection devices. Solid-phase extraction techniques using chromatographic grade silicas with chemically modified surfaces were examined as an alternative to liquid–liquid partition and evaluated using three natural water samples. Solid-phase extraction was found to provide rapid and efficient concentration with quantifiable recovery.

Jahda and Marha [160] investigated the isolation of s-triazines from water using continuous steam–distillation extraction prior to gas–liquid chromatography. Recoveries of seven triazine herbicides, Propazine, Terbutylazine, Atrazine, Prometryne, Terbutryne, Desmetryne and Simazine from water at pH values of 5, 7 and 9 are reported. Recovery rates were independent of pH but generally improved with increase in time of steam–distillation extraction from 1 to 3h. Low recovery rates were obtained for Simazine. Atrazine only gave good recovery rates after 3h steam–distillation extraction.

Lee and Stokker [161] have developed a multi-residue procedure for the quantitative determination of 11 triazines in non saline waters by a gas chromatographic method using a nitrogen–phosphorus detector. Ametryne, Atraton, Atrazine, Cyanazine, Prometron, Prometryne, Propazine, Simazine, Simetone and Simetryne were used. All of them could be successfully quantified on both the Ultrabond 20m and 3% OV–1 columns. Extraction was by methylene chloride and clean-up on Florasil. Recoveries of triazines at 10, 1.0 and 0.1µg L^{-1} were between 87 and 108% except for Simetone and Simetryne which were only 80% at 0.1µg L^{-1}. The method was validated with Ontario lake water at three experimental concentrations and with two other non saline waters at 1µg L^{-1}. The detection limit was 0.025µg L^{-1}.

Other applications of gas chromatography to the determination of triazine herbicides in non saline waters are discussed in Table 15.9.

15.1.1.20 Organophosphorus insecticides

Much of the development work on the determination of organophosphorus insecticides which has been carried out in recent years has hinged on the development of suitable detectors which are ultra-sensitive and which are specific for phosphorus in the presence of other elements such as carbon, hydrogen, oxygen, halogens, nitrogen and sulphur and indeed in some cases, can be used to determine compounds containing these other elements. Several types of organophosphorus insecticides also contain halogens, nitrogen, or sulphur. Electron capture, flame ionisation, flame photometric, microcoulometric, thermionic and electrolytic conductivity detectors have all been studied in this application.

The Karmen–Guiffrida detector (the so-called thermionic detector) has proved extremely useful. It is basically a flame ionisation detector with an alkali salt ring placed on the flame tip. Not only does this enhance the sensitivity but also the selectivity. This may be of the order of 10^4–10^5 for phosphorus-containing compounds as compared with the equivalent carbon compounds. The selectivity for halides and nitrogen is between 10^2 and 10^3 and 10^2 for sulphur and about 10 for arsenic. The responses for phosphorus increases with increasing hydrogen flow but some increase in the background current is also observed. A variety of alkali salts are used including potassium chloride, caesium bromide and rubidium sulphate. The type and shape of the tip and anode, together with the flow rates, are very much a matter of personal choice. De Loach and Hemphill [162,163] have discussed the design of a rubidium sulphate detector and with optimum conditions claim a sensitivity of 1pg. Often a charcoal column clean-up is used before injection to prevent the column becoming contaminated.

Brazhnikov et al. [164] developed a highly stable thermionic detector (the thermaerosol detector) to avoid the limitations imposed by modifications of existing detectors. The thermaerosol detector, which combines a conventional flame ionisation detector with a generator of an aerosol of alkali metal salt was applied to the analysis of organophosphorus insecticides. The detector avoids limitations of conventional thermionic detectors such as a considerable dependence of the sensitivity and reproducibility of the detector operation on the flow rate of hydrogen, air and carrier gas, rapid exhaustion of the alkali metal salt source and the difficulty of replacing one salt by another. The important advantages of this detector are the simplicity of its design, the possibility of rapid replacement of the salt without dismantling the detector and stability of its operation for a long period (several thousand hours) without the need to replace the salt reservoirs.

Kawahara [165] described a procedure for the determination of phosphorothioate insecticides including Parathion (O-(4-nitrophenyl phosphorothiate) and Parathion–methyl (dimethyl-p-nitrophenyl phosphorothioate). It consists of solvent extraction, clean-up by thin layer chromatography on silica gel G (0.25mm layer), and identification by gas chromatography on an aluminium column (1.2m × 0.6cm od) packed with equal portions of acid-washed Chromosorb P supporting 5% of DC 200 silicone oil, and unwashed Chromosorb W supporting 5% of Dow–11 silicone: the column is operated at 180°C, with argon–methane (9:1) as carrier gas (120mL min $^{-1}$) and electron capture detection. If the sample volume is sufficient, identification can be confirmed by infrared spectrometry.

Dale and Miles [166] showed that Abate (O,O,O',O'-tetramethyl-O, O'-(thiodi-p-phenylene) diphosphorothioate) and its sulphoxide were well separated from various other organophosphorus insecticides on a column of XE–60 on silanised Chromosorb W at 240°C, using nitrogen as carrier gas and a flame photometric and electron capture detector. Shafik [167] has described a method for the determination of this insecticide in water. Abate is first converted to its hydrolysis product, 4,4'-thiodiphenol, which is then silated by reaction with chlorotrimethylsilane and hexamethyldisilazane. Separation was achieved on an aluminium column (1.2m × 6mm od) containing 2.5% E–301 or 0.25% Epon 1001 on Chromosorb W (80–100 mesh) operated at 190°C with nitrogen as carrier gas (100mL min $^{-1}$) and a flame photometric detector equipped with a sulphur filter. Miller and Funes [168] used alkali flame gas chromatography to determine Abate. Separation was achieved on a column packed with 2.5% of E–301 plus 0.25% of Epon 1001 on Gas–Chrom W (AW–DMCS)HO, the column and detector being operated at 235°C. The column was conditioned by injecting 20ng standards for up to 200ng Abate. The recovery of 0.01 to 1ppm of added Abate to water was 97%, after extraction by the method of Dale and Miles [166].

A gas chromatographic procedure using electron capture detection has been described for the determination of Dursban (O,O-diethyl-O-(3,5,6-trichloro-2-pyridyl phosphorothioate) in water and silt [169]. In this method, water samples are extracted with dichloromethane, the extract is evaporated, and a solution of the residue is cleaned up on a column of silicic acid, Dursban being eluted with hexane. The eluate is evaporated to dryness under reduced pressure, and a solution of the residue in hexane is subjected to gas chromatography on a glass column (122 × 0.63cm od) packed with 5% of SE–30 on Anakrom ABS (80–90 mesh) operated at 200°C (or 215°C) with methane–argon (1:19) (60mL min $^{-1}$) as carrier gas, or, 5% SF–96 on Chromosorb W (80–100 mesh), operated at 215°C with nitrogen (90mL min $^{-1}$) as carrier gas. Dried silt samples, finely powdered, are blended with dichloromethane and Celite, the filtered extract is evaporated to dryness and a solution of the residue in hexane is extracted with acetonitrile. The concentrated solution plus added hexane is evaporated to dryness under reduced pressure, and Dursban in a solution of the residue in hexane is subjected to clean-up and gas chromatographed as for water samples. Down to 10^{-4}ppm of Dursban in water and down to 5×10^{-3} ppm in silt could be determined; average recoveries from water and silt were 92% and 83% respectively.

Askew *et al.* [138] have developed an early general method for the

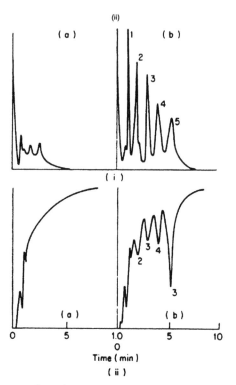

Fig. 15.5 Comparison of gas chromatogram obtained with (i) a phosphorus detector and (ii) an electron capture detector: (a) an extract equivalent to 1 L of Thames water and (b) the same extract fortified with pesticides. Peaks correspond to 5ng of each compound: 1, Demeton-S-methyl; 2, Dimethoate; 3, Pyrimithate; 4, Parathion; 5, Chlorfenvinphos
Source: Reproduced by permission from the Royal Society of Chemistry, London [138]

determination of organophosphorus insecticide residues and their metabolites in river waters and sewage effluents utilising gas chromatography. The organophosphorus pesticides vary greatly in their polarity, and the extent of their extraction from aqueous samples is markedly dependent on the nature of the solvent used. The following insecticides are exceptional in being retained by nuclear carbon: Azinophos–ethyl, Azinophos–methyl, Coumaphos, Dichlorvos, Maloxon, Menazon, Phosalone and Vamidothion. When these insecticides are encountered a clean-up on alumina [170] or magnesium oxide [171] would be preferable.

The general screening technique developed by Askew *et al.* [138] is described in full in Fig. 15.5. Stage (a) is applied as a general screening technique, and stage (b) is incorporated only when suspected pesticides are countered.

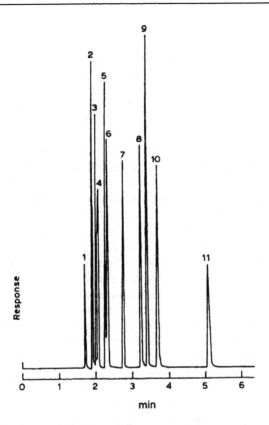

Fig. 15.6 Chromatogram of a standard mixture containing 11 organohalides. Stationary phase, SE–52; 33m × 0.3mm id, inj. temp. 200°C, column temp. 50°C, interface 250°C, detector temp. 250°C. Helium carrier gas flow rate 36ml s^{-1}, scaveriger gas flow 30mL min^{-1}. Split ratio 1:20. 1 = CH_2Cl_2, 2 = $CHCl_3$, 3 = CH_3CCl_3, 4 = CCl_4, 5 = $CHClCCl_2$, 6 = $CHBrCl_2$, 7 = $CBrCl_3$, 8 = $CHBr_2Cl$, 9 = CCl_2CCl_2, 10 = $CHCl_2$, 11 = $CHBr_3$
Source: Reproduced by permission from the Royal Society of Chemistry, London [138]

Fig. 15.6 shows gas chromatographs obtained by Askew *et al.* [138] on solvent extracts of River Thames water unspiked and spiked.

Daughton *et al.* [172,173] have described a procedure for isolating and determining, in large volumes of aqueous media, ionic diethyl phosphate, diethyl thiophosphate, dimethyl phosphate and dimethyl thiophosphate suitable for application in environmental monitoring. Procedures for eliminating interference due to inorganic phosphate are also discussed. In this approach the aqueous sample containing ionic dialkyl phosphates and thiophosphates are passed down a column of Amberlite XAD–4 resin. Recoveries for diethylphosphoric acid and diethylthiophosphoric acid at 0.01–0.1ppm in 500–400mL of aqueous media were 100 and 85%, respectively; recoveries for dimethylphosphoric acid and dimethylthio-phosphoric acid at 0.1ppm in 500mL of aqueous media were 50 and 97% respectively. Following a clean-up procedure the effluents are gas chrom-atographed using a gas chromatograph equipped with a phosphorus thermionic detector and a glass column (1.8m × 2mm id), packed with equal parts of 15% QF–1 and 10% DC–200 on Gas–Chrom Q (80–100 mesh) at 140°C; injector, 200°C; detector, 250°C; nitrogen, 22mL min^{-1}; air 23mL min^{-1}; hydrogen, 55mL min^{-1}.

Thompson et al. [113] have developed a multi-class, multi-residue gas chromatographic method for the determination of insecticides (organo-phosphorus, organochlorine, carbamate types) and herbicides in water samples. The compounds are extracted from water with methylene chloride, and the extract is concentrated by an evaporative technique utilising reduced pressure and low temperature. Compounds are segregated into groups using a column of partially deactivated silica gel and sequential elution with four different solvent systems. Carbamate residues, converted to their 2,4-dinitrophenyl ether derivatives, are gas chromatographed via electron capture detection as the parent compounds of the organohalogen compounds. Organophosphorus compounds are determined by gas chromatography using a flame photometric detector.

Thompson et al. [113] emphasise that even the use of two dissimilar gas chromatographic columns does not ensure irrefutable compound identification. For example, if the retention characteristics of a given peak obtained from two dissimilar columns suggest the possibility of the presence of a compound which appears wholly out of place in a specific sample, further confirmation is clearly indicated by such techniques as specific detectors, coulometry, p values, or gas chromatography–mass spectrometry or thin layer chromatography.

Mallet et al. [174] used an automated gas chromatographic system which consisted of a gas chromatograph mounted with an automatic sample interfaced to an integrator. A Melpar flame photometric detector (phosphorus mode) was connected with the flame gas inlets in the reverse configuration to prevent solvent flame-out. The detector was maintained at 185°C and flame gases were optimised with flow rates (mL min^{-1}) as follows: hydrogen, 80; oxygen, 10; air, 20. A 2.8m × 4.0mm id U-shaped glass column packed with 4% (w/w) OV–101 and 6% (w/w) OV–210 on Chromosorb W AW DCMS, 80–100 mesh was used. Nitrogen was used as carrier gas at a flow rate of 70mL min^{-1}. A column temperature of 195°C sufficiently resolved the Fenitrooxon from its parent compound. The injection port temperature was set at 225°C. The water sample was passed through an XAD–2 column, which was subsequently eluted with ethyl acetate. The ethyl acetate extract was examined by gas chromatography.

Verwej et al. [175] have described a procedure for the determination of PH_3-containing insecticides in surface water. In this procedure the insecticide is hydrolysed to methylphosphonic acid, and the acid is concentrated by anion exchange and converted to the dimethyl ester. After clean-up on a microsilica gel column the ester is analysed by gas chromatography using a thermionic phosphorus-specific detector. Detection limit is 1nmol L $^{-1}$.

Other applications of gas chromatography to the determination of organophosphorus insecticides in non saline waters are reviewed in Table 15.9.

15.1.1.21 Carbamate insecticides

Crosby and Bowers [176] have described a method for determination of Carbaryl (Sevin) (1-naphthyl-N-methylcarbamate) in which the sample (0.5g) is heated under reflux for 1h with 2-chloro-α,α,α-5–nitroluene (1mmol) or 4-chloro-α,α,α-trifluoro-3,3-dinitroluene (0.7mmol), acetone (20mL) and 0.1M $Na_2B_4O_7$) (20mL) to convert its amine moiety into an N-substituted nitro-(trifluoromethyl)aniline derivative. The derivative is subjected to gas chromatography on a stainless steel column 3m × 3mm od packed with 3% SE–30 or 3% FFAP on HMDS treated Chromosorb G at a temperature from 150–250°C, with nitrogen (30mL min $^{-1}$) as carrier gas and electron capture or flame ionisation detection. Down to 50pg of the more volatile or 200pg of the less volatile derivatives can be determined by electron capture detection. Methods also described are the determination of Molinate (5-ethyl hexamethylenecarbamate) in water.

Thompson et al. [113] have devised a multi-residue scheme of analyses based on silica gel chromatography followed by gas chromatography for the analyses of mixtures of organochlorine, organophosphorus and carbamate types of insecticides. The relevant work on seven carbamate insecticides is discussed below. The mixture of the three types of compounds is fractionated into groups on a partially deactivated silica gel column with three sequential elutions. Final determinations were made by gas chromatography using the carbon capture detector for the halogenated compounds and derivatised carbamates, and the flame photometric detector for the organophosphorus compounds. Derivatisation of the carbamate fraction is carried out with 1-fluoro-2,4,-dinitrobenzene as follows. To the tubes containing the 0.1mL concentrates of the carbamate fractions is added 0.5mL of 1-fluoro-2,4-dinitrobenzene (1% in acetone) and 5mL of sodium borate buffer solution ($Na_2B_4O_7.10H_2O$, 1M solution at pH0.4). The reagents are added to an empty tube to serve as a reagent blank. The tubes are tightly capped and heated at 70°C for 1h in a water bath. The tubes are cooled to room temperature, and 5mL of hexane added to each tube. The tubes are shaken vigorously for 3min, either manually or on a wrist action shaker. The layers are allowed to separate and 4mL of the hexane (upper) layer are transferred carefully to a tube and stoppered tightly.

Table 15.8 Substituted urea herbicides with different substituents

Name	Substituents			
	X_1	X_2	Y_1	Y_2
Buturon	H	4-Chlorophenyl	CH_3	$CH(CH_3)C{\equiv}CH$
Chlorobromuron	H	3-Chloro-4-bromophenyl	CH_3	OCH_3
Chlorooxuron	H	4-(Chlorophenoxy)phenyl	CH_3	CH_3
Chlorotoluron	H	3-Chlorotoluyl	CH_3	CH_3
Diuron	H	3,4-dichlorophenyl	CH_3	CH_3
Fenuron	H	Phenyl	CH_3	CH_3
Isoproturon	H	Cumenyl	CH_3	CH_3
Linuron	H	3,4-Dichlorophenyl	CH_3	OCH_3
Monolinuron	H	4-Chlorophenyl	CH_3	OCH_3
Monouron	H	4-Chlorophenyl	CH_3	CH_3
Metabenzthiozuron	CH_3	2-Benzothiazolyl	CH_3	H
Metoxuron	H	3-Chloro-6-methoxyphenyl	CH_3	CH_3
Neburon	H	3,4-Dichlorophenyl	CH_3	CH_3
Siduron	H	2-Methylcyclohexyl-3	CH_3	C_6H_5

Key:
Buturon, 3-(4-chlorophenyl)-methyl-1-(1-methylprop-2-ynyl)-urea
Chlorobromuron, 3-(4-bromo-3-chlorophenyl)-1-methoxy-1-methylurea
Chlorooxuron, 3-(4,4-chlorophenoxy)phenyl)-1, 1-dimethylurea
Chlorotoluron, 3-(3-chlorotoluyl)-1, 1-dimethylurea
Diuron, 3-(3,4-dichlorophenyl-1)-1, 1-dimethylurea
Fenuron, 1,1-dimethyl-3-phenylurea
Isoproturon, 3-(cumethyl)-1, 1-dimethylurea
Linuron, 3-(3,4-dichlorophenyl)-1-methoxy-1-methylurea
Metabenzthiozuron, 3-(2-benzothiazolyl)-1,1-dimethylurea
Metoxuron, 3-(3-chloro-6-methoxyphenyl)-1, 1-dimethylurea
Monolinuron, 3-(4-chlorophenyl)-1-methoxy-1-methylurea
Monouron, 3-(4-chlorophenyl)-1, 1-dimethylurea
Neburon, 1-butyl-3-(dichlorophenyl)-1-methylurea
Siduron, 1-(2-methylcyclohexyl-3)-phenylurea

Source: Own files

Other applications of gas chromatography to the determination of carbamate insecticides in non saline waters are discussed in Table 15.9.

15.1.1.22 Substituted urea herbicides

Substituted urea herbicides with different substitutes are shown in Table 15.8.

Gas chromatography of phenylurea herbicides is difficult because of their ease of thermal decomposition. Procedures have been reported in which careful control of conditions allows these compounds to be gas chromatographed intact [177,178]. Alternatively, the phenylurea herbicides can be hydrolysed to the corresponding substituted anilines (see below) which are then determined either by gas chromatography directly [179], or as derivatives [180], or colorimetrically after coupling with a suitable chromophore [181]. For example, monolinuron is converted to p-chloroaniline, formation of the bromide derivative produces the following:

Herbicide	Structure		Aniline		Derivative

In an electron capture method for estimating Diuron (3-(3,4-dichloro-phenyl-1)-1,–dimethylurea) in surface waters, McKone and Hance [182,183] extract the water sample (100mL) with dichloromethane (2 × 25mL) and the lower layers are combined and washed with water (5mL). After filtering through cotton wool, the organic solvent is evaporated under reduced pressure at 35°C. Saturated aqueous sodium chloride is added to the residue, the mixture is shaken, 2,2,4-trimethyl–pentane (5mL) is added, and the mixture is shaken again. An aliquot of the organic layer is then subjected to gas chromatography in a stainless steel column packed with 5% of E301 (methyl silicone) on Gas Chrom Q (60–80 mesh). Recoveries from controls and from pond, canal and river waters containing 0.001–1ppm of Diuron were about 94%, the coefficient of variation at the higher levels being about 70%.

In a method described by Rosales [184] 1-methoxy-1-methyl-3-phenylurea herbicide is hydrolysed by phosphoric acid to give aniline and N-methodymethamine, which together with certain impurities in the commercial product are titrated with sodium nitrite solution. Aniline and certain byproducts are also determined separately by gas chromatography on a glass column (2m × 4mm) containing 10% of silicone oil OV–17 on Chromosorb Q (80–100 mesh) temperature programmed from 100 to 200°C at 4.5°C min^{-1}, with thermal conductivity detection and helium carrier gas.

Cohen and Wheals [185] used a gas chromatograph equipped with an electron capture detector to determine 10 substituted urea and carbamate herbicides in river water, soil and plant materials in amounts down to 0.001–0.05ppm. The methods are applicable to those urea and carbamate herbicides that can be hydrolysed to yield an aromatic amine. A solution of the herbicide is first spotted on to a silica gel G plate together with herbicide standards (5–10µg) and developed with chloroform or hexane-acetone (5:1). The plate containing the separated herbicide or the free amines is sprayed with 1-fluoro-1,4-dinitrobenzene (4% in acetone) and heated at 190°C for 40min to produce the 2,4-dinitrophenyl derivative of the herbicide amine moiety. Acetone extracts of the areas of interest are subjected to gas chromatography on a column of 1% of XE–60 and 0.1% of Epikote 1001 on Chromosorb C (AE–DCMS) (60–80 mesh) at 215°C.

Kongovi and Grochowski [186] discussed the problems arising during the analysis of pesticides and herbicides by gas chromatography and electron capture detection. During a routine run of pesticide standards (Lindane, Endrin and Methoxychlor) five peaks were obtained, and this led to a study of the Endrin molecule contaminants, and more specifically, to the decomposition of the Endrin molecule in relation to temperature (220–235°C) and nitrogen flow rate. Conclusions were eventually reached that the retention periods of compounds generally provide good criteria for identification, but that this was not always the case, particularly with esters of low-molecular-weight acids. Also the retention period alone does not serve as an absolute identifying criterion; other confirmation, eg mass spectroscopy, is required.

Lee *et al.* [187] developed a multi-residue method with a low detection limit for 10 commonly used acid herbicides in non saline waters. The herbicides were Dicamba, MCPA, 2,4–DP, 2,3,6–TBA, 2,4–D, Silvex, 2,4,5–T, MCPB, 2,4,5–DB and Picloram. The method used solvent extraction and the formation of pentafluorobenzyl esters. The derivatives were quantified by capillary column gas chromatography with electron capture detection. The detection limit was 0.05µg L $^{-1}$. Recoveries of herbicides from spiked Ontario lake water (0.5–1.0µg L $^{-1}$) were 73–108% except for Picloram recovery which was 59% at 0.1µg L $^{-1}$.

Other applications of gas chromatography to the determination of substituted urea herbicides in non saline water are discussed in Table 15.9.

15.1.1.23 Phenoxyacetic acid herbicides

- 2,4-D ester and sodium salts (2,4-dichlorophenoxyacetic acid)
- 2,4-DP (Dichlorprop; 2,4-dichlorophenoxypropionic acid)
- 2,4-DB (4-(2-dichlorophenoxy)–butyric acid)
- MCPA (4-chloro-2-methylphenoxyacetic acid)
- MCPB (4-(4-chloro-2-methylphenoxy)–butyric acid)
- Silvex (2(2,4,5-trichlorophenoxy)–propionic acid)
- MCPP (Mecoprop; mixture of Mecoprop and 2-(2-chloro-4-methyl-phenoxy)–propionic acid)
- 2,4,5-T (2,4,5-trichlorophenoxyacetic acid)
- Dicamba
- Trifluralin (2-methoxy-3,6-dichlorobenzoic acid)
- Methoxychlor
- Fenoprop

Devine *et al.* [188] adjust the water sample (1L) to pH2 with hydrochloric acid and extract it with benzene (100, 50 and 50mL). The extract is dried over sodium sulphate, concentrated to 0.1mL and methylated by the addition of diazomethane in ethyl ether (1mL). After 10min, the volume is reduced to about 0.1mL, acetone is added and an aliquot is analysed by gas chromatography on one of three columns: (1) 5% SE–30 on 60–80 mesh Chromosorb W at 175°C, (2) 2% QF–1 on 90–100 mesh Anakron ABS at 175°C or (3) 20% Carbowax 20m on 60–80 mesh Chromosorb W at 220°C. In each instance nitrogen is the carrier gas and detection is by electron capture. The minimum detectable amount of pesticide in water was 2 parts per 10^9 for MCPA (4-chloro-2-methyl–phenoxyacetic acid) and 0.01–0.05 parts per 10^9 for 2,4–D (2,4-dichloro–phenoxyacetic acid) and its esters, 2,4,5–T (2,4,5–trichlorophenoxyacetic acid), Dicamba, Trifluralin (2-methoxy-3,6-dichlorobenzoic acid), and Fenoprop. Recoveries were 50–60% for MCPA and Dicamba and 80–95% for the other compounds.

Larose and Chau [189] state that owing to the similar retention times of several common phenoxyacetic acid type herbicides; the alkyl esters are subject to incorrect identification if several herbicides are present. Also, the sensitivity obtainable by means of electron capture detection of the alkyl esters by some herbicides, such as MCPA and MCPB is very poor and therefore the method is generally not suitable for the determination of these compounds in water. In addition, the methyl ester of MCPA has a very short retention time close to the solvent front and is prone to interference from sample coextractives, which usually appear in this region. In fact the MCPA methyl ester often cannot be detected even at higher levels because of overlapping with coextraction peaks when the same gas chromatographic parameters as for the determination of organochlorine pesticides are used. Hence other derivatives have been considered.

Agemian and Chau [190] have reported a method for determining low levels of 4-chloro-2-methylphenoxyacetic acid and 4-(4-chloro-2-methyl-phenoxy)–butyric acid in non saline and waste waters by derivatisation with pentafluorobenzyl bromide. The increased sensitivity of the penta-fluorobenzyl esters of these two herbicides over the 2-chloroethyl methyl esters as well as their longer retention times make pentafluoro-benzyl bromide the preferred reagent.

These workers [190] used a gas–liquid chromatograph equipped with a nickel detector, a 1.8m × 6mm id coiled glass column and an automatic sampler connected to a computing integrator for data processing. The column used was 3.5% w/w OV–101 and 5.5% w/w OV–210 on 80–100 mesh Chromosorb W, acid washed and treated with dimethylchlorosilane. The operating conditions were as follows: injector temperature 220°C, column temperature 220°C, detector temperature 300°C, carrier gas argon–methane (9:1) at a flow rate of 60mL min^{-1}. Agemian and Chau [190] found that the few organochlorine pesticides that are eluted in the same fraction as the pentafluorobenzyl derivatives of the phenoxyacetic acid herbicides do not interfere because they have distinct retention times. Organophosphorus pesticides do not interfere. Twenty-four of the most widely used phenols either are eluted with the PCBs and organochlorine pesticides or have distinct retention times from those of the pentafluoro-benzyl esters of the two herbicides. The whole of the above procedure, at a level of 0.5µg L^{-1} MCPA in 1L of distilled water, gave an average recovery of 75–80% with a coefficient of variation between 9 and 15%.

Agemian and Chau [192] using the above method compared the pentafluorobenzyl bromide reagent for forming esters and the boron trichloride-2-chloroethanol and dicyclohexylcarbondiimide-2-chloro-ethanol reagents for forming 2–chloroethyl esters of phenoxyacetic acid type herbicides, coupled with a complete solvent extraction system to obtain multi-residue methods for determining these compounds in non saline waters at sub-microgram per litre levels.

The nine herbicides studied by these workers were Dicamba, MCPA, MCPB, 2,4–DB, Picloram (4-amino-3,5,6-trichloropicolinic acid), 2,4–D, 2,4,5–T, Silvex and 2,4–DP.

Other applications of gas chromatography to the determination of phen-oxyacetic acid herbicides in non saline waters are discussed in Table 15.9.

15.1.1.24 Miscellaneous herbicides/pesticides

Paraquat (1, 1'-dimethyl-4,4-bipyridium chloride and Diquat (1, 1'-ethylene-2,2-bipyridylium bromide)

Soderquist and Crosby [193] added to the water sample (100mL), sulphuric acid (3mL) and platinum dioxide (25mg) and hydrogen was bubbled through for 1h, whereby Paraquat is converted into 1,1'-dimethyl-4,4'-bipiperidyl. This is extracted with dichloromethane (3 × 50mL) in the presence of 11ml of 50% sodium hydroxide solution and the combined extract is treated with 0.01N hydrochloric acid (4mL) and evaporated in a rotary evaporator at 50–55°C. The aqueous residue is transferred with 1mL of 0.01N hydrochloric acid to a 15mL screw-cap tube and shaken with 50% sodium hydroxide solution (0.5mL) and carbon disulphide (1mL). Aliquots of the carbon disulphide phase (1–10µL) are injected on to a glass column (20m × 3mm) packed with 10% Triton X–100 and 1% potassium hydroxide on AW–DCMS Chromosorb G (70–80 mesh), and operated at 150°C with nitrogen as the carrier gas (30–40mL min^{-1}) and flame ionisation detection. The calibration graph is rectilinear for up to 1ppm of Paraquat. The limit of detection is 0.1ppm but recovery is only 36–43%, although reproducible.

To determine Paraquat in agricultural run-off water Payne [194] separated the sediment from the sample (2L) by adding calcium chloride to aid flocculation, leaving the mixture overnight in a refrigerator for the sediment to settle. A 1L aliquot of the filtrate is extracted with dichloromethane. The dichloromethane extracts are concentrated by evaporation and the Trifluralin and Diphenamid are determined by direct injection, without further purification on to a glass column 1.8 × 6mm od packed with 10% DC 200 on Gas–Chrom Q and operated at 220°C with helium as carrier gas (100mL min $^{-1}$) and a Coulson electrolytic–conductivity detector (N mode). Paraquat is determined in the filtrate by a modification of a conventional colorimetric method. Recoveries of the three substances were between 82 and 95% from water.

Cannard and Criddle [195] have described a rapid pyrolysis–gas chromatography method for the simultaneous determination of Paraquat and Diquat in pond and river waters in amounts down to 0.001ppm.

Piclorom (4-amino-3,5,6-trichloropicolinic acid)

Abbott *et al.* [196] described a pyrolysis unit for the determination of Picloram and other herbicides in water. The determination is effected by electron capture–gas chromatography following thermal decarboxylation of the herbicide. Hall *et al.* [197] reported further on this method.

Dicamba (2-methoxy-3,6-dichlorobenzoic acid)

This herbicide is discussed in methods for its codetermination with phenoxyacetic acid herbicides. Norris and Montgomery [198] described a procedure for the determination of traces of Dicamba and 2,4–D in streams after forest spraying. Dicamba and its metabolites (3,6-dichlorosalicylic acid and 5-hydroxydicamba) were determined gas chromatographically.

The application of gas chromatography to the determination of other miscellaneous insecticides is reviewed in Table 15.9.

15.1.1.25 Growth regulators

Getzendaner's method [199] for determining Dalapon (2,2'dichloropropionic acids is applicable to plant tissues and body fluid and doubtlessly to water samples. The sample was extracted with ethyl ether and the residue was analysed by gas chromatography on a glass column (1.2m × 2mm) of 4% LAC–2R plus 0.5% of phosphoric acid on Gas Chrom S (60–80 mesh) at 100°C with nitrogen as carrier gas (85mL min $^{-1}$) and electron capture detection. Recoveries of about 90% were obtained for 10ppm of the herbicide.

The Frank and Demint [200] method is directly applicable to water samples. After addition of sodium chloride (340g L $^{-1}$) and aqueous hydrochloric acid (1:1) to bring the pH to 1, the sample was extracted with ethyl ether and the organic layer was then extracted with 0.1M sodium bicarbonate (saturated with sodium chloride and adjusted with sodium hydroxide to pH8). The aqueous solution adjusted to pH1 with hydrochloric acid was extracted with ether and after evaporation of the ether to a small volume, Dalapon was esterified at room temperature by addition of diazomethane (0.5% solution in ether) and then applied to a stainless steel column (1.5m × 3mm) packed with Chromosorb P (60–80 mesh) pretreated with hexamethyldisilazane and then coated with 10% FFAP. The column was operated at 140°C, with nitrogen carrier gas (30mL min^{-1}) and electron capture detection. The recovery of Dalapon ranged from 91 to 100%; the limit of detection was 0.1ng. Herbicides of the phenoxyacetic acid type did not interfere; trichloroacetic acid could be determined simultaneously with Dalapon.

In a later method [201] for the determination of Dalapon in non saline waters and plant tissues the herbicide is first esterified with 3-phenol-propanol-1 then determined by electron capture–gas chromatography. As little as 0.001mg Dalapon per litre of water can be determined by this method. These workers used a gas chromatograph with two ^{63}Ni electron capture detectors. The detectors were operated in the pulse mode at 50V. Two columns, both glass, were used to determine the ester. Column A was packed with 3% OV–1 and 2.7% OV–210 on Gas Chrom Q (80–100 mesh), Column B with 1.8% OV–1 and 2.7% OV–210 on Gas Chrom Q (80–100 mesh). The carrier gas was nitrogen, flow rate 70mL min^{-1} on both columns. The temperatures of the column oven, injectors and detectors were 160, 205 and 275°C respectively. Recoveries of between 94 and 103% were obtained. The response of the detector was linear up to nanogram amounts of the 3–phenylpropyl ester of Dalapon injected. As little as 0.001mg Dalapon per litre non saline water can be determined by this method.

15.1.1.26 Other applications

As well as those mentioned above in sections 15.1.1.1 to 15.1.1.25 gas chromatography has been applied to the determination of various other organic compounds in non saline waters including: glycols, alcohols, aldehydes, ketones, esters, humic and fulvic acids, non ionic surfactants, carbohydrates, aromatic halogen compounds, acrylamide, aromatic amines, nitrophenols, aminophenols, acrylonitrile, ethylene diamine tetracetals, diazo compounds, azarenes, dioxins, alkyl and aryl phosphates, growth regulators, volatile organic compounds, mestranol, squoxin, coprostanol, glyphosphate and pyrethins.

These are reviewed in Table 15.9.

Table 15.9 Gas chromatography of organics in non saline water

Organic	Extraction solvent from water	GLC column details	Detector	Comments	LD	Ref
Oil	Amberlite XAD4	–	–	Biodegradation study, comparison of GLC, IR, UV and mass spectrometry	–	[191, 202]
Gasoline	Stripping pre-concentration	–	–	95–100% recovery of gasolines	10–500 ppb	[203]
Jet fuel, JP4, diesel fuel	–	Capillary column GLC	–	–	µg L^{-1}	[204]
Naphtha	–	–	–	Headspace GLC	–	[205]
Total petroleum hydrocarbons	Freon	–	Flame ionisation	–	–	[206]
Aromatic hydrocarbons	–	–	–	–	–	[207]
Benzene, 1,4 dimethyl napthalene	–	–	Flame ionisation	Steam chromatography	10µg L^{-1}	[208]
Aromatic hydrocarbons	–	–	–	Capillary column GLC	–	[209]
Dimethyl benzene	–	Apiezon	–	Preconcentration on Apiezon L coated nickel wire then Curie point pyrolysis–GC	–	[210]
Polyaromatic hydrocarbons	Dimethyl sulphoxide	–	–	Partition coefficients determined	–	[211]
Polyaromatic hydrocarbons	Methyl chloride, n-pentane, ethyl ether, acetone, acetonitrile	–	–	7 PAHs studied	–	[212]

322

Table 15.9 continued

Organic	Extraction solvent from water	GLC column details	Detector	Comments	LD	Ref
Polyaromatic hydrocarbons and aliphatic hydrocarbons	—	—	—	HPLC used to extract these from water, then GLC of extracts from polychlorobiphenyl	—	[213]
Pairs of polyaromatic hydrocarbons eg benzo(a)pyrene and benzo(e)pyrene	—	—	—	—	—	[214]
Polyaromatic hydrocarbons	—	—	—	—	—	[35,215]
Polyaromatic hydrocarbons	—	—	Fluorescence	—	ng	[216–219]
Polyaromatic hydrocarbons	—	—	—	Study of effect of suspended solids on extraction efficiency of PAH from water and sampling techniques	<0.1 µg L^{-1}	[35]
Polyaromatic hydrocarbons	Dimethyl sulphoxide	—	—	—	—	[220, 221]
Methanol, glycols	—	—	—	—	—	[222] [223]
Mono, di and tri-ethylene glycols	Mixtures of water and dioxane or acetone	Nitrile– siloxane rubber	Flame ionisation	Solvent extract acetylated in presence of BF_4	2mg L^{-1}	[224]
Benzoic, oleic linoleic and undecanoic acids	Chloroform	—	Electron capture	Acids converted to pentafluoro benzyl esters	—	[225, 226]

Table 15.9 continued

Organic	Extraction solvent from water	GLC column details	Detector	Comments	LD	Ref
Oxalic, malonic, succinic, malic, glutaric, adipic, fumaric, aconitic, lactic, tarturic, citric, pyruvic and gallic acids	Ethylether: butanol (1:1) and isobutyl alcohol:ethyl acetate	–	–	–	–	[227]
Normal and iso carboxylic acids	3% FFAP on Chromosorb 101 at 180°C	–	Dual flame ionisation	–	–	[228]
Organic acids	Anion exchange chromatography	–	–	Acids converted to methyl derivatives then GLC	–	[229]
Acrylic acid	Tri-n-phosphine oxide	–	Electron capture	Acid converted to penta-fluorobenzyl ester	–	[230]
Acetic acid	–	–	–	–	–	[231]
Free fatty acids	Chloroform	–	–	–	–	[232]
Malonic, hexabenzene carboxylic, octanoic and octane dioic acids	–	–	–	Combination of isotype dilution gas chromatography and Fourier transform infrared spectroscopy	–	[233]
Sodium mono-fluorocetate	–	–	–	–	$0.6\mu g\ L^{-1}$	[234]
Aldehydes and organic acids	–	–	–	Aldehydes and ketones determined directly, oxoacids converted to methyl esters with diazomethane	–	[235]

Table 15.9 continued

Organic	Extraction solvent from water	GLC column details	Detector	Comments	LD	Ref
Acetone, methyl ethyl ketone methyl isobutyl ketone, tetrahydrofuran	–	–	–	Direct injection GC	–	[236]
Formaldehyde, acetaldehyde	Gas solid chromatography	–	N.P. detection (for cyano-hydrins)	Aldehydes converted to cyano-hydrins with hydrogen cyanide	–	[237]
Monosaccharides	–	–	–	Monosaccharides converted to trimethyl silyl derivatives	–	[238, 239]
Acrolein	–	Electron capture	–	Acrolein brominated to acrolein-o-methyloxime	0.4μg L^{-1}	[240]
Phenols	–	–	–	Converted to pentafluoro-benzyl esters prior to GLC	–	[241]
Phenols	–	Acid washed graphitised carbon black, modified with trimesic acid and PEG 20M	–	Polar volatile analysis	–	[242]
Phenols	–		–	–	–	[243]
Phenols	–	Electron capture	–	Phenols first brominated to tribromophenol	–	[244–245]
2-methyl phenol, phenolic disinfectants, phenolic priority pollutants and 2,5 xylenol	–	–	–	–	–	[246–248]

Table 15.9 continued

Organic	Extraction solvent from water	GLC column details	Detector	Comments	LD	Ref
Phenols, hydroxy-phenols	None or various solvents	–	–	–	–	[39,65, 241,245, 250–274]
Phthalate esters	–	Flame ionisation	–	–	20µg L^{-1}	[275]
Fulvic and humic acids (bromination and chlorination products of)	–	Microwave emission	–	–	–	[276]
Carbon tetrachloride, haloforms, halomethanes	–	–	–	–	–	[277]
Trichloroethylene	–	–	–	Hollow fibre membrane directly interfaced with GLC	1ppb	[278]
Chloroparaffins	–	–	Micro-coulometric	–	–	[279–281]
Halogenated organic compounds	–	–	Atomspheric pressure induced helium micro-wave induced plasma emission spectrometry	Uses a heated discharge tube for pyrolysis	–	[282]
Iodinated aliphatic compounds MeI, EtI	–	Electron capture (negative ion hydration and photodetachment)	–	–	–	[283]

Table 15.9 continued

Organic	Extraction solvent from water	GLC column details	Detector	Comments	LD	Ref
Trihalomethanes, chloroethanes, dichloroethanes	Adsorption on XAD–4 resin and elution with ethanol	–	–	Study of adsorption of chlorocompounds on XAD4 resin	–	[284, 285]
Trihaloforms	Adsorption on XAD–2 resin and elution with pyridine	–	–	–	–	[286]
Ethylene dibromide	Extraction with XAD–4 resin	Electron capture	–	–	$<1\,\mu g\,L^{-1}$	[287]
$CF_2\,Br_2$, $CH_2\,Br_2$	–	Modulated electron capture	–	–	–	[288]
Chlorinated benzenes	Macroreticular resin or pentane extraction	Coated open tubular glass capillary column	^{60}Ni electron capture	–	$1ng\,L^{-1}$ (pentane extraction) dichlorobenzene $0.01\,mg\,L^{-1}$ (hexachlorobenzene extraction)	[289]
Chlorinated humic acid products	–	–	–	–	–	[290]
Hexachlorobenzene	–	–	Electron capture	Hexachlorobenzene partially dechlorinated in carrier gas doped with hydrogen, over a nickel catalyst, then passed to separation column	–	[291]

Table 15.9 continued

Organic	Extraction solvent from water	GLC column details	Detector	Comments	LD	Ref
Volatile haloaromatics	–	–	–	–	20pg p-di bromobenzene	[292, 293]
2,4 dichlorotoluene	.	–	–	–	sub ppb	[294]
Chlorobenzenes	–	–	–	–	sub ppb	[295]
Pentachlorophenol and other chlorophenols	–	–	–	Recoveries 80–95%	0.2ppb	[296]
Chlorophenols	–	Capillary column coated with SE–30	–	Comparison with other capillary columns (eg OV–1)	–	[297]
Chlorophenols	–	–	–	Chlorophenols derivitivised with acetic anhydride or pentafluorobenzoyl chloride	–	[298]
Chlorinated phenols and catechols	–	SE–30 quartz capillary column	–	–	–	[299]
Pentachlorophenol	–	–	Electron capture	Pentachlorophenol converted to trimethyl silyl ether	0.01 µg L^{-1}	[300]
Pentachlorophenol	Adsorption on ion exchange column then extracted with acidified methanol	–	Flame ionisation	–	0.001– 0.1µg L^{-1}	[65]
Pentachlorophenol	–	–	Flame ionisation	–	–	[301]
Pentachlorophenol	n-hexane	–	Electron capture	Pentachlorophenol acetylated with acetic anhydride, then n-hexane extraction, then GLC	0.1µg L^{-1}	[302]

Table 15.9 continued

Organic	Extraction solvent from water	GLC column details	Detector	Comments	LD	Ref
Pentachlorophenol	–	–	–	Pentachlorophenol brominated then GLC	1 µg L^{-1}	[303]
Polychlorobiphenyls Arochlor 1242, 2143, 1252, 1260	–	–	–	Simca pattern recognition analysis performed	–	[304]
27 polychlorobiphenyl isomers	–	Open tubular column SE-30 at 190°C	Flame ionisation	Preparative gas chromatography	–	[305]
Polychlorobiphenyls 2',2',3,3'6,6' hexa-chlorobiphenyl	–	–	–	Identification	–	[306]
Polychlorobiphenyls	–	–	Electron capture	–	–	[307–309]
Polychlorobiphenyls	–	–	Flame ionisation	–	–	[310]
Polychlorobiphenyls	–	–	–	Separation on column of pyrolytically deposited carbon	–	[311]
Polychlorobiphenyls	–	–	–	Mechanism of transport of polychlorobiphenyl in rivers	–	[312]
Polychlorobiphenyls	–	–	–	Selective analysis of penta-chlorobiphenyl isomers	–	[313]
Polychlorobiphenyls	–	–	–	Computer-assisted production of GLC retention indexes of polychlorobiphenyls	–	[314]
Polychlorobiphenyls	–	–	–	Prediction of GLC retention characteristics of polychloro-biphenyls	–	[315]

329

Table 15.9 continued

Organic	Extraction solvent from water	GLC column details	Detector	Comments	LD	Ref
Polychlorobiphenyls	–	Silica gel	–	–	–	[316, 317]
Polybromobiphenyls	–	–	–	UV irradiation, then GLC of irradiation products	–	[318–321]
Aniline, toluidenes, monochloroanilines	Iso octane	Capillary column	Electron capture	Extracted amines brominated	5–15ng L^{-1}	[322]
Amino compounds	–	–	–	Amino compounds acetylated in presence of BF$_3$	–	[323]
Acrylamide	–	–	–	–	0.1μg L^{-1}	[324, 325]
Acrylamide	–	–	–	–	–	[71–73]
Acrylamide	Ethyl acetate	–	Electron capture	Acrylamide brominated	–	[326]
Aromatic amines	Toluene	–	–	Amines converted to amides by reaction with pentafluorobenzoyl chloride in 1% aqueous sodium bicarbonate, then extraction with toluene, then GLC	20–50ppt	[327]
o-,m,p-aminophenols	–	–	–	–	–	[328]
Aminophenols	–	–	–	–	0.1mole	[329]
Nitrophenol	Adjust to pH11 extract with benzene diethyl ether (4:1)	5% DC–200 on Gas-Chrom Q or 1.5% OV–17 and 1.95% QF–1 on Chrom-V at 140°C	Electron capture	Solvent extraction then phenol derivativised with hexamethyl disilazine, then GLC	0.05mg L^{-1}	[330]

Table 15.9 continued

Organic	Extraction solvent from water	GLC column details	Detector	Comments	LD	Ref
Nitrophenol and dinitrophenols	Toluene	–	–	Phenol derivativised with heptafluorobutyranhydride then toluene extraction then GLC	0.01 μg L⁻¹ (nitrophenol) 1.6μg L⁻¹ (dinitrophenol)	[331]
Amino acids	Concentration by ion and ligand exchange chromatography	0.7% XE-60 0.5% OV-101 0.2% QF-1 on Diatoport S	Flame ionisation	After preliminary concentration prepare N-trifluoroacetyl methyl esters	–	[332]
Amino acids	Concentrated on Amberlite IR-120	OV-175	–	Amino acids converted to n-butyl-N-trifluro-acetyl esters. Argenine, cysteine and histidine not eluted from OV-175	–	[333]
Acrylonitrile	–	–	–	–	μg L⁻¹	[334]
Nitriloacetic acid	–	2% poly(ethanediol) on Chromosorb W	–	Nitroacetic acid converted to trimethyl ester prior to GLC	25μg L⁻¹	[335]
Nitriloacetic acid	Concentration on ion exchange column	–	–	Nitriloacetic acid esterified with formic acid, conversion to butyl esters then GLC	0.71μg L⁻¹	[336]
Nitriloacetic acid	–	–	–	Nitriloacetic acid converted to trimethyl silyl ester	1μg L⁻¹	[337]
Nitrosamines	–	–	–	–	–	[338]
Diazocompounds	–	–	Atmospheric pressure ionisation mass spectrometry	–	–	[339]

Note: The LD values use LaTeX: 0.01 μg L^{-1}, 1.6 μg L^{-1}, 25 μg L^{-1}, 0.71 μg L^{-1}, 1 μg L^{-1}.

Table 15.9 continued

Organic	Extraction solvent from water	GLC column details	Detector	Comments	LD	Ref
Azarenes	—	—	—	—	—	[340]
Dimethyl sulphide	—	—	—	—	—	[341]
Reduced sulphur compounds methyl mercaptan, dimethyl sulphide, H$_2$S, CS$_2$	—	—	Flame ionisation	—	0.2mg L^{-1} CS$_2$, 0.6ng L^{-1} (methyl mercaptan)	[342]
Reduced organo-sulphur compounds	—	—	Flame photometric	Cryogenic trapping used to minimise losses, sulphur compounds then revolatilised by controlled heating and injected into GLC column	—	[343]
Ethylene thiourea	—	—	—	—	—	[344]
Ethylene thiourea	—	—	Negative ion chemical ionisation	Derivativisation with 3,5 bis (trifluoromethyl) benzyl bromide	50ppt	[345]
Tetrahydrothiophen	—	—	—	—	—	[346]
Linear alkylbenzene sulphonates (C$_5$–C$_{15}$ homologues)	LAS as 1 methyl-heptyl amine extracted into hexane	—	—	Microsulphonation–GLC technique	10µg L^{-1}	[347]
Organosulphur compounds	Purge and trap analysis	—	GC–microwave induced atomic emission spectrometry	—	ppt	[348]

332

Table 15.9 continued

Organic	Extraction solvent from water	GLC column details	Detector	Comments	LD	Ref
2,3,7,8 tetrachloro-dibenzo-p-dioxin	–	–	–	–	3–5ng kg^{-1}	[349]
Organophosphorus compounds	–	–	Supported copper-cuprous oxide island film	–	mg L^{-1}	[350, 351]
Organophosphorus compounds	–	–	Chemiresist sensors	–	–	[352]
Organophosphorus compounds	–	–	Surface acoustic wave	–	0.01mg L^{-1}	[353]
Parathion, Malathion	–	–	–	Study of lyophilisation and co-crystallisation preconcentration methods	–	[354]
Organophosphorus sulphur compounds	XAD resin	–	Flame photometric 526nm (PH$_3$) 384nm (H$_2$S)	After extraction on to XAD resin, compounds reduced with hydrogen at 1100°C resulting in production of phosphine and hydrogen sulphide	0.1ng (P) 1ng (S)	[355]

Chlorinated insecticides

Organic	Extraction solvent from water	GLC column details	Detector	Comments	LD	Ref
Organochlorine insecticides and photodegradation products	–	–	–	UV irradiation, then GLC	–	[356]

Table 15.9 continued

Organic	Extraction solvent from water	GLC column details	Detector	Comments	LD	Ref
Organochlorine insecticides	-	-	-	Practical advice on methodology	-	[357, 358]
Organochlorine insecticides	-	-	-	Clean up of water with deactivated alumina silica gel and Florasil prior to GLC	-	[359]
Organochlorine insecticides	-	-	-	Removal of sulphur from water samples prior to GLC using copper or a column containing 4% OV–17, 47% QF–1 and 1% DC–200 on Gas Chrom G	-	[360–361]
Dieldrin, Endrin	Derivitivization then hexane extraction	OV–17 and QF–1 on Gas Chrom Q	Electron capture	Derivitivised with BF_3 in 2-chloroethanol prior to GLC	0.01 ppm	[362]
DDT, δBHC	Hexane	-	-	Clean up on silica gel modified with trichlorooctadecyl silicone or dichlorodimethyl silane	-	[363]
Aldrin	Hexane	3%OV–17 on Chromosorb W–HP at 180°C	Electron capture	Removal of sulphur interference	-	[364]
DDT, Dieldrin	-	-	Electron capture	Discusses interferences	-	[365]
DDT	Light petroleum	5% DC–11 on Chromosorb NAW DMCS	Tritium electron capture	Petroleum extracted clean up on Florasil	-	[102]
Organochlorine insecticides	-	-	-	Silica gel clean up	-	[366]

Table 15.9 continued

Organic	Extraction solvent from water	GLC column details	Detector	Comments	LD	Ref
Organochlorine insecticides and polychlorobiphenyls	–	–	Dual detection system	–	–	[367]
Organochlorine insecticides and polychlorobiphenyls	–	–	–	Water sampling apparatus	–	[368]
25 organochlorine insecticides and polychlorobiphenyls	–	–	Electron capture	Discussion on sample preservation clean up and recovery from water	–	[111]
DDT, HCHs, HCB	Hexane	–	Electron capture	Discusses clean up with sulphuric acid	low ppt	[369]
Organochlorine insecticides	–	–	Electron capture	Study of analysis conditions	–	[370]
Hexachlorocyclohexane hexachlorobenzene	–	–	–	Study of analysis conditions	Ultra low level	[118]
Trichlorphon, Dichlorvos	–	–	–	Discusses Trichlorophon degradation on storage of water samples	–	[371]
Trichlorphon	–	16% XF-1150 on Chromosorb W–AW at 125°C	Flame photometer in phosphorus	–	0.002ppm	[372]
Toxaphene	–	–	–	–	–	[373]
Hexachlorophane	1-chlorobutane	Sil–x silica	UV absorption	–	<20ng	[374]
Hexachlorophane	–	–	Electron capture	Hexachlorophane methylated then GLC	0.005μg L^{-1}	[375]

Table 15.9 continued

Organic	Extraction solvent from water	GLC column details	Detector	Comments	LD	Ref
α and β Endosulphan	–	–	–	Endosulphan converted to an ether by heating on Al$_2$O$_3$-concentrated sulphuric acid column, then GLC	0.003ng	[376]
Endosulphan	–	–	–	Study of occurrence in River Rhine	–	[377]
α and β Endosulphan	–	–	–	–	0.18pg	[378]
Chlorosulphuron	–	–	Electron capture	Chlorosulphuron converted to N,N bis(pentafluorobenzyl) 2-chlorobenzene sulphonamide, then GLC	0.1ppb	[379]
Chlorosulphuron	–	–	–	–	ppt	[380]
Organophosphorus insecticides						
Dursban	–	3% Carbowax 20M on Gas Chrom	Thermionic and flame photometric	Good separation of Dursban from other organophosphorus insecticides	0.01ppm	[381]
Nitrogen and phosphorus containing pesticides	–	–	–	Solid phase microextraction of water samples	–	[382]
Dichlorvos, omethoate, methamidophos, dimethoate, methyl parathion	–	–	–	–	0.5ng	[383]
Parathion	–	–	–	–	–	[384–386]

Table 15.9 continued

Organic	Extraction solvent from water	GLC column details	Detector	Comments	LD	Ref
Parathion, paraoxon	–	–	Flame ionisation	–	0.2ppg (parathion) 2ppb (para-oxon)	[387]
Parathion, Malathion	–	–	Phosphorus sulphur	Comparison of GLC and thin layer chromatography	–	[354]
Phosphorus–sulphur type halogen	–	–	Flame ionisation	–	–	[388, 389]
Sulphur–phosphorus type	–	–	Flame ionisation	–	–	[390]
Phosphorus–halogen type	–	–	Flame ionisation	–	–	[391]
Phosphorus–sulphur type	–	–	Helium plasma microwave	–	0.05–0.5 mg kg^{-1}	[392]
Phosphorus type	–	–	Electron capture (flame burns above Cs tip)	–	3×10^{-12}g	[390,393, 394]
Phosphorus–sulphur type	–	–	Flame photometric	–	P 0.005ppm S 1ppm	[395]
Phosphorus–sulphur, halogen type	–	–	Micro coul-ometric titration	Insecticides separated on GLC column, converted at 950°C to phosphine, hydrogen sulphide and hydrogen chloride prior to titration with silver	–	[389,395, 396]

Table 15.9 continued

Organic	Extraction solvent from water	GLC column details	Detector	Comments	LD	Ref
Phosphorus–nitrogen arsenic–chlorine type	–	–	Alkali metal salt therm-ionic type	Potassium chloride source for phosphorus compounds, rubidium source for nitrogen compounds	–	[397]
Phosphorus–chlorine, nitrogen–sulphur carbon type	–	–	Alkali metal thermionic type	Alkali metal salt molecular sieve source, eg CsCl for nitrogen compounds	–	[398]
Phosphorus type	–	–	–	Caesium nitrate aspirated into flame	–	[399]
Phosphorus–chlorine type	–	–	Alkali metal thermionic type	–	–	[400]
Phosphorus–sulphur type	–	–	Alkali metal thermionic	–	–	[401]
Phosphorus–nitrogen halogen type	–	–	Alkali metal thermionic	–	–	[402]
Phosphorus type	–	–	Alkali metal thermionic	Caesium bromide vapour detection	–	[403]
Phosphorus–sulphur type	–	–	Alkali metal thermionic	Melpar (rubidium–caesium) detection	–	[404]
Phosphorus–sulphur type	–	–	Alkali metal thermionic	Melpar detection	S 200pg P 40pg	[405]
Phosphorus–nitrogen type	–	–	Alkali metal thermionic	Alkali metal flame detection, RbSO$_4$ and KBr at top of flame	–	[406]

Table 15.9 continued

Organic	Extraction solvent from water	GLC column details	Detector	Comments	LD	Ref
Phosphorus type	–	–	Combined flame therm-ionic and photometric	–	–	[407]
Phosphorus–nitrogen type	–	–	Combined RbCl flame thermionic and electro-lytic con-ductivity detector	–	–	[408]
Phenoxy acetic acid herbicides						
Phenoxy carboxylic acids	Solid phase extraction with C_{18} resin	–	–	Phenoxyacetic acids derivativised with pentafluorobenzyl bromide, then GLC	ppt	[409]
Phenoxy alkanoic acids	–	–	–	Conversion to alkyl ester, then GLC	–	[409–417]
Phenoxy alkanoic acids	Chloroform or dichloromethane	Silicone DOW 710 on Chromosorb W–AW	–	Derivativisation to methyl esters with diazomethane	1 ppm	[188, 418]
2,4D and other phenoxy alkanoic acids	–	–	–	Study of conditions for nitration (using N,O-bis-trimethyl silyl-acetamide), bromination (using 1-2,3,4,5,6 pentafluorobenzene) and silation of phenoxy alkanoic acids prior to GLC	–	[419, 420]

Table 15.9 continued

Organic	Extraction solvent from water	GLC column details	Detector	Comments	LD	Ref
Phenoxy alkane alkanoic acids	–	–	Flame ionisation	–	5–10µg L^{-1}	[421]
Phenoxy alkane alkanoic acids	–	–	Electron capture	Study of back flushing conditions	–	[370]
Phenoxy alkane alkanoic acids	–	–	–	Conversion of phenoxy alkanoic acids to methyl esters with diazomethane, then GLC or derivitivisation with 2-chloro-ethyl and pentafluoro benzyl esters then GLC	–	[419, 422]
Azine herbicides						
Atrazine, simazine demetryn, prometryn	–	–	Nitrogen specific detector	–	10^{-8}M	[423]
Substitute phenyl urea herbicides						
15 phenyl urea type	–	–	–	Discussion of analysis schemes, including catalytic hydrolyses to anilines and derivitivisation with hexafluorobutyric anlydride	–	[424]
Phenyl urea type	–	–	Nitrogen specific	–	–	[122, 425–427]

Table 15.9 continued

Organic	Extraction solvent from water	GLC column details	Detector	Comments	LD	Ref
Carbamate pesticides/insecticides						
Carbamate type	–	–	–	Methylation, then GLC	25–50ppb	[428]
Carbamate type	Hexane	1% GE-XE-60 and 0.1% Epikote 1001 on AW–DMCS Chromosorb G at 211°C	Electron capture	Conversion to 2,4 dinitro phenyl hydiazones by reaction with 1-fluoro-2,4 dinitrobenzene, then GLC	0.005ppm	[429]
Aldicarb, Aldecarb oxime and Aldecarb nitrile	–	Short capillary column	–	–	0.3–1.3ng	[430]
Aldicarb and oxidative metabolites	–	Open tubular column	Nitrogen–phosphorus type	Recoveries 95% to 105%	1µg L^{-1}	[431]
m-S-butyl–phenyl methyl (phenylthio–) carbamate (RE 11775))	Acetonitrile or dichloromethane or chloroform	5% OV-225 on Gas Chrom Q at 242°C	Flame photometric in S mode, or electro-lytic conductivity detector	Clean up on Florasil, silica gel or alumina	0.01ppm	[432]
Methomyl (S-methyl -N-(methyl carbamoxyl) oxythioacetamide	Dichloromethane	10% DC 200M Chromosorb W-HP at 140°C	–	–	0.01ppm	[433]
Carbaryl	–	–	–	–	–	[434–437]

Table 15.9 continued

Organic	Extraction solvent from water	GLC column details	Detector	Comments	LD	Ref
Carbaryl and its hydrolysis product	Benzene	Gas Chrom G at 145°C	–	–	–	[438]
N-methyl carbamates, incl. Propoxur, Carbofuran, 3-ketocarbofuran Met mercapturon, carbaryl and Mobam	Methylene chloride	–	Electron capture	Methylene dichloride extract of water hydrolysed with methanolic potassium hydroxide to produce phenols, derivitivisation of phenols with pentaflurobenzenzyl bromide to produce ethers, then GLC	–	[439]
N-methyl carbamates	Silica phase extraction	–	–	–	–	[440]
Aldicarb	–	–	Flame photometric	–	$<1\mu g\,L^{-1}$	[441]
Miscellaneous anions, insecticides, herbicides, pesticides						
Misc.	–	–	–	Interference effects of filter paper on GLC chromatogram	–	[442]
Misc. pesticides	Membrane disc extraction on-line GC	–	–	Fully automated procedure	–	[443]
Misc. pesticides	Continuous liquid–liquid extraction	–	–	On-line GC	ppb	[444]
Catoran, Fenuron, Dicuron	–	–	Electron capture	Derivitivisation with polyfluoro-carboxylic anlydride, then GLC	$0.004mg\,L^{-1}$	[445]

Table 15.9 continued

Organic	Extraction solvent from water	GLC column details	Detector	Comments	LD	Ref
Fenitrothion	–	–	Flame photometric	–	–	[446–447]
Fenitrothion, Fenitrooxon, Aminofenitrothion	–	SE–30 plus OF–1 column	Flame photometric in phosphorus mode	–	–	[448]
Fenitrothion, Aminocarb and phenolic hydrolyses products	–	–	Nitrogen–phosphorus selective detector	Phenols derivitivised to esters with aceticanhydride then methylene chloride extraction, then GLC	1 µg L^{-1}	[449]
Diazinon, Parathion, Malathion, Fenthion and oxygen analyses and hydrolysis products	–	Reoplex–400	Electron capture flame ionisation	–	–	[450]
Miscellaneous organics						
Chloromethoxynil, Bifenox, Butachlor	Extraction on XAD–2 resin	–	Electron capture	–	–	[451]
Methanol, acetaldehyde, hexafluoroacetone, acetic acid	–	–	Chromo-sorb 104	–	–	[452]
Miscellaneous organics	–	–	–	–	–	[453]
Miscellaneous organics	Diethyl ether carbon disulphide, methylene dichloride	Capillary column	Electron capture nitrogen–phosphorus flame ionisation	–	–	[454]

Table 15.9 continued

Organic	Extraction solvent from water	GLC column details	Detector	Comments	LD	Ref
Miscellaneous organics	SE-30, OV-M on bentone didecylphthalate	-	-	-	$\mu g\ L^{-1}$	[455]
Miscellaneous organics in boiling point range 77–238°C	-	-	-	Steam distillation gas chromatography	-	[456]
Miscellaneous organics	-	Capillary column	Flame ionisation	-	-	[457]
Organic trace analysis	-	-	-	Automated extraction procedure	-	[458]
Miscellaneous organic pollutants	Solid phase extraction	-	-	-	sub ppt	[459]
Miscellaneous organic pollutants	Solid phase extraction	-	-	-	Trace	[460]
Pesticides, nitro-aromatics, poly-aromatic hydrocarbons	Direct injection of water	-	-	-	-	[461, 462]
Mestranol, ethynyl-oestradiol	-	-	-	Water cleaned on Florasil, then GLC	$0.1mg\ L^{-1}$	[463]
Squoxin piscicide (1'1'-methylene di-2-naphthol)	-	-	-	-	$0.1mg\ L^{-1}$	[464]
Coprostanol	-	-	-	-	$1\ \mu g\ L^{-1}$	[465]
Glyphocate (amino-methyl) phosphoric acid and glufosinate	-	-	-	Glyphosphate derivitivised to N-isopropoxy carbonyl methyl esters, then GLC	10ppb	[466]

Table 15.9 continued

Organic	Extraction solvent from water	GLC column details	Detector	Comments	LD	Ref
Pyrethrin insecticides and piperonyl butoxide N-(zethylhexyl) Norborn-5-enedicarboxamide	Hexane	5% SE–30 on AW–DMCs Chromosorb W at 190°C	Flame ionisation	93–94% recovery	0.2µg	[467]
Pyrethroils	Solid phase extraction	–	–	–	–	[468]
Fluoridone aquatic herbicide and photo-lysis product N-methyl formamide	–	–	Hall electrolytic detector in N mode	–	–	[469]
Miscellaneous organics	–	–	Electrolytic conductivity and photoion-isation	–	–	[470]
Miscellaneous organics	–	–	–	Correlations of GLC with water vapour carrier	µg L^{-1}	[471]
Miscellaneous organics	Solid phase extraction coupled to GLC	–	–	–	–	[472]
Volatile organics	–	–	–	Coupling of purge and trap with GLC, investigation of parameters, (purge gas flow rate, purge gas and sample volume)	–	[473]

Table 15.9 continued

Organic	Extraction solvent from water	GLC column details	Detector	Comments	LD	Ref
Miscellaneous organics	Miscellaneous solvents	–	–	Comparison of direct solvent extraction GLC and head-space analysis	–	[474]
Miscellaneous organics	–	–	Hall electrolytic and photoionisation	Purge and trap technique	0.1–0.9ppb	[475]
Volatile organic compounds	Capillary column	Volatiles purged from water	–	Capillary column cryogenically cooled (–80° to –90°)	–	[476–478]
Volatile organic compounds	–	–	–	Evaluation of Environmental Protection Agency GLC–MS method	–	[479]
Halogen, sulphur and nitrogen compounds, aromatic and unsaturated hydrocarbons	–	–	–	Study of use of polycomplexonates of copper, mercury and silver as vinyl derivatives of pyridine as adsorbents for organics prior to GLC	–	[480]

Source: Own files

15.1.2 Aqueous precipitation

15.1.2.1 Chlorinated insecticides

Engst and Knoll [481] used hexane extraction gas chromatography to study the occurrence in rain water of down to 0.001µg L^{-1} of p,p'-DDT, p,p'-DDE and pp'-TDE. Samples (500mL) were extracted by shaking with hexane (100, 50 and 50mL) for 1min each time. The combined hexane phases were dried with sodium sulphate and evaporated to between 1 and 5mL in a rotary evaporator. The extract (1mL) was injected, without further clean-up, on to a gas chromatographic assembly consisting of one column (1.8m × 6mm) packed with 5% QF-1 on Varaport 30 (100–120 mesh) and a second column (0.9m × 6mm) packed with 4% of OV-17 on AW–DMCS Chromosorb W (60–80 mesh) both operated at 180°C with nitrogen as the carrier gas (30mL min^{-1}) and a tritium electron capture detector.

15.1.2.2 Carboxylates

Kawamura and Kaplan [482] have described a sensitive method for measuring volatile acids (C_1–C_7) in rain and fog samples using p-bromophenacyl esters and a high resolution capillary gas chromatograph employing fused silica columns. Experiments showed that the measured concentrations of volatile acids in spiked rain samples increased linearly in proportion to the concentrations of volatile acids added. Relative standard deviations were less than or equal to 18% for C_1, C_2 and C_3 acids. The distributions of volatile acids in Los Angeles rain and fog samples are discussed.

15.1.3 Seawater

15.1.3.1 Petroleum products

In the marine environment gas chromatography has been employed to identify petroleum products [12,483–493]. Here the pollutants are crude oil, marine residual fuel oil and crude oil sludge consisting of a concentrated suspension of high melting point paraffin wax in crude oil. Although weathering of marine oil pollutants can be such that the oil is rendered unrecognisable, the time required to achieve this was found to be so long as to be insignificant in regard to pollutant identification.

Occasionally the mound of unresolved components on the chromatogram supporting the superimposed n-alkane peaks confuse the true n-alkane profile. This has been overcome by separating off the n-alkanes using molecular sieves, prior to gas chromatography [493]. However, separation of n-alkanes in this way, or by urea complex formation

[492], is reported as being more applicable to distillates rather than residual materials, and also the separation is not entirely specific [12]. In the case of marine pollutants it has been found advantageous to chromatograph a distilled residue (bp > 343°C) [493], or fraction (bp 254–370°C) [483], which avoids problems caused by evaporation of lower ends by weathering.

Zafiron and Oliver [494] have developed a method for characterising environmental hydrocarbons using gas chromatography. Solutions of samples containing oil were separated on an open tubular column (15.2m × 0.05cm) coated with OV–101 and temperature programmed from 75 to 275°C at 6°C per minute; helium (50mL min $^{-1}$) was used as carrier gas and detection was by flame ionisation.

Rasmussen [495] has described gas chromatography methods for the identification of hydrocarbon oil spills. The spill samples are analysed on a 30.5m Dexsil–300 support coated open tube (SCOT) column to obtain maximum resolution.

15.1.3.2 Chlorinated insecticides and polychlorobiphenyls

Girenko et al. [496] noted that it was difficult to analyse samples of sea water because they are severely polluted by various co-extractive substances, chiefly chlorinated biphenyls. To determine organochlorine insecticide residues by gas chromatography with an electron capture detector, the chlorinated biphenyls were eluted from the column together with the insecticides. They produce inseparable peaks with equal retention times, thus interfering with the identification and quantitative determination of the organochlorine insecticides. The presence of chlorinated biphenyls is indicated by additional peaks on the chromatographs of the water samples and aquatic organic organisms. Some of the peaks coincide with the peaks of the o,p' and p,p' isomers DDE, DDD and DDT and some of the constituents are eluted after p,p'-DDT.

Work by Wilson et al. [497,498] has indicated that organochlorine pesticides were not stable in water.

Since petroleum ether was the solvent used in these earlier studies for extracting DDT from sea water, Wilson and Forester [499] initiated further studies to evaluate the extraction efficiencies of other solvent systems, viz. petroleum ether, 15% ethyl ether in hexane followed by hexane or methylene chloride.

Wilson and Forester [499] discussed the determination of Aldrin, Chlordane, Dieldrin, Endrin, Lindane, o,p and p,p' isomers of DDT and its metabolites, Mirex and Toxaphene in sea water. The concentrated solvent extracts were analysed by electron capture gas chromatography using columns of different liquid phases. The following columns were used: DC–200, QF–1, EGS, OV–101, mixed DC–200–QF–1 and mixed OV–101–

Table 15.10 Gas chromatography of organic compounds in seawater

Organic	Extraction from water	GLC column conditions	Detector	Comments	LD	Ref.
Carboxylic acids	–	–	–	–	–	[500]
Monosaccharides	–	–	Electron capture	Monosaccharides derivativised with trifluoroacetyl chloride	–	[501]
Trichlorofluoromethone, dichlorofluoromethane	–	–	Electron capture	Headspace analysis–GLC technique	0.005×10^{-12} mol kg^{-1}	[502]
Polychlorobiphenyls	Adsorption on XAD–2 resin	–	Electron capture	–	–	[503]
Organosulphur compounds	–	Open tabular column	Chemiluminescence	–	1 ppb	[504, 505]
Atrazine	–	–	–	Sampling technique, analysis of microsurface of estuary waters	–	[506, 507]
Hydrocarbons	–	–	–	High resolution GLC	–	[508]
Hydrocarbons	–	–	–	–	–	[509]

Source: Own files

OV–17. Just prior to extraction, all samples were fortified with *o,p′*-DDE to evaluate the integrity of the analysis. The recovery rates of *o,p′*-DDE in all tests were greater than 89%, indicating no significant loss during analyses.

15.1.3.3 Other applications

Table 15.10 presents a summary of the applications of gas chromatography to the determinations of other organic compounds, including chlorofluoro-paraffins, carboxylic acids, monosaccharides, organosulphur compounds, polychlorobiphenols and azine herbicides in seawater.

15.1.4 Groundwaters

15.1.4.1 Hydrocarbons

Roe *et al.* [510] determined volatile aromatic compounds in groundwaters using a headspace gas chromatography method.

Solid phase micro-extraction coupled with gas chromatography has been used to identify jet fuel components in groundwaters [511].

15.1.5 Potable waters

15.1.5.1 Polyaromatic hydrocarbons

Saxena *et al.* [512] used polyurethane foams to concentrate trace quantities of six representatives of polynuclear aromatic hydrocarbons (fluoranthene, benzo(k)fluoranthene, benzo(j)fluoranthene, benzo(a) pyrene, benzo(ghi) perylene, and indeno(1,2,3–cd)pyrene) prior to regular screening of these compounds in US raw and potable waters. Final purification and resolution of samples was by gas chromatography and two-dimensional thin layer chromatography, followed by fluoro-metric analysis and quantification.

In this method the PAHs are collected by passing water through a polyurethane foam plug. Water is heated to $62 \pm 2°C$ prior to passage and flow rate is maintained at approximately 250mL min^{-1} to obtain quantitative recoveries. The collection is followed by elution of foam plugs with organic solvent, purification by partitioning with solvents and column chromatography on Florasil, and analysis by two-dimensional thin layer chromatography on cellulose–acetate–alumina plates followed by fluorometry and gas–liquid chromatography using flame ionisation detection. The latter method was less sensitive than thin layer chromatography. Employing this method and a sample volume of 60L, PAHs were detected in all the water supplies sampled. Although the sum of the six representative PAHs in drinking waters was small (0.9–15ppt), the values found for raw waters were as high as 600ppt.

In further work Saxena *et al.* [513] and Basu and Saxena [36] showed that the polyurethane–foam plug method had an extraction efficiency for PAHs of at least 88% from treated waters and 72% from raw water.

15.1.5.2 Chlorophenols

Chriswell and Cheng [65] have shown that chlorophenols and alkylphenols in the milligram to microgram per litre range in potable water can be determined by sorption on macroporous A–26 anion exchange resin followed by elution with acetone and measurement by gas chromatography.

The gas chromatography of chlorophenols is also discussed in Table 15.11.

15.1.5.3 Vinyl chloride

The Bellar [514] purge and trap method has been applied to the determination of vinyl chloride in potable water. Chloroform, bromidichloromethane and dibromochloromethane are common to chlorinated drinking waters and result from the chlorination process. Low levels of methylene chloride are often observed in samples analysed by this technique. These are attributed to method background.

15.1.5.4 Haloforms and halogenated aliphatic compounds

Numerous methods based on gas chromatography have been described for the determination of haloforms in potable water:

- direct injection of aqueous samples into the gas chromatograph;
- gas chromatography on conventional columns of organic solvent extract of water sample;
- adsorption of ion exchange resins followed by gas chromatography of extracts; and
- capillary column gas chromatography of organic solvent extract of water sample.

Direct injection gas chromatography

Nicolson and Meresz [515] directly injected the drinking water sample into a gas chromatograph equipped with a scandium tritide electron capture detector. Glass columns (1.2mm × 6mm) packed with Chromosorb 101 (60–80 mesh) were used for the analysis.

The conditions for operating the instruments were as follows.

- Varian 2400: injector temperature 230°C; detector temperature 230°C; oven temperature 130°C; nitrogen flow rate 50mL min $^{-1}$.
- Varian 2100: injector temperature 220°C; detector temperature 225°C; oven temperature 150°C; nitrogen flow rate 60mL min $^{-1}$.

The detection limits for the trihalogenated compounds are all below the 10µg L $^{-1}$ level.

The detection limit for dichloropropane is only 60µg L $^{-1}$, and the dichlorobenzenes cannot be detected below 500µg L $^{-1}$. This analysis is not, therefore, suitable for detecting trace levels of some of the dichlorinated hydrocarbons in water.

Renberg [284] has reported an ion exchange resin adsorption method for the determination of trihalomethanes and chloroethanes and dichloroethane in water. In this method halogenated hydrocarbons are determined by adsorption on to XAD–4 polystyrene resin and eluted with ethanol. The extract is analysed by gas chromatography and is sufficiently enriched in hydrocarbon to be suitable for other chemical analysis or biological tests. Volatile hydrocarbons yielded recoveries of 60–95%. By using two series-connected columns Renberg [284] was able to study the degree of adsorption and the chloroethanes were found to be more strongly adsorbed than the haloalkanes.

Eklund et al. [516,517] have developed a method for the determination of down to 1µg L $^{-1}$ volatile organohalides in waters which combines the resolving power of the glass capillary column with the sensitivity of the electron capture detector. The eluate from the column is mixed with purge gas of the detector to minimise band broadening due to dead volumes. This and low column bleeding give enhanced sensitivity. Ten different organohalides were quantified in potable water. Retention times were measured on two columns with different stationary phases, ie SE–52 and Carbowax 400.

Trussell et al. [518] used glass capillary gas chromatography for the precise and rapid analysis of trihalomethanes in n-pentane extracts of potable water. These workers claim that the use of glass capillary gas chromatography has several advantages over conventional gas chromatography. First, the glass capillary column provides better resolution of individual components; the typical packed column has 2000–10000 theoretical plates, whereas capillary columns range from 15000 to 50,000 theoretical plates. Second, the high quality resolution allows shorter gas chromatographic runs without overlapping peaks and thus brings about time savings. Finally, the trihalomethanes can be resolved well on general usage liquid phases, which give versatility in analytical capability in that a wider range of types of organic compounds can be analysed without requiring time-consuming column changes.

The gas chromatography of haloforms is also discussed in Table 15.11.

15.1.5.5 Nitrosamines

Fine *et al.* [519] have described a gas chromatographic method for the determination of N-nitroso compounds in potable water.

A gas chromatograph equipped with a flame ionisation detector and a 3.6m × 3mm stainless steel column packed with 10% diethyleneglycol succinate on 80–100 mesh Chromosorb W (HO) was used. Recoveries obtained by this procedure were 77% or higher and were particularly good for the nitrosamine levels of about $10\mu g$ L^{-1}. Two different concentration and extraction procedures were used by these workers; one based on liquid–liquid extraction and the other based on the adsorption of the organic fraction on carbon and its subsequent extraction with chloroform and alcohol. In both cases, final quantitative analysis and identification were carried out on a single-column gas chromatograph equipped with N-nitroso compound specific thermal energy analyser.

15.1.5.6 Chlorinated insecticides

Sackmauereva *et al.* [520] have described the method for the determination of chlorinated insecticides (BHC isomers, DDE, DDT and hexachlorobenzene) in potable waters.

The water sample (1–3L) is extracted with three portions of petroleum ether (boiling point 30–40°C). The petroleum ether layer is then concentrated to a volume of about 0.5mL using a vacuum and purified on an alumina column (Woelm, neutral, activated by heating at 300°C for 3h and deactivated by adding 11% water). Thereafter, insecticides were eluted with 15% dichloromethane in petroleum ether. The eluate was concentrated in a vacuum rotary evaporator to a volume of 1mL and then used for gas liquid chromatography. When the individual insecticides are present in the solution in such a concentration range, the electron capture responds nearly uniformly to all insecticides. A column filled with 1.5% silicone OV–17 plus silicone oil (fluoralkylsiloxane) on Chromosorb W (80–100 mesh) is used for separation of the BHC, alpha, beta, gamma and delta isomers (hexachlorocyclohexane), o,p'-DDT, p,p'-DDE, p,p'-DDD and p,p'-DDT, α-BHC and hexachlorobenzene (HCB) have a common peak. They can be separated on a column filled with 2.5% silicone oil XE–60 (β-cyanoethylmethysilicone) on Chromosorb W (18–800 mesh).

The gas chromatography of chlorinated insecticides is also discussed in Table 15.11.

15.1.5.7 Polychlorobiphenyls

Le'Bel and Williams [521,522] found that low procedural blank values, equivalent to 0.05ng L^{-1} for a 200L potable water sample were attainable only by using doubly distilled solvents and by exhaustive washing of all reagents and glassware with these solvents.

Gas chromatographic analysis of concentrated extracts of potable water samples without Florasil column clean-up gave off-scale peaks at instrument settings suitable for low nanogram per litre PCB analysis. However, fractionation of the extract by Florasil column chromatography gave a PCB fraction sufficiently clear of interfering organics to permit PCB analysis at 1–10ng L^{-1}.

The detection limit was estimated to be ca. 0.04ng L^{-1} of Arochlor 1016 from this source of water with Arochlors 1232, 1242 and 2154 having similar levels of detection. When the method was applied to potable water from a river source, the interference in the gas chromatogram from other organic compounds present in the sample made quantitation difficult at the 1ng L^{-1} level. The gas chromatography of polychloro-biphenyls is also discussed in Table 15.11.

15.1.5.8 Other applications

Gas chromatography has also been applied to the determination of the following types of organic compounds in potable waters (see Table 15.11): aliphatic hydrocarbons, phenols, nitrophenols, chlorinated dioxins, anticholisterinase insecticides, chlorinated humic acid and chloroligno-sulphonic acids.

15.1.6 Sewage effluents

15.1.6.1 Carboxylates

Gas chromatography is a very attractive possibility for volatile acids determination, since it makes the separation of the individual acids for qualitative and quantitative determination *in situ* possible. In practice, many difficulties in analysing volatile acids in aqueous systems, resulting mainly from the presence of water have been reported [568]. The volatile acids' high polarity as well as their tendency to associate and to be adsorbed firmly on the column require esterification prior to gas chromatography determination. The presence of water interferes in esterification so that complex drying techniques and isolation of the acids by extraction, liquid–solid chromatography, distillation and even ion exchangers have to be used [569–572].

The introduction of the more sensitive hydrogen flame ionisation detector has made possible the analysis of dilute aqueous solutions of organic acids by gas–liquid chromatography. Problems, such as 'ghosting' at high acid concentrations and an excessive tailing effect of the water in dilute solutions, masking the components, have been reported for aqueous solutions [573].

Table 15.11 Gas chromatography of organic compounds in potable waters

Organic	Extraction from water	GLC column conditions	Detector	Comments	LD	Ref.
C_1–C_6 hydrocarbons	Stripped from water with helium	–	–	Helium stripped hydrocarbons can be adsorbed on to carbon to increase sensitivity	<mg L^{-1}	[53, 523]
Volatile hydrocarbons and chlorohydrocarbons	Air stripped from water	–	Flame ionisation	–	–	[27]
Phenols and chloro-phenols	–	–	Electron capture	Phenols reacted with bromine to produce bromophenols and determined by GLC	1 µg L^{-1}	[303]
Haloforms	–	–	Simultaneous flow ionisation and electron capture	–	Traces	[524]
Haloforms	–	–	–	Inject water directly into gas chromatograph	–	[525–536]
Chloroform, dichloro-bromomethane, bromoform, trihalo-methane	n-pentane	Capillary column OV–101	Electron capture	–	0.1–1	[518, 537]
Trihalomethanes	n-pentane	Fused silica column	Electron capture	–	–	[538]
Trihalomethanes	Liquid–liquid extraction	–	Hall electrolytic conductivity in hydrogen mode	Recovery 88–98.6%	Low µg L^{-1}	[539–541]

Table 15.11 continued

Organic	Extraction from water	GLC column conditions	Detector	Comments	LD	Ref.
Chloroform, bromoform, dichlorobromomethane, dibromochloromethane	—	1% SRI100 on Carbopak 3	Hall electrolytic conductivity	—	—	[286, 540]
Non volatile organic chlorine compounds	—	—	—	Studies of formation of non volatile organic chlorine compounds formed during chlorination of potable water	—	[542]
Chloroalkanes	—	—	—	Closed loop stripping gas chromatography	—	[543]
Chloroform, carbon tetrachloride, di-bromochloromethane, bromoform	—	—	—	—	$0.9–2.6\mu g\ L^{-1}$	[544]
Lower halogenated hydrocarbons, trichloromethane, dichloro-bromomethane, di-chloromethane, tri-chloromethane, tri-bromomethane, tri-chloroethylene, tetra chloromethane, tetra-chloroethylene	Xylene	—	—	—	$\mu g\ L^{-1}$	[545]
Trihaloforms	Pyridine	High resolution capillary column	—	Trihaloforms adsorbed on acetylated XAD-2 resin	$<1\mu g\ L^{-1}$	[546]

Table 15.11 continued

Organic	Extraction from water	GLC column conditions	Detector	Comments	LD	Ref.
Haloforms	Cryogenic refocuser	Capillary column	—	Liquid nitrogen cryogenic refocuser interfaced with capillary column	sub µg L^{-1}	[547]
Haloforms, halo-methanes, haloethanes	Vacuum distillation–cryogenic trapping	—	Electron capture	—	1ng L^{-1}	[548]
Trihalomethanes	Methyl cyclohexane	Short capillary column	Electron capture	—	1µg L^{-1}	[549]
Chloroform, dichloro-bromomethane, di-bromochloromethane, chloroform	Pentane or iso-octane	10% squalene on Chromosorb	Electron capture	72–83% recovery	0.1–0.5µg L^{-1}	[550, 551, 820]
Chloroform, dibromo-chloromethane, chloro-dibromomethane, bromoform	Pentane or hexane or methyl-cyclohexane	3% SP-1000 on Supelcoport or 10% Squalane on Chromosorb WA–10	Electron capture	Recovery: CHCl$_3$, 106–110% BrCl$_2$CH$_2$, 108–125% ClBrCH$_2$ 74–94% CHBr$_3$, 75–114%	—	[545, 553]
Chloroform, chloro-dibromomethane, dibromochloromethane, bromoform	Pentane	10% Carbowax 20M on Chromosorb WHP	Electron capture	—	0.1–10µg L^{-1}	[552]
Chloroform, bromo-dichloromethane, bromoform	Petroleum ether	5% FFAP on Chromosorb HPW	Electron capture	83–96% recovery	0.02–0.05µg L^{-1}	[554]
Chloroform, dichloro-bromomethane, di-chloromethane bromoform	Methyl cyclo-hexane	10% Squalane on Chromosorb WAW	Electron capture	85–94% recovery	0.2–0.8µg L^{-1}	[531, 550]

Table 15.11 continued

Organic	Extraction from water	GLC column conditions	Detector	Comments	LD	Ref.
Bromoform, chloroform, bromodichloromethane, dibromochloro methane	Petroleum ether	—	Electron capture	90–94% recovery	0.1–0.8µg L⁻¹	[555]
Tetrachloromethane, chloroform, dichloroethane, tetrachloroethylene, bromodichloro methane, chlorodibromomethane, bromoform	Hexane	—	Electron capture	60–111% recovery	0.1–10µg L⁻¹	[556]
Hexachlorocyclopentadiene	—	—	—	Solvent extract procedure	—	[557]
Polyhalogenated phenols	—	—	—	—	<1µg L⁻¹	[558]
Chlorophenols	Liquid nitrogen trapping	—	Fourier transform infrared	—	—	[559]
Nitrobenzene, 2,4 dinitrophenol	—	—	—	—	—	[560]
Polychlorobiphenyls	—	—	Flame ionisation	—	—	[310]
Polychlorobiphenyls	—	—	Electron capture	—	—	[561]
Chlorinated insecticides and polychlorobiphenyls	—	—	—	—	—	[562]
Chloroligno sulphonic acids	—	—	—	Pyrolysis–GLC technique	—	[563]

Table 15.11 continued

Organic	Extraction from water	GLC column conditions	Detector	Comments	LD	Ref.
Toxaphene insecticide, polychlorobiphenyl	–	–	Flame ionisation, electron capture	–	–	[564]
2,3,7,8 tetrachloro-dibenzo-p-dioxin	–	–	–	–	1 pg L^{-1}	[565]
Anticholinesterase pesticides	Carbon disulphide	–	–	Sample treated with bromine and resulting 3,3 dimethyl-butylacetate extracted with CS$_2$, then GLC	–	[566. 567]
Chlorination products of fulvic acid	–	–	–	–	–	[276]

Source: Own files

Subsequently phosphoric [574] or metaphosphoric acids [575] were added to the liquid phase, resulting in more reproducible column performance and reduced 'ghosting'. Addition of formic acid to the carrier gas was recommended by Cochrane [576] to overcome all the problems normally associated with analysing free fatty acids by gas chromatography.

Baker [577] used an FFAP column for direct injection of dilute aqueous solutions of acids (FFAP is a reaction product of polyethylene glycol 20,000 and 2-nitrophthalic acid developed by Varian Aerograph). The acetic acid peak was not clear and the ability of this column to separate normal and iso fatty acids was not reported. Van Huyssteen [608] successfully used a Chromosorb 101 column coated with 3% FFAP for separation of volatile acids by direct injection of synthetic aqueous solutions and anaerobic digester samples, which were first centrifuged and acidified to pH1-2 with hydrochloric acid. His column affected complete separation of the C_2–C_6 straight and branched short-chain fatty acids from synthetic aqueous solutions, but less sharpened peaks were obtained from anaerobic digester samples. The response with acetic acid approximated that of the other acids; additional peaks, probably alcohols, appeared between the acid peaks. To eliminate 'ghosting' 1–2µL water was injected between samples.

Narkis and Henfield-Furie [578] have described a direct method for the identification and determination of volatile water-soluble C_1–C_5 acids in municipal waste water and raw sewage. The method involves direct injection of the sewage into a gas chromatograph equipped with a Carbowax 20m on acid-washed Chromosorb W column and a flame ionisation detector. Preliminary preparation of the sample is limited to the addition of solid metaphosphoric acid to the sewage and removal of precipitated proteins and suspended solids by centrifuging.

Results for individual volatile acid concentrations in raw sewage determined by the direct injection procedure of Narkis and Henfield-Furie [578] and by that of Standard Methods [579]. The results were also expressed as acetic acid for comparison with the collective total amount of organic acids determined by the Standard Method [579]. The total amount of organic acids determined according to the Standard Method is higher than that found by the Narkis and Henfield-Furie [578] method. On average between 85 and 98% of the organic acids determined by the Standard Methods procedure were found to be volatile acids by the direct injection method.

The determination of carboxylic acids is also discussed in Table 15.13.

15.1.6.2 Organochlorine insecticides and polychlorobiphenyls

Mattson and Nygren [580] devised a solvent extraction method for extracting PCBs from sewage sludge containing lipids. They point out

Table 15.12 Effect of treatment of a solution of chlorinated hydrocarbons and the internal standard hexabromobenzene with fuming sulphuric acid (I), fuming sulphuric acid plus potassium cyanide (II), and potassium hydroxide (III) expressed as percentages of the compounds in an untreated solution

Compound name (no.)	Concn. (ng mL⁻¹)	Method of treatment		
		I	II	III
α-BHC(1)	0.022	94	59	0
β-BHC(2)	0.072	94	70	6
Lindane (3)	0.026	92	55	11
Heptachlor (4)	0.026	94	90	104
Aldrin (5)	0.020	92	87	104
Heptachlor epoxide (6)	0.033	83	79	100
p,p'-DDMU (7)	0.090	100	91	190
Dieldrin (8)	0.048	0	0	103
p,p'-DDE (9)	0.022	104	100	437
p,p'-DDD (10)	0.085	100	95	0
o,p'-DDT (11)	0.096	98	95	0
p,p'-DDT (12)	0.103	100	95	0
Hexabromobenzene (13)	0.57	(100)	(100)	(100)

Source: Own files

that lipids and some other impurities in the crude extracts of sewage sludge can be destroyed by treatment with fuming sulphuric acid, either by shaking the acid [581], or by eluting on a fuming sulphuric acid–Celite column [582,583]. Dieldrin is decomposed by this treatment but DDT and its metabolites, DDD and DDE, are not. Extracts of sewage sludges often contain large amounts of elemental sulphur, particularly after treatment with sulphuric acid. These interfere with early eluting compounds in the gas chromatographic step.

Sulphur was removed by the Bartlett and Skoog [584] method in which the sulphur is reacted with cyanide in acetone solution to produce thiocyanate. BHCs are decomposed to some extent, probably to penta-chlorocyclohexane. An alternative procedure for the removal of sulphur utilising barium hydroxide is also described. Alkali hydroxides should not be used as they cause dehydrochlorination of BHCs [581]. Lindane and its isomers are dehydrochlorinated to trichlorobenzenes [585] and are eluted together with the solvent. Cochrane and Maybury [586] have used the reaction with sodium hydroxide in methanol for the identification of BHCs. Dieldrin is not decomposed in the potassium hydroxide treatment and can thus be detected in the chromatogram of that aliquot. Some common chlorinated hydrocarbon pollutants and the internal standard hexabromobenzene were treated, according to the general procedure described above with sulphuric acid, potassium cyanide and potassium hydroxide. The results of the recovery experiments are shown in Table 15.12. When using packed columns, a precolumn of sodium and potassium hydroxides will give the same effect as the potassium hydroxide treatment described above [587]. Mattson and Nygren [580] have also tested a column with a packed alkaline post-column to remove the sulphur peak from the chromatogram. In the post-column, DDT and DDD are dehydrochlorinated but this does not effect their retention times. This method has good reproducibility and has a detection limit for the total amount of PCBs in the dried sample of at least 0.1mg kg $^{-1}$ and for DDT, DDD and DDE limits of 0.01, 0.005, and 0.005mg kg $^{-1}$, respectively.

Jensen et al. [588] have described a procedure for the determination of organochlorine compounds including PCBs and DDT in sediments and sewage sludge in the presence of elemental sulphur. The method can also be used for a search for both volatile and/or polar pollutants. The sulphur interfering in the gas chromatographic determination is removed in a non-destructive treatment of the extract with tetrabutylammonium sulphite. This lipophilic ion pair rapidly converts the sulphur to thiosulphate in an organic phase. The recovery of added organochlorines was above 80% and the detection limit in the range of 1–10ppb organochlorines from a 10g sample. Elemental sulphur present in most sediment and digested sludge has caused significant problems in residue analysis [140,589]. If the sulphur level is high, the electron capture detector will be saturated for a considerable period of time, and if the level of sulphur is low, it gives three or more distinct peaks on the chromatogram which can interfere with BHC isomers and Aldrin. Treatment of the crude extract with potassium hydroxide in ethanol [581] or Raney nickel [590] will quantitatively destroy all sulphur, but will at the same time convert DDT and DDD to DDE and DDMU (1-chloro-2,2-bis(4-chlorophenyl)ethane), respectively, and most BHC isomers are lost. Metallic mercury has also been used for removal of sulphur [591]. Jensen et al. [588] described an efficient, rapid, non-destructive method to remove the sulphur according to the reaction

$$(TBA^+)_2SO_3^{2-} + S(s) \rightarrow 2TBA^+ + S_2O_3^{2-}$$

where TBA$^+$ is the tetrabutylammonium ion.

The sample extracts were run with an electron capture detector (^{63}Ni). The 2.4m × 0.18cm (id) glass column was filled with a mechanical mixture of two parts of 8% QF–1 and one part of 4% SF–96 on acid-washed silanised Chromosorb W (100–120 mesh). Typical chromatograms obtained by this procedure are shown in Fig. 15.7.

The determination of chlorinated insecticides and of polychlorobiphenyls is also discussed in Table 15.13.

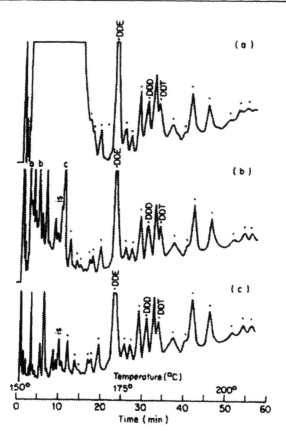

Fig. 15.7 (a) Typical digested sewage sludge chromatogram, severely contaminated with sulphur. (b) The same sample after a normal TBA–sulphite treatment, showing that most of the sulphur has disappeared. A number of peaks (a, b, c) originating from traces of sulphur appear in the BHC–Aldrin region. (c) The final chromatogram after additional treatment with sodium sulphite.
* = PCB components. IS = internal standard

Source: Reproduced by permission from the American Chemical Society [588]

15.1.6.3 Other applications

Other organic compounds that have been determined in sewage effluents include the following (see Table 15.13): hydrocarbons, alcohols, carboxylic acids, esters, chlorobenzenes, nitrosamines, ethylene diamine tetraacetic acid, nitriloacetic acid, organophosphorus compounds, linear alkyl benzene sulphonates, methyl mercaptan, polychlorobiphenyls and chlorinated insecticides.

15.1.7 Waste waters

15.1.7.1 Polyaromatic hydrocarbons

CONCAWE [621] recommended methods described by the Environmental Protection Agency for the determination of PAHs in oil refinery effluents. The method involves extraction of the effluent with methylene dichloride followed by clean-up procedures followed by gas chromatography or liquid chromatography.

For the gas chromatographic method no detector is specified; flame ionisation seems to be the best choice. Lack of selectivity in the method can lead to interference by compounds that are not completely removed by the clean-up. Higher selectivity can be achieved by use of a photoionisation detector. In addition some pairs of the PAH isomers are incompletely separated by the 15cm column used in this method, while the heavier PAHs often show tailing peaks.

The liquid chromatographic procedure [622] uses reversed phase liquid chromatography with fluorescence detection to separate all 16 PAHs completely. The method is sensitive and so selective as often to allow the method to be applied without clean-up procedure. For gas chromatographic methods, detection limits are about $1\mu g \, L^{-1}$, whereas for the liquid chromatographic methods, limits are between 1 and $100\mu g \, L^{-1}$ for two- and three-ring aromatics and below $1\mu g \, L^{-1}$ for the four-, five- and six-ring compounds.

For PAH analysis the extraction with methylene chloride may successfully be substituted by cyclohexane extraction allowing high recoveries by one single extraction.

15.1.7.2 Organosulphur compounds

Vitenberg et al. [623] have described a gas chromatographic method for the determination of traces (down to $1-10\mu g \, kg^{-1}$) of sulphur compounds, such as hydrogen sulphide, mercaptans, sulphides and disulphides, in industrial waste waters (kraft paper mill effluents) by a combination of headspace analysis and microcoulometry. This method increases the analytical sensitivity 10^2-10^3 times without any preliminary concentration of the sample.

At a concentration level of 1ppm sulphur compounds, the analytical error does not exceed 8% for the flame ionisation detector and 12% for the microcoulometric detector, of the given amount of the compound. When analysing solutions with concentrations of not more than 0.1ppm, the analytical error in the region of the highest sensitivity reaches 15% and 20% for the flame ionisation and the microcoulometric detector, respectively.

Table 15.13 Gas chromatography of organic compounds in sewage effluents

Organic	Extraction from sample	GLC column conditions	Detector	Comments	LD	Ref.
C$_{16}$–C$_{26}$ hydrocarbons	–		–	Recovery 95–102%	<20mg L^{-1}	[592]
Methanol	–	Tenax GC at 70°C preheated with 85% phosphoric acid	Nitrogen flame ionisation	–	0.5mg L^{-1}	[593]
2-propanol, acetic acid, oxallic acid	–		Dual flame ionisation	Effect of ozonisation on breakdown of these compounds	–	[594, 595]
Formic, acetic, propionic acids	–		–	Acids converted to benzyl esters, then GLC	–	[49]
Oxalic acid	–		–	Oxalic acid converted to ethyl ester with diazomethane, then GLC	–	[596, 597]
Formaldehyde, acetaldehyde, propionaldehyde, methyl glyoxylic acid, pyruvic acid	–		–	Converted to benzyl esters, then GLC	–	[598]
C$_2$–C$_5$ carboxylic acids	–	Chromosorb W pretreated with phosphoric acid	–	–	–	[599]
Votalite fatty acids	–		Flame ionisation	Sample steam distilled, then direct aqueous injection into GLC, recovery 97.6–99%	–	[600]
Salicyclic acid	–		–	–	–	[601]

Table 15.13 continued

Organic	Extraction from sample	GLC column conditions	Detector	Comments	LD	Ref.
Phthalate esters	Dichloromethane, and dichloro-methane:hexane (15:85)	–	Electron capture, flame ionisation	–	–	[602]
Chlorobenzenes, organochlorine insecticides, poly-chlorobiphenyls, halogenated solvents	–	–	–	–	–	[603]
Nitrosamines	–	–	Chemifluorescent	–	–	[604]
Ethylene diamine tetraacetic acid	–	–	–	EDTA, converted to methyl ester, then GLC	–	[79, 605]
Nitriloacetic acid	NTA separated from sewage on an anion exchange column	2% GF–1 on Varoport 30	Electron capture	Nitriloacetic and derivitivised with boron trifluoride in 2-chloroethanol	0.1 µg L^{-1}	[606]
Nitriloacetic acid, ethylene diamine tetraacetic and diethylene triamine pentaacetic acid	Sample at pH2 chloroform	5% VT on Aeropak, 150 285°C	Electron capture	Compounds converted to methyl esters using methanolic boron trifluoride	–	[606]
Nitriloacetic acid, citric acid	–	Carbowax on Celite 545 at 183°C	–	Compounds converted to butyl esters	–	[607]

Table 15.13 continued

Organic	Extraction from sample	GLC column conditions	Detector	Comments	LD	Ref.
Nitriloacetic acid and amino poly-carboxylic acids	–	0.65%R ethane–diol adipate on Chromosorb W 80 to 220°C	Flame ionisation	Compounds converted to butyl or N-trifluorylacetyl esters	–	[80]
Organophosphorus compounds	–	–	Flame photometry	–	–	[608]
Linear alkyl benzene sulphonates	Tetrabutyl ammonium hydrogen sulphate in chloroform	–	Flame ionisation	Gas in sewage preconcentrated on C_{18} Empone disc	–	[609]
Methyl mercaptan	Adsorbed on traps containing Tenax GL	–	–	Headspace analysis	–	[610]
Methyl mercaptan, hydrogen sulphide	–	FEF Teflon column, Poropak QS	Sulphur specific flame ionisation	–	0.25ng H_2S 0.5ng mesh	[611]
Chlorinated insecticides polychlorobiphenyls	Hexane	1.5% OV–17 plus 1.95% QF–1 on Supelport and 1% NPGG on Gas Chrom Q	Electron capture	Hexane extract cleaned up on alumina plus silver nitrate	–	[612–614]

Table 15.13 continued

Organic	Extraction from sample	GLC column conditions	Detector	Comments	LD	Ref.
Polychlorobiphenyls and chlorinated insecticides	–	–	–	–	mg L⁻¹	[615]
Polychlorobiphenyls and chlorinated insecticides	–	–	–	–	–	[562]
Miscellaneous organics	Solvent extraction, freeze drying	–	–	Reviews	–	[616–620]

Source: Own files

15.1.7.3 Organophosphorus insecticides

Gas chromatography has been used [624] to determine the following organophosphorus insecticides at the microgram per litre level in water and waste water samples: Azinophos–methyl, Demeton–O, Demeton–S, Diazinon, Disolfoton, Malathion, Parathion–methyl and Parathion–ethyl. This method is claimed to offer several analytical alternatives, dependent on the analysts' assessment of the nature and extent of interferences and the complexity of the pesticide mixtures found. Specifically, the procedure uses a mixture of 15% v/v methylene chloride in hexane to extract organophosphorus insecticides from the aqueous sample. The method provides, through use of column chromatography and liquid– liquid partition, methods for the elimination of non-pesticide interference and the pre-separation of pesticide mixtures. Identification is made by selective gas chromatographic separations and may be corroborated through the use of two or more unlike columns. Detection and measurement are best accomplished by flame photometric gas chromatography using a phosphorus-specific filter. The electron capture detector, though non-specific, may also be used for those compounds to which it responds. Confirmation of the identity of the compounds should be made by gas chromatography–mass spectrometry when a new or undefined sample type is being analysed and the concentration is adequate for such determination.

15.1.7.4 Organochlorine insecticides and polychlorobiphenyls

A method has been described [625] for the determination of the following PCBs (Arochlors) at the nanogram level in 15:85 methylene chloride–hexane extracts of water and waste water: PCB–1016, PCB–1221, PCB–1232, PCB–1242, PCB–1248, PCB–1254 and PCB–1260. This method is an extension of the method for chlorinated hydrocarbons in water and waste water (described by Goerlitz and Law [626]). It is designed so that determination of both PCBs and the following organochlorine insecticides can be made on the same sample: Aldrin, DDT, Mirex, BHC, Heptachlor, pentachloronitrobenzene, Chlordane, Heptachlor epoxide, Strobane, DDD, Lindane, Toxaphene, DDE, Methoxychlor and Trifluralin.

The determination of polychlorobiphenyls and chlorinated insecticides in waste waters is also reviewed in Table 15.14.

15.1.7.5 Other applications

Other organic compounds that have been determined in waste waters include petroleum hydrocarbons, carboxylic acids, aliphatic and aromatic chlorocompounds and phenoxyacetic acid herbicides. See Table 15.14.

15.1.8 Trade effluents

15.1.8.1 Miscellaneous organic compounds

Gas chromatography has been applied to the determination of a wide range of organic compounds in trade effluents including the following types of compounds which are reviewed in Table 15.15: aromatic hydrocarbons, carboxylic acids; aldehydes, non ionic surfactants (alkyl ethoxylated type) phenols; monosaccharides; chlorinated aliphatics and haloforms; polychlorobiphenyls; chlorlignosulphonates; aliphatic and aromatic amines; benzidine; chloroanilines; chloronitroanilines; nitro-compounds; nitrosamines; dimethylformamide; diethanolamine; nitriloacetic acid; pyridine; pyridazinones; substituted pyrrolidones; alkyl hydantoins; alkyl sulphides; dialkyl suphides; dithiocarbamate insecticides; triazine herbicides and miscellaneous organic compounds.

15.1.9 High purity waters

15.1.9.1 Nitrosamines

Kimoto et al. [681] extracted nitrosamines from 1L water samples with 3 × 150mL portions of dichloromethane with shaking for 5min per extraction. The extracts were combined, dried over anhydrous sodium sulphate and concentrated in a Kuderna–Danish evaporator to 1.0mL. These workers determined nitrosamines in dichloromethane extracts by a gas chromatographic procedure.

15.1.10 Swimming pool waters

15.1.10.1 Aliphatic halogen compounds

A gas chromatographic procedure for determination of volatile organo-chlorine compounds in swimming pool water has been described [682].

A column packed with Carbopack B with 0.8% SE–30 as the mobile phase is used, the sample being extracted by n-pentane in the sampling container. Retention times are given for a range of chlorinated and bromochloro derivatives.

15.2 Organometallic compounds

15.2.1 Non saline waters

15.2.1.1 Organoarsenic compounds

Andreae [683] described a sequential volatilisation method for the sequential determination of arsenate, arsenite, mono-, di- and tri-methylarsine, monomethylarsonic and dimethylarsinic acid, and

Table 15.14 Gas chromatography of organic compounds in waste waters

Organic	Extraction from waste effluents	GLC column conditions	Detector	Comments	LD	Ref.
Petroleum hydro-carbons	–	SE–52 on silanised Chromosorb W, 50–300°C	Flame ionisation also infrared spectroscopy	–	–	[627]
Higher carboxylic acids	Tertbutyl methyl ether	–	Flame ionisation	Carboxylic acids derivitivised with diazomethane	5µg L^{-1}	[628]
Higher carboxylic acids		Capillary column	Electron capture flame ionisation	Carboxylic acids derivitivised to p-bromophenacyl esters	–	[629]
C$_1$–C$_4$ carboxylic acids		3% FFAP on Chromosorb 101 at 180°C	Dual flame ionisation	–	–	[228]
Acetic acid	–	–	–	–	–	[630]
Aliphatic and aromatic chlorocompounds	–	–	Halogen specific	–	–	[631]
Chlorophenols	–	Capillary column	Electron capture	Chlorophenols derivitivised by alkylation with pentafluoro-benzyl chloride, then GLC, recovery 70%	ng L^{-1}	[632]
Pentachlorophenol	Benzene	–	Electron capture	Benzene extract extracted with 0.1M potassium carbonate, acetylation of potassium carbonate extract which is extracted with benzene	10pg L^{-1}	[633]
Substituted phenols	Methylene dichloride	–	–	Phenols acetylated, then extracted with methylene dichloride, then GLC	–	[634]

Table 15.14 continued

Organic	Extraction from waste	GLC column conditions	Detector	Comments	LD µg L^{-1}	Ref.
Polychlorobiphenyls and chlorinated insecticides	–	–	–	–	–	[615]
Polychlorobiphenyls and 25 chlorinated insecticides	–	–	–	–	–	[111]
Polychlorobiphenyls and chlorinated insecticides	Hexane	–	Electron capture	Sample clean-up on alumina–silver nitrate and silica gel	–	[635]
Polychlorobiphenyls	–	–	–	Pattern recognition	–	[304]
Phenoxyacetic acid herbicides	–	–	–	–	–	[636]
Miscellaneous organic compounds	–	–	–	Pyrolysis–GLC	–	[637]
Miscellaneous organic compounds	–	–	–	–	–	[638]

Source: Own files

Table 15.15 Gas chromatography of organic compounds in trade effluents

Organic	Extraction from trade effluents	GLC column conditions	Detector	Comments	LD	Ref.
Aromatic hydrocarbons	–	–	–	Headspace analysis	μg L^{-1}	[639]
Aromatic hydrocarbons	Silver membrane filler, cyclohexane extraction	–	–	Steam carrier GLC	–	[640]
Polyaromatic hydrocarbons		Dimethysilated Chromosorb–9, 175–275°C	Split flame ionisation and ultraviolet detection	Combination of GLC and ultraviolet spectroscopy	–	[641]
Ethylbenzene styrene	Carbon disulphide	5% nitrile silicone rubber on Celite 545 at 135°C	Flame ionisation	–	10mg L^{-1}	[642]
Acetic acid	–	–	–	–	–	[630]
Carboxylic acids	–	–	–	–	–	[643]
Phenols	–	–	–	–	–	[644–646]
Phenols, carboxylic acids	–	–	Electron capture	Phenols and carboxylic acids alkylated with pentafluorobenzyl bromide, then GLC	1–10μg L^{-1}	[647]
Syringaldehyde	–	Apiezon non-granulated Teflon at 220°C	Hydrogen flame ionisation	2,4 dinitrophenyl hydrazone	–	[648]
2-furfuraldehyde	–	10% Polysorbate on Chromosorb W at 130°C	Flame ionisation	–	40μg L^{-1}	[649]
Alkyl ethoxylated surfactants	Anion exchange resin	–	–	Surfactants eluted from anion exchange resin with ethanolic hydrochloric acid hydrolysis and conversion to alkyl bromides	–	[650]

Table 15.15 continued

Organic	Extraction from trade effluents	GLC column conditions	Detector	Comments	LD	Ref.
Mono- and di-saccharides	–	10% FFAP on AW–DMCS Chromosorb W at 150–220°C	Dual flame ionisation	–	0.05mg L⁻¹	[651]
Chlorinated aliphatic hydrocarbons	–	–	–	–	–	[652]
Organohalides	–	Capillary column	Electron capture	–	1µg L⁻¹	[516, 517]
Chloroform, bromo dichloromethane, dibromochloro-methane, bromoform	Pentane	–	Electron capture	–	<2µg L⁻¹	[653]
Total organohalogen	–	–	Hall electrolytic	–	–	[654]
1,1,3 tribromopropane	–	–	–	–	–	[655]
Polychlorobiphenyls	–	–	–	–	2µg L⁻¹	[656]
Polychlorobiphenyls	–	–	–	Effect of suspended solids on determination of PCB	–	[656, 657]
Chlorolignosulphonates	–	–	–	Pyrolysis–GLC	–	[563]
Aromatic amines	–	–	–	–	–	[658, 659]
Aniline	–	–	Thermionic N–P detector	Recoveries in excess of 75%	µg L⁻¹	[660]

Table 15.15 continued

Organic	Extraction from trade effluents	GLC column conditions	Detector	Comments	LD	Ref.
Dimethylamine, dimethylformamide, propylamine, di-isopropylamine	–	5% amine 220 on Chromosorb W at 25°C or 28% Pennwalt 223 and 4% potassium hydroxide on Gas Chrom R at 134°C	Flame ionisation	–	–	[71–73]
Benzidine	–	–	–	–	µg L^{-1}	[661]
Chloroanilines, chloronitroanilines	–	–	N–P specific and Hall electrolytic conductivity	–	–	[662]
Nitrocompounds	Select phase extraction on XAD–4 resin ethyl acetate extraction	–	–	–	µg L^{-1}	[663]
2,4 trinitrotoluene, teryl, 2,4 dinitrotoluene, 2,6 dinitrotoluene	Toluene	Capillary column	Electron capture	Recovery 96–103%	µg L^{-1}	[664]
2,4 dinitrotoluene	–	–	–	–	4–12µg L^{-1}	[665]
Nitrosamines	–	–	–	–	–	[604]
Dimethylformamide	–	–	Flame ionisation	–	–	[72]
Diethanol amine	Dimethyl-acetamide	–	–	Derivitivisation GLC	0.5µg L^{-1}	[666]
Nitriloacetic acid	–	–	–	–	10µg L^{-1}	[667]
N-methyl pyrrolidine	–	–	–	–	1µg L^{-1}	[668]

Table 15.15 continued

Organic	Extraction from trade effluents	GLC column conditions	Detector	Comments	LD	Ref.
Pyridasinones	–	–	–	–	–	[669]
1-methyl-2-pyrrolidones	–	–	–	–	mg L^{-1}	[670]
5,5 dialkyl and 5-alkyl hydantoins	–	–	–	–	–	[671]
Pyridine	–	–	–	–	40μg L^{-1}	[672]
Methane, diol, dimethyl sulphide	–	–	–	–	0.1mg L^{-1}	[623]
Mercaptans, disulphides	–	–	Microcoulo-metric and flame ionisation	Headspace sampling of trade effluent to increase sensitivity by 10^2 to 10^3	Traces	[673, 674]
Dithiocarbamate insecticides	–	–	Hall electrolytic conductivity in S mode	Conversion of insecticide to carbon disulphide	–	[675]
Triazine herbicides	–	–	–	–	–	[676]
Carboxylic acids, nitriles, aldehydes, ketones	–	–	–	Steam distillation, solvent extraction, ion exchange chromatography, GC	–	[677]
Miscellaneous organic compounds	–	Capillary	–	Capillary chromatography with simultaneous headspace injection on two columns of different polarity	–	[678]
Solvent residues	–	–	–	Gas chromatography on various porous polymer packed columns	–	[679]
Miscellaneous low carbon number organic compounds	–	–	–	–	–	[680]

Source: Own files

trimethylarsine oxide in non saline waters with detection limits of several ng L $^{-1}$. The arsines are volatilised from the sample by gas stripping; the other species are then selectively reduced to the corresponding arsines and volatilised. The arsines are collected in a cold trap cooled with liquid nitrogen. They are then separated by slow warming of the trap or by gas chromatography and measured with atomic absorption, electron capture and/or flame ionisation detectors. He found that these four arsenic species all occurred in non saline water samples.

The gas chromatograph is equipped with a ^{63}Ni electron capture detector mounted in parallel with a flame ionisation detector and an auxiliary vent by the use of a column effluent splitter. The separation is performed on a 4.8mm od, 6m long stainless steel column packed with 16.5% silicone oil DC–550 on Chromosorb W AW DMCS.

To isolate the arsine species the sample 50mL is introduced into the gas stripper with a hypodermic syringe through the injection port.Any volatile arsines in the sample are stripped out by bubbling a helium stream through the sample. Then 1mL of the Tris buffer solution for each 50mL sample is added, giving an initial pH of about 6. Into this solution 1.2mL of 4% sodium borohydride solution is injected while continuously stripping with helium. After about 6–10min the As(III) is converted to arsine and stripped from the solution. The pH at the end of this period is about 8. Then 2mL of 6N hydrochloric acid is added, which brings the pH to about 1. The addition of three aliquots of 2mL of 4% sodium boro-hydride solution during 10min reduces As(IV), monomethylarsonic acid, dimethylarsinic acid, and trimethylarsine oxide to the corresponding arsines, which are swept out of the solution by the helium stream coming from the reaction vessel stream.

The applications of gas chromatography to the determination of organo-arsenic compounds in non saline waters are reviewed in Table 15.17.

15.2.1.2 Organolead compounds

Chau et al. [684,685] have described a simple and rapid extraction procedure to extract the five tetraalkyllead compounds (Me$_4$Pb, Me$_3$EtPb, Me$_2$Et$_2$Pb, MeEt$_3$Pb and Et$_4$Pb) in hexane extracts of water samples. The extracted compounds are analysed in their authentic forms by a gas chromatographic–atomic absorption spectrometry system. Other forms of inorganic and organic lead do not interfere. The detection limit for water (200mL) is 0.50µg L $^{-1}$. An average recovery of 89% was obtained by this procedure for the aforementioned alkyllead compounds.

The application of a combination of gas chromatography and atomic absorption spectrometry to the determination of tetraalkyllead compounds has also been studied by Segar [686]. In these methods the gas chromatography flame combination showed a detection limit of about 0.1µg lead. Chau *et al.* [685] have applied the silica furnace in the atomic absorption unit and have also shown that the sensitivity limit for the detection of lead can be enhanced by three orders of magnitude.

Chau *et al.* [687] developed a gas chromatographic method for the determination of ionic organolead species and applied it to samples of Lake Ontario water. When 1L of sample water is taken the following species can be determined in amounts down to 0.1µg L $^{-1}$: Me_2Pb^{2+} Me_3Pb^+, Et_2Pbt^{2+} Et_3Pb^+ and Pb^{2+}.

The highly polar ions are quantitatively extracted into benzene from aqueous solution after chelation with dithiocarbamate. The lead species are butylated by a Grignard reagent to the tetraalkyl form, $R_nPbBu_{(4-n)}$ (R = CH_3 or C_2H_5) and Bu_4Pb, all of which can be quantified by a gas chromatography–atomic absorption spectrometry method. Molecular covalent tetralkyllead species, if present in the sample, are also extracted and quantified simultaneously. Other metals co-extracted by the chelating agent do not interfere.

Chau *et al.* [688,689] have described a simple and rapid extraction procedure to extract the five tetraalkyllead compounds (Me_4Pb, Me_3Et_2Pb, $MeEt_3Pb$, Me_2Et_2Pb and Et_4Pb) from non saline waters. The extracted compounds are analysed in their authentic forms by a gas chromatographic–atomic absorption spectrometry system. Other forms of inorganic lead do not interfere. The detection limit for water (200mL) is 0.50µg L $^{-1}$.

The main interest of Chau *et al.* [689] was in the determination of organically bound lead produced by biological methylation of inorganic and organic lead compounds in the aquatic environment by micro-organisms. The gas chromatographic–atomic absorption system used by Chau *et al.* (used without a sample injection trap) for this procedure has been described [690]. The extract was injected directly into the column injection port of the chromatograph. Instrumental parameters were identical. A Perkin–Elmer electrodeless discharge lead lamp was used; peak areas were integrated.

Other applications of gas chromatography to the determination of organolead compounds in non saline waters are discussed in Table 15.17.

15.2.1.3 Organomanganese compounds

Aue *et al.* [691] used gas chromatography with electron capture detection to determine methylcyclopentadienyl manganese tricarbonyl in non saline waters.

15.2.1.4 Organomercury compounds

Nishi and Horimoto [692,693] determined trace amounts of methyl-, ethyl- and phenylmercury compounds in river waters and industrial effluents. In this procedure, the organomercury compound present at less that 0.4ng L $^{-1}$ in the sample (100–500mL) is extracted with benzene (2 × 0.5 vol. relative to that of the aqueous solution). The benzene layer is then back-extracted with 0.1% L–cysteine solution (5mL), and recovered from the complex by extracting with benzene (1mL) in the presence of hydrochloric acid (2mL) and submitted to gas chromatography using a stainless steel column (197cm × 3mm) packed with 5% of diethylene glycol succinate on Chromosorb W (60–80 mesh) with nitrogen as carrier gas (60mL min $^{-1}$) and an electron capture detector. The calibration graph is rectilinear for less than 0.1µg of mercury compound per millilitre of the cysteine solution. This method is capable of determining mercury down to 0.4µg L $^{-1}$ for the methyl and ethyl derivatives and 0.86µg L $^{-1}$ for the phenyl derivative.

The above method has been modified [693] for the determination of methylmercury(II) compounds in aqueous media containing sulphur compounds that affect the extractions of mercury. The modified method is capable of handling samples containing up to 100mg of various organic and inorganic sulphur compounds per 100mL. The aqueous test solution (150mL) containing 100mg of methylmercury ions per 100mL is treated with hydrochloric acid until the acid concentration is 0.4%, then 0.3–1g of mercuric chloride is added (to displace methyl mercury groups bonded to sulphur), and the mixture is filtered. The filtrate is treated with aqueous ammonia in excess to precipitate the unconsumed inorganic mercury which is filtered off; this filtrate is made 0.4% in hydrochloric acid and extracted with benzene. The benzene solution is shaken with 0.1% L–cysteine solution, the aqueous phase is acidified with concentrated hydrochloric acid and then shaken with benzene for 5min and this benzene solution is analysed by a gas chromatography as described above.

Dressman [694] used the Coleman 50 system in his determination of dialkylmercury compounds in river waters. These compounds were separated in a glass column (1.86m × 2mm) packed with 5% of DC–200 plus 3% of QF–1 on Gas Chrom Q (80–100 mesh) and temperature programmed from 70 to 180°C at 20°C min $^{-1}$, with nitrogen as carrier gas (50mL min $^{-1}$). The mercury compound eluted from the column was burnt in a flame ionisation detector, and the resulting free mercury was detected by a Coleman mercury analyser MAS–50 connected to the exit of the flame ionisation instrument; down to 0.1mg of mercury could be detected. River water (1L) was extracted with pentane–ethyl ether (4:1) (2 × 60mL). The extract was dried over sodium sulphate, evaporated to 5mL and analysed as above.

Ealy et al. [695] have discussed the determination of methyl-, ethyl- and methoxymercury(II) halides in water. The mercury compounds were separated from the sample by leaching with 1M sodium iodide for 24h and then the alkylmercury iodides were extracted into benzene. These iodides were then determined by gas chromatography with 5% of cyclohexanesuccinate on Anakron ABS (70–80 mesh) and operated at 299°C with nitrogen (56mL min $^{-1}$) as carrier gas and electron capture detection. Good separation of chromatographic peaks was obtained for the mercury compounds as either chlorides, bromides or iodides. The extraction recoveries were monitored by the use of alkylmercury compounds labelled with ^{208}Hg.

Zarnegar and Mushak [696] have described a gas chromatographic procedure for the determination of organomercury compounds and inorganic mercury in non saline water. The sample is treated with an alkylating or arylating reagent and the organomercury chloride is extracted into benzene. Gas chromatography is carried out using electron capture detection. The best alkylating or arylating reagents were penta-cyano(methyl)cobalt(III) and tetraphenylborate. Inorganic and organic mercury could be determined sequentially by extracting and analysing two aliquots of sample, of which only one had been treated with alkylating reagent. The limits of detection achieved in the method were 10–20ng.

Cappon and Crispin-Smith [697] have described a method for the extraction, clean-up and gas chromatographic determination of organic (alkyl and aryl) and inorganic mercury in water. Methyl-, ethyl- and phenylmercury are first extracted as the chloride derivatives. Inorganic mercury is then isolated as methylmercury upon reaction with tetramethyltin. The initial extracts are subjected to thiosulphate clean-up and the organomercury species are isolated as the bromide derivatives. Total mercury recovery ranges between 75 and 90% for both forms of mercury, and is assessed by using appropriate ^{203}Hg-labelled compounds for liquid scintillation spectrometric assay. Specific gas chromatographic conditions allow detection of mercury concentrations of 1µg L $^{-1}$ or lower. Mean deviation and relative accuracy average 3.2 and 2.2% respectively.

Bowles and Apte [698] have described a method for the determination of methylmercury compounds in non saline waters using steam distillation followed by gas chromatography with an atomic fluorescence spectrometric detector. These workers evaluated steam distillation as a technique for the separation of methylmercury compounds from water and obtained recoveries in spiking experiments ranging from approximately 100% in fresh waters and estuaries to 80% in sea water.

The addition of ammonium pyrrolidine dithiocarbamate improved recoveries from sea water to 85%. The co-distillation of inorganic mercury was prevented by the addition of ammonium pyrrolidine dithiocarbamate.

The precision achieved in this method on Milli–Q water was 0.2% rsd at the 0.2ng L $^{-1}$ methylmercuric chloride level and 1.56% rsd at the 2ng

L $^{-1}$ level. A detection limit of 0.24ng L $^{-1}$ was achieved for a 50mL water sample.

Other applications of gas chromatography to the determination of organomercury compounds in non saline water are reviewed in Table 15.17.

15.2.1.5 Organoselenium compounds

The determination of dimethylselenides and diethylselenide in scrubbed air samples has been discussed [699]. This method is based on a gas chromatographic–atomic absorption procedure and could, no doubt, be adapted for the analysis of waters.

15.2.1.6 Organosilicon compounds

Mahone et al. [700] have given details of a procedure to determine water-borne organosilicon substances such as silanols or silanol–functional materials by conversion to trimethylsilylated derivatives which can be determined by gas–liquid chromatography. The method gives good accuracy and precision in the milligram per litre range and with suitable precautions can be extended to the low microgram per litre range.

15.2.1.7 Organotin compounds

Soderquist and Crosby [701] have described a gas chromatographic procedure for the determination of down to 0.01µg L $^{-1}$ triphenyltin hydroxide and its possible degradation products (tetraphenyltin, diphenyltin, benzene stannoic acid and inorganic tin) in water. The phenyltin compounds are detected by gas–liquid chromatography using an electron capture detector after conversion to their hydride derivatives, while inorganic tin is determined by a novel procedure which responds to tin dioxide and aqueous solutions of tin salts. The basis for the method involves extraction of the phenyltin species from water followed by their quantitation as phenyltin hydrides by electron capture gas chromatography and analysis of the remaining aqueous phase for inorganic tin (Sn^{4+} plus SnO_2) by colorimetry.

$$Sn^{4+}aq$$
$$\nearrow$$
$$Ph_3Sn^+aq \rightarrow Ph_2Sn^{+2}aq \rightarrow PhSn^{+3}aq \qquad \text{Possible degradation}$$
$$\searrow$$
$$SnO_2aq$$

$$Ph_3SnH, Ph_2SnH_2, PhSnH, \qquad Sn\text{–}PCV \text{ complex} \qquad Analysis$$
$$ec\text{–}glc \qquad\qquad colorimetry$$

Table 15.16 Method sensitivity

Species	Method	Minimum detectable amount	Method sensitivity $(\mu g\ mL^{-1})^a$
Ph_4Sn	FID–GLC	5.0ng as Ph_4Sn	0.015
Ph_3Sn^{1+}	EC–GLC	0.2ng as Ph_3SnH	0.003
Ph_3Sn^{2+}	EC–GLC	0.2ng as Ph_2SnH_2	0.003
$PhSn^{3+}$	EC–GLC	0.2ng as $PhSnH_3$	0.003
Total extractable organotins	Colorimetry	1.0µg as Sn	0.01
Sn^{4+}	Colorimetry	1.0µg as Sn	0.007
$SnO_3 + Sn^{4+}$	Colorimetry	1.0µg as Sn	0.007

Note
a For 200ml samples

Source: Reproduced by permission from the American Chemical Society [701]

Gas–liquid chromatography was performed on a dual-column–dual-detector Varian Model 2400 instrument equipped, on one side, with a flame ionisation detector and 0.7m × 2mm (id) glass column containing 3% OV–17 on 60–80 mesh Gas Chrom Q. Column, injector and detector temperatures were 265, 275 and 300°C, respectively; the carrier gas was nitrogen. Tetraphenyltin eluted within 8min under those conditions. The second side of the chromatograph was equipped with a tritium electron capture detector and a 1.1m × 2mm (id) glass column containing 4% SE–30 on 60–80 mesh Gas Chrom Q. The injector and detector temperature were 210°C, the carrier gas (nitrogen) flow rate was 20mL min^{-1}, and column temperatures which eluted the following compounds within 6min were: Ph_3SnH, 190°C; Ph_2SnH_2, 135°C; $PhSnH_3$, 45°C. Combined gas–liquid chromatography–mass spectrometry was performed on a Finnigan Model 1015 utilising a 1.0m × 2mm (id) glass column containing 3% OV–17 on 60–80 mesh Gas Chrom Q. Infrared spectra were obtained in hexane solution.

The sensitivity of the method for each of the tin species examined is summarised in Table 15.16. These limits could be decreased either by increasing the sample size, avoiding some of the sub-analyses, or both. None of the non saline water samples analysed by Soderquist and Crosby [701] contained materials which interfered with the determination of any of the tin compounds of interest.

Braman and Tompkins [702] have developed methods for the determination of trace amounts (ppb) of inorganic tin and methyltin compounds in river waters and, indeed, these workers were the first to confirm the presence of methyltin compounds in non saline waters. Tin compounds are converted to the corresponding volatile hydride (SnH_4, CH_3SnH_3, $(CH_3)_2SnH_2$ and $(CH_3)_3SnH$) by reaction with sodium borohydride at pH6.5 followed by a separation of the hydrides and then atomic absorption spectrometry using a hydrogen-rich hydrogen–air flame emission type detector (Sn–H band). The technique described has a detection limit of 0.01ng as tin and hence parts per trillion of organotin species can be determined in water samples.

Braman and Tompkins [702] found that stannane (SnH_4) and methyl-stannanes (CH_3SnH_3, $(CH_3)_2SnH_2$, and $(CH_3)_3SnH$) could be separated very well on a column comprising silicone oil OV–3 (20% w/w) supported Chromosorb W.

Chau et al. [703] described an improved extraction procedure for the polar methyltin compounds, using benzene containing tripolone, from water saturated with sodium chloride. Tetramethylbutyltin derivatives were prepared in the extracts and were separated by gas chromatography in well-defined peaks. The difference in sensitivity of the different tin species is attributed to difference of behaviour in the atomic absorption furnace. A large number of samples can be analysed in large volumes of water. The overall recovery is satisfactory, coefficients of variation using six replicate samples was 5–11% and a detection limit of 0.04µg L^{-1} was achieved.

Other applications of the application of gas chromatography to the determination of organotin compounds in non saline waters are reviewed in Table 15.17.

15.2.2 Aqueous precipitation

15.2.2.1 Organolead compounds

Lobinski et al. [747] carried out speciation analysis of organolead compounds in Greenland snow at the femtogram per gram level by capillary gas chromatography using an atomic absorption detector. In this procedure the snow sample was mixed with EDTA buffered to pH8.5 and extracted with hexane to preconcentrate organolead compounds. The extract was propylated using propyl magnesium chloride and the product analysed by capillary gas chromatography.

Et_3Pb^+ and Et_2Pb^{2+} could be detected in amounts down to 0.02 and 0.02–0.5µg kg^{-1} respectively. Neither Me_3Pb^+ or Me_2Pb^{2+} were found in the snow samples analysed.

15.2.2.2 Organotin compounds

Braman and Tompkins [748] have developed methods for the deter-mination of microgram per litre amounts of inorganic tin and methyltin

Table 15.17 Gas chromatography of organometallic compounds in non saline waters

Organometallic	Extraction from water	GLC column details	Detector	Comments	LD	Ref.
Organoarsenic compounds						
Alkyl arsenic compounds	–	–	Electron capture or flame ionisation	Use of sodium borohydride as reducing agent prior to GLC	0.21 µg L^{-1}	[683, 704–720]
			Atomic absorption	as above		[683, 706]
			Emission spectrometric	as above		[711–717]
			Spectrophotometric	as above		[716]
			Mass specific	as above		[706]
			Neutron activation	as above		[712]
Hydroxidimethyl arsine oxide	–	10% DC–200 on GasChrom Q	Electron capture	Hydroxydimethyl arsine oxide converted to iododimethyl arsine, then GLC, 92.3% recovery	–	[719]
Trimethyl arsine	–	–	Atomic absorption	–	5µg	[720]
Organoarsenic	–	Poropak Q	Electrooxidation potential	–	0.21	[709]
Organo antimony compounds						
Organoantimony	–	Poropak Q	Electro oxidation potential	–	0.2µg L^{-1}	[709]

Table 15.17 continued

Organometallic	Extraction from water	GLC column conditions	Detector	Comments	LD	Ref.
Organolead compounds						
Alkyllead compounds	Hexane–pentane	–	–	Derivitivisation of organolead with ammonium in tetraborate, solvent extraction, then GLC	–	[721]
Alkyllead salts	–	–	–	–	–	[722]
Methyl lead ions	–	–	Atomic absorption	Conversion of lead hydride with sodium tetraethyl borate, then purge and trap GLC	0.2ng L^{-1}	[723]
Dialkyl and tri-alkyllead salts	–	–	Atomic absorption	90% recovery of di- and trialkyl lead salts from water	1.25mg Pb L^{-1} (500ml sample)	[724]
EtMe$_2$PhPb, Et$_2$MePhPb, EtMePh$_2$Pb	Benzene:hexane (1:1)	Capillary column fused silica DB–1	Electron capture	Extraction from water with dithizone in benzene:hexane (1:1), derivitivised with Grignard reagent (phenyl magnesium bromide) in tetrahydrofuran	1–20µg L^{-1}	[725]
Alkyllead compounds and alkyllead salts	–	–	–	Results compared to thin layer chromatography results. Study of influence of sediments on results	0.1–0.5µg alkyllead salt	[726]
Organomercury compounds						
Methylmercury	–	–	Atomic absorption	–	sub ppt	[727]

Table 15.17 continued

Organometallic	Extraction from water	GLC column conditions	Detector	Comments	LD	Ref.
Dimethyl, dipropyl and dibutyl mercury	–	5% DC200 and 3% QF–I on GasChrom Q programmed 60 to 180°C	Coleman 50 mercury analyser	–	–	[728]
Dimethyl mercury	–	–	Fourier transform infrared	Reaction with sodium tetra-borohydride to convert to mercury hydrides	50–100pg	[729]
Aryl mercury compounds	–	10% Dexil 300 on Anakom SD and Durapak Carbo-wax–400 on Parasil F	–	70.5% recovery	–	[730–732]
Methyl and ethyl mercury compounds	Preconcentrated on sulphydril cotton; then benzene extraction	–	Electron capture	Mercury compounds eluted from sulphydril cotton with hydro-chloric acid–sodium chloride, benzene extraction, then GLC 42–68% recovery of methy mercury	0.04ng L^{-1}	[733]
			Organotin compounds			
Mono, di, tri and tetraalkyl tin compounds	Pentane	Capillary column	Helium microwave induced plasma emission spectrometry	Ionic alkyltins extracted as diethyldithio-carbamates into pentane then converted to pentyl magnesium bromide with Grignard reagent, then GLC	0.5pg	[734]

Table 15.17 continued

Organometallic	Extraction from water	GLC column conditions	Detector	Comments	LD	Ref.
Organotin compounds	–	High resolution capillary column	Flame photometric	Butyltin compounds ethylated prior to GLC	–	[735]
Tributyl, dibutyl tin species	Cation exchange column	–	–	Compounds concentrated on cation exchange column, desorbed into diethyl ether–hydrogen chloride, then methylated with Grignard reagent, then GLC	–	[736, 737]
Trialkyl tin and triphenyl tin compounds	Benzene	–	Electron capture	Silica gel clean-up of benzene extract, then conversion to tin hydrides with sodium borohydride	0.8μg L^{-1}	[738]
Butyltin chlorides	–	–	–	–	0.4ng	[699]
Butyl phenyl and cyclohexyl tin compounds	–	–	Flame photometric	Organotin compounds ethylated with sodium ethyl borate, collected on C$_{18}$ resin disc and eluted with supercritical carbon dioxide prior to GLC	–	[739]
Organotin compounds	–	–	Flame photometric	Organotin compounds ethylated and extracted with sodium tetraethyl borate	0.4–400 ng dm^3	[740]
Organotin compounds	–	–	Mass spectrometry and atomic emission spectrometry	–	–	[741]
Tin tetraalkyls	–	–	–	Determination of solubility of tin tetraalkyls in water	–	[742]

Table 15.17 continued

Organometallic	Extraction from water	GLC column conditions	Detector	Comments	LD	Ref.
13 organotin compounds	Supercritical fluid extractor	–	Atomic emission spectrometry	Speciation study. Organotin compounds reacted with pentyl magnesium bromide Grignard reagent	–	[743, 744]
Organotin compounds	Benzene–tropalone	–	–	Solvents for preconcentration organotin	–	[745] [746]
	n-hexane tropalone	–	–	as above		
	Methylene dichloride	–	–	as above		[701]
	benzene	–	–	as above		[738]

Source: Own files

compounds in rain and other waters. Tin compounds are con-verted to the corresponding volatile hydride (SnH_4, CH_3SnH_3, $(CH_3)_2SnH_2$ and $(CH_3)_3SnH$) by reaction with sodium borohydride at pH6.5 followed by gas chromatographic separation of the hydrides and then atomic absorption spectroscopy using a hydrogen-rich hydrogen–air flame emission type detector (Sn–H band). The technique described has a detection limit of 0.01ng as tin and hence parts per trillion of organotin species can be determined.

Braman and Tompkins [748] found that stannane (SnH_4) and methyl-stannanes (CH_3SnH_3, $(CH_3)_2SnH_2$ and $(CH_3)_3SnH$) could be separated very well on a column comprising silicone oil OV–3 (20% w/w) supported Chromosorb W.

An average total tin content of rain water was found to be 25ng L^{-1} and the methyltin form constitutes 24% of that total.

15.2.3 Seawater

15.2.3.1 Organomercury compounds

Chiba *et al.* [749] used atmospheric pressure helium microwave induced plasma emission spectrometry with the cold vapour generation technique combined with gas chromatography for the determination of methylmercury chloride, ethylmercury chloride and dimethylmercury in sea water following a 500-fold preconcentration using a benzene–cysteine extraction technique.

The analysis system consisted of a Shimsdzu QC–6A gas chromato-graph, a chemically deactivated four-way valve for solvent ventilation, a heated transfer tube interface, a Beenakker-type TM_{010} microwave resonance cavity, and an Ebert-type monochromator (0.5m focal length).

The chromatograms were detected with the thermal conductivity and the microwave-induced plasma detectors in series. When the microwave-induced plasma detector was used as a detector, the emission signals were monitored at 253.7nm, and the solvent was vented through a four-way valve before reaching the microwave induced plasma detector.

Detection limit and standard deviation date obtained for these organo-mercury compounds without preconcentration are shown in Table 15.18.

The total extraction efficiency involved in the three stages of the extraction of methylmercury chloride from sea water by the cysteine–benzene extraction technique was reproducible at 42% for a 500-fold concentration giving a detection limit of 0.4ng L^{-1} and a relative standard deviation of 6% at the 20ng L^{-1} level.

Summarising, it can be seen that in non saline and sea waters organomercury compounds can be speciated and determined in amounts down to 10ng L^{-1}, thereby meeting even the most searching present day requirements.

Table 15.18 Analytical figures of merit in the determination of alkylmercury compounds by the GLC–MIP system

Compound	Detection limit $\mu g\ L^{-1}$	Rel. Std. deviation[a]	Dynamic range decades
Methylmercury chloride	0.09	2.0	5
Ethylmercury chloride	0.12	2.0	5
Dimethylmercury	0.40	3.0	4.5

[a]Measured with $1\ \mu g\ L^{-1}$ mercury for each compound

Source: Reproduced by permission from the American Chemical Society [749]

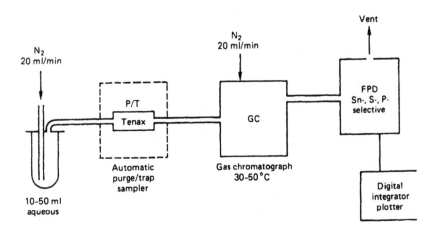

Fig. 15.8 The purge/trap GC–FPD system and operating conditions
Source: Reproduced by permission from Kluwer Academic, Plenum, Amsterdam [750]

15.2.3.2 Organotin compounds

Brinckmann and co-workers [750] used a gas chromographic method with or without hydride derivatisation for determining volatile organotin compounds (eg tetramethyltin) in seawater. For non-volatile organotin compounds, a direct liquid chromatographic method was used. This system employs a 'Tenax–GC' polymeric sorbent in an automatic purge and trap (P/T) sampler coupled to a conventional glass column gas chromatograph equipped with a flame photometric detector (FPD). Fig. 15.8 is a schematic of the P/T–GC–FPD assembly with typical operating conditions. Flame conditions in the FPD were tuned to permit maximum response to SnH emission in a H-rich plasma, as detected through narrow band–pass interference filters (610 ± 5nm) [751]. Two modes of analysis were used: (1) volatile stannanes were trapped directly from sparged 10–50ml water samples with no pretreatment and (2) volatilised tin species were trapped from the same or replicate water samples following rapid injection of aqueous excess sodium borohydride solution directly into the P/T sparging vessel immediately prior to beginning the P/T cycle [752].

Brinckmann [750] generated calibration curves by the P/T gas chromatography–PFD method for borohydride reductions of tin(IV), tin (II) and Me_2Sn^{2+} species to SnH_4 and SnH_2 Me_2 SnH_3Me, respectively in distilled water, 0.2M sodium chloride and bay water. All three analytes showed a substantial increase in their calibration slopes in going from distilled water to 0.2M sodium chloride solution, the latter approximating the salinity and ionic strength common to estuarine waters. Presumably these effects could arise from formation of chlorohydroxyl tin species favouring more rapid hydridisation (see equations 15.1 and 15.2) as well as the more propitious partition coefficients for dynamic gas stripping of the volatile tin hydrides from saline solutions. In the typical laboratory distilled water calibration solutions, only 16% of tin(II) was recovered as SnH_4, compared with tin(IV), though this sensitivity ratio can probably be altered somewhat with pH changes. However, in spiking anaerobic pre-purged Chesapeake Bay water with these three tin species, a striking reversal occurred in overall relative sensitivities, ie calibration slopes. Brinckmann [750] found that not only was Me_2SnH_2 generation repressed by 50% but that, very significantly, SnH_4 formation from tin(IV) was reduced by 15-fold as compared with the sodium chloride medium.

$$RnSn_4-^{n+}(aq) + excess\ BH_4^- \rightarrow RnSNH_{4-n} \qquad (15.1)$$

$$4HSnO_2^- + 3BH_4^- + 7H+ + H_2O \rightarrow 3H_3BO_3 + 4SnH_4 \qquad (15.2)$$

The overall effect of estuarine water on the hydridisation process is thus one of reducing yields of the three tin species tested. It is expected that not only the dissolved and particulate organics and chloride influence formation of Sn–H bonds, but that other aquated metal ions play an important role, too. Several workers have reported that, for example, arsenic(III), arsenic(V), copper(II), nickel(II), mercury(II), lead(II) and silver(I) interfere by unknown means at low concentrations [702,753].

The hydride generation method cannot adequately differentiate between aquated tin(IV) and tin(II) which may coexist in certain, especially anaerobic, environments found in marine waters. Previous reports [699,702] or inorganic tin, speciated as 'tin(IV)', should probably

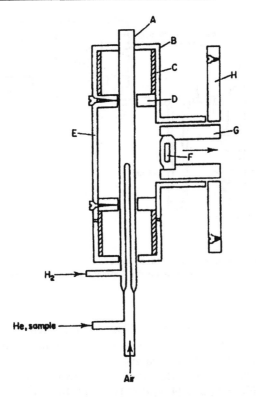

Fig. 15.9 Quartz burner and housing. A, quartz burner; B, PVC cap; C, PVC tubing; D, mounting ring; E, PVC T–joining, 1.25 inch; (METRIC); F, filter and holder; G, PVC coupling; H, connection to photomultiplier housing (PM)
Source: Reproduced by permission from the American Chemical Society [702]

be regarded as 'total reducible inorganic tin' until more discriminatory techniques become available.

Takahasi [754] converted tributyltin and triphenyltin compounds in seawater to the corresponding organotin hydrides prior to determination by packed column gas chromatography with electron capture detection.

Valkirs *et al.* [755,756] compared two methods for the determination of µg L^{-1} levels of dialkyltin and tributyltin species in marine and estuarine waters. The two methods studied were hydride generation followed by atomic absorption spectrometry and gas chromatography with flame photometric detection. Good agreement was obtained between the results of the two methods. Down to 0.01mg kg^{-1} of butyltin compounds, including tri-*n*-butyl tin and tri-*n*-butyl tin oxide, could be detected.

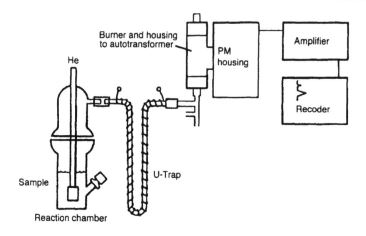

Fig. 15.10 Apparatus arrangement for tin analysis
Source: Reproduced by permission from the American Chemical Society [702]

Studies by Braman and Tompkins [702] have shown that non volatile methyltin species $Me_nSn_{(4-N)}$ ($n = 1 - 3$) are ubiquitous at ng L^{-1} concentrations in marine and freshwaters.

Tin compounds are converted to the corresponding volatile hydride (SnH_4, CH_3, SnH_3, $(CH_3)_2$ SnH_2 and $(CH_3)_3SnH$) by reaction with sodium borohydride at pH6.5 followed by separation of the hydrides by gas chromatography and then detection by atomic absorption spectroscopy using a hydrogen-rich hydrogen–air flame emission type detector (Sn–H band). The apparatus used is shown in Figs. 15.9–15.11.

The technique described has a detection limit of 0.01ng as tin and hence parts per trillion of organotin species can be determined in water samples.

Braman and Tompkins [702] found that stannane (SnH_4) and methylstannanes (CH_3SnH_3, $(CH_3)_2SnH_2$ and $(CH_3)_3SnH$) could be separated well on a column comprising silicone oil OV–3 (20% w/w) supported Chromosorb W. A typical separation achieved on a coastal water sample is shown in Fig. 15.11. Average tin recoveries from seawater are in the range 96–109%.

A number of esturial waters, from in and around Tampa Bay, Florida, area were analysed by this method for tin content. All samples were analysed without pre-treatment. Samples which were not analysed immediately were frozen until analysis was possible. Polyethylene bottles, 500mL volume, were used for sample acquisition and storage. The results of these analyses appear in Table 15.19, and the average total tin content of estuarine waters was 12ng L^{-1}. Approximately 17–60% of

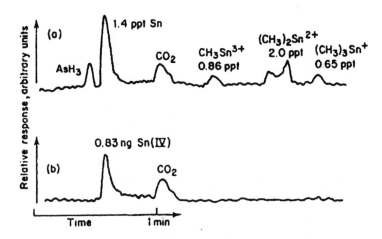

Fig. 15.11 Environmental sample analysis and blank: separation of methyl stannanes (a) environmental analysis, Old Tampa Bay; (b) typical blank
Source: Reproduced by permission from the American Chemical Society [702]

the total tin present was found to be in methylated forms. This procedure, although valuable in itself, is incomplete in that any monobutyltin present escapes detection. Excellent recoveries of monobutyltin species are achieved with tropolone.

15.2.4 Potable waters

15.2.4.1 Organolead compounds

Estes et al. [722] described a method for the measurement of triethyl- and trimethyllead chloride in potable water, using fused silica capillary column gas chromatography with microwave excited helium plasma lead specific detection. Element specific detection verified the elution of lead species, a definite advantage to the packed column method. The method involved the initial extraction of trialkyllead ions from water into benzene, which was then vacuum reduced to further concentrate the compounds. Direct injection of the vacuum concentrated solutions into the gas chromatography–microwave excited helium plasma system gave delectability of triethyllead chloride at the 30mg L^{-1} level and trimethyl-lead chloride at the mg L^{-1} level, but the method was time consuming and only semiquantitative.

Estes et al. [757] and Chau et al. [758] have also reported the n-butyl Grignard derivatisation of the trialkyllead ions extracted into benzene as the chlorides from spiked tap water which has been saturated with

Table 15.19 Analysis of estuarine water samples*

Sample	Tin(IV) ng L⁻¹	Tin(IV) %	Methyltin ng L⁻¹	Methyltin %	Dimethyltin ng L⁻¹	Dimethyltin %	Trimethyltin ng L⁻¹	Trimethyltin %	Total tin ng L⁻¹
Sarasota Bay	5.7	47	3.3	27	2.0	16	1.1	9.1	12
Tampa Bay	3.3	27	8.0	66	0.79	6.5	nd		12
McKay Bay	20	88	nd		2.2	9.6	0.45	2.0	23
Hillsborough Bay,	nd		dd		1.8	71	0.71	29	2.5
Hillsborough Bay, Seddon Channel North	12	86	0.74	5.3	0.91	6.6	0.35	2.5	14
Hillsborough Bay, Seddon Channel South	13	83	nd		2.4	15	0.31	1.9	16
Manatee River	4.8	61	1.4	1.7	1.1	14	0.65	8.2	7.9
Alafia River	3.4	73	nd		0.75	16	0.55	12	4.7
Palm River‡	567	98	nd		4.6	0.80	4.0	0.69	576
Bowes' Creek	8.6	42	8.5	42	3.3	16	nd		20
Average	7.9	63	2.4	19	1.7	14	0.46	3.7	12

* Data are average of duplicates
† nd less than 0.01ng L⁻¹ for methyltin compounds and 0.3ng L⁻¹ for inorganic tin
‡ This set of values was not used in computing the average

Source: Reproduced by permission from the American Chemical Society [702]

sodium chloride. A precolumn trap enrichment technique is substituted to replace solvent extract vacuum reduction. Final measurement of the lead compounds, now as n-butyltrialkylleads, is undertaken with the gas chromatography–microwave emission detector system. Precolumn Tenax trap enrichment of the derived trialkylbutylleads enables determination to low µg L $^{-1}$ levels to be carried out. In this procedure the water sample (100mL) is adjusted to pH7, saturated with sodium chloride and extracted with a small volume of Specpure benzene.

15.2.4.2 Organotin compounds

The hydride conversion gas chromatographic method of Braman and Tompkins [702] and discussed in section 15.2.1.7 has been applied to the determination of methyltin in compounds in potable water.

Maguire and Huneault [759] developed a gas chromatographic method with flame photometric detection for the determination of bis(tri-n-butyltin) oxide and some of its possible dialkylation products in potable and non saline waters. This method involves extraction of bis(tri-n-butyltin) oxide, Bu_2Su^{2+}, $BuSn^{3+}$ and Sn^{4+} from the water sample with 1% tropolone in benzene, derivatisation with a phenyl Grignard reagent to form the various $Bu_n(pentyl_{(4-n)}Sn$ species followed by analysis of these by flame photometric gas chromatography–mass spectrometry. The pentyl derivatives are all sufficiently non-volatile compared with benzene that none are lost in the solvent stripping yet they are volatile enough to be analysed by gas chromatography.

15.2.5 Trade effluents

15.2.5.1 Organomercury compounds

The method discussed in section 15.2.1.4 for the determination of organo-mercury in non saline waters has also been applied to industrial effluents [692,693].

15.3 Cations

15.3.1 Non saline waters

15.3.1.1 Arsenic, antimony, selenium and tin

Kusaka et al. [760] generated the gaseous hydrides of antimony(III), arsenic(III) and tin by sodium borohydride reduction. The hydrides were swept from solution onto a Porapak Q column where they were separated and detected at a gold gas-porous electrode by measurement of the respective electro-oxidation currents. Detection limits for 5ml samples were: As(III) (0.2µg L $^{-1}$); Sn(II) (0.8µg L $^{-1}$); Sb(III) (0.2µg L $^{-1}$). The order of elution is hydrogen, arsine, stannane, stibine and mercury, ie the order of increasing molecular weight.

Most of the detectors commonly used for gas chromatography have been applied to the detection of the hydrides, among them the thermal conductivity, flame ionisation and the electron capture detector [761].

Vien and Fry [762] have reported a gas chromatographic determination of arsenic, selenium, tin and antimony in natural waters. The gaseous hydrides are generated, concentrated on a cold trap, and then injected into the gas chromatograph with the use of drying agents or carbon dioxide scrubbing. A specially conditioned Tenax column suppresses unwanted byproduct elution and separates the volatile hydrides at room temperature. A photoionisation detector was used and the authors reported a detection limit as low as 0.001µg L $^{-1}$.

Cutter *et al.* [763] have described a method for the simultaneous determination of inorganic arsenic and antimony species in non saline waters using selective hydride generation with gas chromatography–photoionisation detection.

These workers showed that dissolved arsenic and antimony in natural waters can exist in the trivalent and pentavalent oxidation states, and the biochemical and geochemical reactivities of these elements are dependent upon their chemical forms. They developed a method for the simultaneous determination of arsenic (III) + antimony (III + V) + antimony (III + V) that uses selective hydride generation, liquid nitrogen cooled trapping, and gas chromatography–photoionisation detection. The detection limit for arsenic is 10pmol L $^{-1}$, while that for antimony is 3.3pmol L $^{-1}$; precision (as relative standard deviation) for both elements is better than 3%.

The application of gas chromatography to the determination of selenium in non saline waters is also discussed in Table 15.20.

15.3.1.2 Beryllium

Beryllium has been determined [764] in non saline waters and in sea water at oceanic levels of 2.30pM. Two ml of 0.1M EDTA, 2ml of 1.0M sodium acetate, 1.0ml of benzene and 100µl of 1,1,1-trifluoro-2-4-pentanedione were added sequentially to 150ml samples. Following liquid–liquid extraction using detailed handling procedures, the organic phase was mixed with 1.0ml of 1.0M sodium hydroxide (de-emulsifier), washed several times with distilled water and the resultant beryllium 1,1,1-tri-fluoro-2,4-pentanedione complex analysed by gas chromatography with electron capture detection.

Tao *et al.* [765] describe a unique method combining gas chromatography and inductively coupled plasma emission spectrometry. Beryllium is extracted from a non saline water sample with acetylacetone into chloroform and concentrated by evaporation.

The beryllium acetylacetonate is separated in a gas chromatograph and injected into the helium plasma emission spectrometer. The detection limit is 10pg in a 30mL water sample and the standard deviation was 4.1% at 10ng of beryllium.

The determination of beryllium in non saline waters is also discussed in Table 15.21.

15.3.1.3 Selenium

Shimoishi and Toei [766] have described a gas chromatographic determination of selenium in non saline waters based on 1,2-diamino-3,5-dibromobenzene with an extraction procedure that is specific for selenium (IV). Total selenium is determined by treatment of non saline water with titanium trichloride and with a bromine–bromide redox buffer to convert selenide, elemental selenium and selenate to selenious acid. After reaction, the 4,6-dibromopiazselenol formed from as little as 1ng of selenium can be extracted quantitatively into 1 ml of toluene from 500ml of natural water; up to 2ng L^{-1} of selenium(IV) and total selenium can be determined. The percentage of selenium(IV) in the total selenium in river water varies from 35 to 70%.

Uchida et al. [793] determined various selenium species in river water and sea water by electron capture detection gas chromatography after reaction with 1,2-diamino-3,5-dibromobenzene. This reagent reacts with selenium(IV) to form 4,6-dibromopiazselenol which is extracted into toluene. After Se(–11) and Se(0) has been reduced by a bromine–bromide solution to selenium(IV) state, total selenium is determined by the same method. The limit of detection is 0.002μg L^{-1}.

Flinn and Aue [767] proposed a photometric detector for selenium analysis which enables determination of 2×10^{-12}g selenium s^{-1}.

Johansson et al. [768] determined selenium in non saline waters by derivitivisation followed by gas chromatography using an electron capture detector.

De la Calle Guntinas et al. [769] volatilised selenium from natural water samples by reaction with sodium tetraethylborate and measured the volatilised selenium by gas chromatography microwave-induced plasma atomic emission spectrometry. The detection limit for a 5mL sample was 8ppt.

Estimation of selenium by gas chromatography is based almost exclusively on measurement of the amount of piazselenol formed in the reaction of selenium(IV) with an appropriate reagent in acidic media. Piazselenols are easily extracted with organic solvents (most frequently toluene) in which they can be subsequently determined by

Table 15.20 Examples of selenium determination in various materials by gas-liquid chromatography

Materials	Detection limit	Reference	Reagent	Determination conditions
Water solutions	4×10^{-8} g	[838]	4-Chloro-phenylenediamine	Glass column, 2 m × 4 mm, with 15% SE-30 on 60–80 mesh Chromosorb W:$t = 200°C$, $V_{N2} = 50$ ml/min; ECD detector
Sea water	2×10^{-9} g	[839]	4-Nitro-o-phenylene-diamine	Glass column 1 m × 4 mm with 15% SE-30 on 60–80 mesh Chromosorb W:$t = 200°C$; $V_{N2} = 53$ ml/min; ECD detector
River water	5×10^{-10} g	[840]	2,3-Diamino-naphthalene	Stainless column 6 ft × 1.8 in. with 3% SE-30 on Chromosorb G. 60–80 m mesh: $t = 165°C$; $V_{He} = 40$ ml/min; ECD detector
River and sea water	2×10^{-9} g	[841]	1,2-Diamino-3,5-dibromobenzene	Glass column 1 m × 4 mm with 15% SE-30 on Chromosorb W 60–80 mesh: $t = 200°C$; $V_{N2} = 28$ ml/min; ECD detector

Form of selenium	Detection limit (μg L^{-1})	Reagent	Reference
Se(IV)	–	4,5-Dichloropiazselenol	[770]
Se(IV)	2	2,3-Diaminopiazselenol	[771]
Se(total)	2	Reduction + 2,3–diaminopiazselenol	[771]
Se(IV)	0.1	5-Nitropiazselenol	[772]
Se(total)	0.1	Reduction + 5-nitropiazselenol	[772]
Se(IV)	0.05	5-Chloro-piazselenol	[772]
Se(total)	0.05	Reduction + 5–chloropiazselenol	[773]
Se(total)	0.04	Reduction + 4-nitropiazselenol	[774]
Se(IV)	0.02	4-Nitropiazselenol	[775]
Se(total)	0.02	Reduction + 4-nitropiazselenol	[775]

Table 15.20 continued

Form of selenium	Detection limit (µg L^{-1})	Reagent	Reference
Se(IV)	0.01	2,3-Diaminopiazselenol	[776]
Se(IV)	0.01	4-Nitropiazselenol	[777]
Se(total)	0.01	Photo-oxidation + 4-nitropiazselenol	[777]
Se(IV)	0.002	4,6-Dibromopiazselenol	[775]
Se(total)	0.002	Reduction + 4,6-dichloropiazselenol	[775]
Se(IV)	0.002	4,6-Dibromopiazselenol	[793]
Se(−II, 0, IV)	0.002	Bromine oxidation + 4,6-dibromopiazselenol	[793]
Se(−II, 0, IV,VI)	0.002	Br$_2$/Br− buffer + 4,6-dibromopiazselenol	[793]
Se(IV)	0.008	5-Nitropiazselenol	[778]
Se(total)	0.0008	Photo-oxidation + 5-nitropiazselenol	[778–780]

Source: Own files

Table 15.21 Gas chromatography of cations and non metallic elements in non saline waters

Cation	Extraction from water	GLC column conditions	Detector	Comments	LD	Ref.
Aluminium	Trifluoroacetyl acetone in toluene	4.6 DC710 and 0.2% Carbowax 20M on GasChrom Z at 118°C	Electron capture	–	6pg in 2μL extract	[781]
Beryllium	1,1,1 trifluoro 2,4 pentane dione in benzene	–	Electron capture	–	2.0pM	[764]
Beryllium	Acetyl–acetone in chloroform	–	Inductively coupled plasma emission spectrometer	–	10pg per 30ml	[765]
Chromium(III), total chromium	Trifluoro acetyl acetone in solvent	–	Electron capture	–	–	[782]
Cobalt	Di(trifluoroethyl) dithiocarbamate in solvent	–	Electron capture	–	0.2ng L^{-1}	[783]
Mercury	–	–	Microwave plasma atomic emission spectrometer	–	ppt	[784]
Mercury	–	–	–	Conversion of inorganic mercury to organomercury compound using arene sulphites, formation of tri-methyl–silyl derivitives then GLC	–	[785, 786]

Table 15.21 continued

Cation	Extraction from water	GLC column conditions	Detector	Comments	LD	Ref.
Selenium	–	–	Inductively coupled plasma mass spectrometer	–	0.2ppb	[787, 788]
Selenium	–	–	Electron capture	Derivativisation GLC	–	[768]
Selenium	–	–	Microwave induced plasma atomic emission	Reaction of selenium with sodium in tetraethylborate to form volatile selenium	8ppg	[769]
Silicon silands	–	–	–	Silenol converted to tri-methyl silated derivatives, then GLC	mg L^{-1} µg L^{-1}	[700]

Source: Own files

spectrophotometric, fluorometric, or chromatographic methods. In gas chromatography, piazselenols are usually estimated with an electron capture detector due to its highest sensitivity and selectivity with respects to these compounds (Table 15.20). Apart from the superior sensitivity and selectivity, the gas chromatography method allows for elimination of interferences from the matrix.

The application of gas chromatography to the determination of selenium in non saline waters is also discussed in Table 15.21.

15.3.1.4 Other applications

Gas chromatography has also been applied to the determination of various other cations in non saline waters including aluminium, chromium, cobalt, mercury and silicon.

See Table 15.21.

15.3.2 Seawater

15.3.2.1 Aluminium

An example of a gas chromatographic method is that of Lee and Burrell [781]. In this method the aluminium is extracted by shaking a 30ml sample (previously subjected to ultraviolet radiation to destroy organic matter) with 0.1M trifluoroacetylacetone in toluene for 1h. Free reagent is removed from the separated toluene phase by washing it with 0.01M aqueous ammonia. The toluene phase is injected directly on to a glass column (15cm × 6mm) packed with 4.6% of DC710 and 0.2% of Carbowax 20M on GasChrom Z. The column is operated at 118°C with nitrogen as carrier gas (285ml per min) and electron capture detection. Excellent results were obtained on 2µL of extract containing 6pg of aluminium.

15.3.2.2 Selenium

Shimoishi [775] determined selenium by gas chromatography with electron capture detection. To 50–100ml of seawater was added 5ml concentrated hydrochloric acid and 2ml 1% 4-nitro-o-phenylenediamine and, after 2h, the product formed was extracted into 1ml of toluene. Wash the extract with 2ml 7.5M hydrochloric acid, then inject 5µL into a glass gas–liquid chromatography column (1 × 4mm) packed with 15% of SE–30 on Chromosorb W (60–80 mesh) and operated at 200°C with nitrogen (53min $^{-1}$) as carrier gas. There is no interference from other substances present in seawater.

Measures and Burton [778] used gas chromatography to determine selenite and total selenium in seawater.

15.3.3 Potable water

15.3.3.1 Beryllium

Kuo et al. [789] determined down to 2×10^{-11}g (0.02μg L^{-1}) of beryllium in potable water by gas chromatography with electron capture detection. After mixing with an ethanolic solution of N-trifluoroacetylacetone the Be(TFA)$_2$ complex formed is extracted with cyclohexane. To the excess of free N-trifluoroacetylacetone is added sodium bicarbonate solution with shaking for at least 15s. The pH must be above 7.5, preferably in the range 8–5–9.0. Deionised water showed no detectable levels of beryllium, while Taiwan tap water had a level of 0.18μg L^{-1}.

The peak height of Be(TFA)$_2$, at a retention time of 1.9 min, was taken for the quantitative determination of beryllium as shown in Fig. 15.12 (a). In the chromatogram (Fig. 15.12 (b)) the solvent peak of cyclohexane (retention time ca. 1min) was a tailing that resulted from incomplete removal of free HTFA, in which a random determination would be obtained. The calibration curve of Be(TFA)$_2$ was obtained from the measurement of standard aqueous solutions or standard solutions of Be(TFA)$_2$ in cyclohexane.

15.4 Anions

15.4.1 Non saline waters

15.4.1.1 Nitrite

Funazo et al. [791] have given details of a procedure for the determination of nitrite in river water by gas chromatography after reaction with m-nitroaniline and hypophosphorous acid to convert nitrite to nitrobenzene.

It is well known that nitrite diazotises aromatic amines in acidic media and that the resulting diazonium ions are reduced by hypophosphorous acid to form benzene derivatives. The reactions are formulated as follows.

An electron capture detector was used to monitor the gas chromatographic effluent. The detection limit and determination range of nitrite are 0.5μg L^{-1} and 1.00–1000μg L^{-1} respectively, which are much

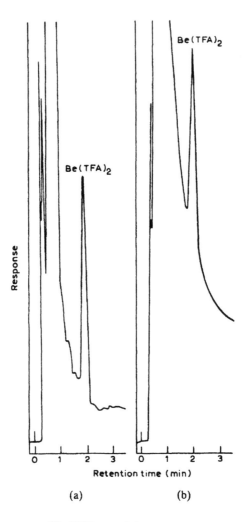

Fig. 15.12 Chromatograms of Be(TFA)$_2$ in cyclohexane medium; (a) peak of Be(TFA)$_2$ with complete washing at 1.9 min retention time; (b) tailing peak for incomplete removal of free HTFA.
Source: Reproduced by permission from Kluwer Academic Plenum, Amsterdam [789]

lower than those of the widely used colorimetric method for the determination of nitrite (detection limit 16μg L^{-1}).

The gas chromatographic method was tested in the presence of several ions normally found in environmental samples. In Table 15.22 the peak area of nitrobenzene produced from the standard solution containing

Table 15.22 Results of interference study the concentration of nitrite was 0.10µg mL^{-1}

Ion	Concentration/ mg L^{-1}	Relative peak area^a	95% Confidence interval
Standard	–	100.0 ± 2.0	97.5–102.5
Cl$^-$	20000	101.2 ± 2.3	98.3–104.1
Cl$^-$	100	99.4 ± 2.1	96.8–102.2
Br$^-$	100	100.4 ± 2.2	97.7–103.1
F$^-$	100	98.8 ± 1.9	96.4–101.2
SO$_4^{2-}$	100	97.8 ± 1.3	96.2–99.4
HCO$_3^-$	100	99.1 ± 1.5	97.2–101.0
H$_2$PO$_4^-$	100	98.6 ± 2.5	95.5–101.7
NO$_3^-$	100	111.9 ± 1.3	110.3–113.5
NO$_3^-$	10	99.6 ± 1.8	97.4–101.8
NH$_1^+$	100	100.4 ± 1.1	99.0–101.8

^{a}Mean ± standard deviation for five replicate analyses (between batch).The peak area of nitrobenzene converted from the standard solution was arbitrarily assigned a value of 100

Source: Reproduced with permission from the Royal Society of Chemistry, London [791]

100µg L^{-1} of nitrite was arbitrarily assigned a value of 100. The concentrations of the ions (100mg L^{-1}) added to the standard nitrite solution were much higher than those in environmental samples. The results indicated that these anions, except for nitrate, do not interfere in this method. Moreover, chloride added as sodium chloride at a concentration of 2.0% to the standard solution has no effect, which suggests the possibility of applying this method to the analysis of seawater. However, nitrate interferes positively at a concentration of 100mg L^{-1} although this interference is not found at 10mg L^{-1}.

Derivatisation–electron capture gas chromatography has been used to determine µg L^{-1} quantities of nitrite in water without interference from halides, nitrate, phosphate, sulphate, bicarbonate, ammonium and alkali metals and alkaline earth metals [792].

15.4.1.2 Selenate and selenite

Uchida *et al.* [793] have described a method for determination of selenium in river and sea water by gas chromatography with electron–capture detection. The specific reaction of 1,2-diamino-3,5-dibromo-benzene with selenium(IV) is used, the product (4,6–dibromopiazselenol) being extracted into toluene and determined from its peak height in the gas chromatogram. Selenium (–II, 0) is oxidised to selenium(IV) by bromine solution and selenium(VI) is reduced to the same state by bromine/ bromide redox buffer solution and determined as the piazselenol as above.

The gas chromatograph used in this study was equipped with ^{63}Ni electron capture detector and the column comprised 15% of SE–30 on 60–80 mesh Chromosorb W. The column and detector temperatures were maintained at 200 and 280°C respectively. The nitrogen flow rate was 28ml min $^{-1}$.

15.4.1.3 Other applications

Further applications of the determination of gas chromatography to the determination of anions in non saline waters are reviewed in Table 15.23. These include: aminoacetates, arsenate, bromide, chloride, fluoride, iodide, cyanide, ethylene diamine tetraacetate, nitrate, phosphate, thiocyanate and sulphide.

15.4.2 Aqueous precipitation

15.4.2.1 Chloride, bromide and iodide

Mack and Grimsrud [790] have described a photochemical modulated pulsed electron capture detector suitable for the gas chromatographic determination of chloride, bromide and iodide in rainwater.

15.4.2.2 Sulphate, nitrate and phosphate

Faiqle and Klochow [805] applied gas chromatography to the determination of traces of sulphate (and nitrate and phosphate) in rainwater. The dissolved salts are freeze dried and converted to the corresponding silver salts. These are then converted to the *n*-butyl esters with the aid of *n*-butyl iodide, the butyl esters being determined by gas chromatography. Sulphate (and phosphate) are determined simultaneously on a column containing 3% OV–17 on Chromasorb, while nitrate is determined separately with 3% of tri-*p*-cresol phosphate on Chromasorb.

15.4.3 Seawater

15.4.3.1 Sulphide

Leck and Bagander [342] determined reduced sulphur compounds (hydrogen sulphide, methyl mercaptan, carbon disulphide and dimethyl sulphide) in seawater by gas chromatography using flame detection. Detection limits ranged from 0.2ng L $^{-1}$ for carbon disulphide to 0.6ng L $^{-1}$ for methyl mercaptan. Hydrogen sulphide was determined at the 1ng L $^{-1}$ level.

Cutter and Oatts [806] determined dissolved sulphide and studied sedimentary sulphur speciation using gas chromatography in conjunction with a photoionisation detector. The determination of dissolved sulphide and sedimentary sulphur is important to studies of trace element cycling in the aquatic environment. A method employing selective generation of hydrogen sulphide, liquid nitrogen-cooled trapping, and subsequent gas chromatographic separation/photoionisation detection has been developed for such studies. Dissolved sulphide is determined via acidification and gas stripping of a water sample, with a detection limit of 12.7nM and a precision of 1% (relative standard deviation). With preconcentration steps the detection limit is 0.13nM. Hydrogen sulphide is generated from sedimentary acid volatile sulphides via acidification, from greigite using sodium borohydride and potassium iodide, and from pyrite using acidic chromium(II). The detection limit for these sulphur species is 6.1μg of S/g, with the precision not exceeding 7% (relative standard deviation). This method is rapid and free of chemical interferences, and field determinations are possible.

15.4.3.2 Selenate and selenite

See section 15.4.1.2.

15.4.4 Potable water

15.4.4.1 Chloride, bromide and iodide

Grandet *et al.* [807] have described a method for the determination of traces of bromide and iodide in potable water which permits determination of 50μg bromide L^{-1} and 5μg iodide L^{-1}, without preconcentration of the sample or 0.5μg bromide and 0.2μg iodide L^{-1} after preconcentration. The method is based on the transformation of the halides into 2-bromo- or 2-iodo–ethanol. These derivatives are extracted with ethyl acetate and determined with an electron capture detector. The difference in the retention periods enables both halides to be determined simultaneously in one sample. The method is substantially free of interferences and is suitable for use on many types of water.

Bachmann and Matusca [808] have described a gas chromatographic method for the determination of μg L^{-1} quantities of chloride, bromide and iodide in potable water. This method involves reaction of the halide with an acetone solution of 7-oxabicyclo–(4.1.0) heptane in the presence of nitric acid to form halogenated derivatives of cyclohexanol and cyclohexanol nitrate. The composition of the reagent is 1% 0.1N nitric acid 7-oxabicyclo–(4.1.0) heptane, 25% water, 75% acetone. The column effluent passes through a pyrolysis chamber at 800°C and then through a conductivity detector. The solution is injected onto a gas chromatographic column (OV–10–Chromosorb W–HP1 80–100 mesh), 150cm long, 0.2cm

Table 15.23 Gas chromatography of anions in non saline waters

Anion	Extraction from water	GLC column conditions	Detector	Comments	LD	Ref.
Amino acetate	Concentration on ion exchange column	–	–	Eluate from ion exchange column converted to N–trifluoroacetyl methyl esters, then GLC	–	[332]
Arsenate, arsenite, monomethyl arsenic acid	Benzene	–	Flame photometric	Arsenic anions converted to 2,3 mercaptopropanol complexes, then GLC	0.2ng (AsO_3 AsO_4) 0.04ng (monomethyl arsenic acid	[794]
Bromide	–	–	–	Headspace gas chromatography, bromide oxidised to $OH_2Br_2COCBr_3$ which is then decomposed to bromoform, then GLC of bromoform	–	[795]
Bromide	Bromide removed from water on strong anion exchange resin	–	–	Bromide eluted from resin with 0.5M sodium nitrate, then GLC	1–2µg L^{-1}	[796]
Bromide	–	–	Electron capture	Bromide converted to cyanogen bromide with cyanide ions and chlorine, then GLC Br^- + $2Cl_2$ + $2CN^-$ = $BrCN$ + $ClCN$ + $3Cl^-$. Aromatic compounds interfere	0.5mg L^{-1}	[797]
Bromide, iodide	n-hexane or benzene	–	Electron capture	Bromide and iodide; converted to acetone derivatives by reaction with chromate, permanganate and acetone, then extracted with benzene to remove bromo and iodo acetones, then GLC	<10^{-7}mg L^{-1}	[798]

Table 15.23 continued

Anion	Extraction from water	GLC column conditions	Detector	Comments	LD	Ref.
Carboxylates, malonate, hexa-benzene carboxy-lates, octonioates octandioate	–	–	–	Acids converted to methyl esters then isotope dilution gas chrom-atography–Fourier transform infrared spectroscopy	–	[233]
Chloride	Carbon tetrachloride	–	–	Chloride reacted with phenyl mercuric nitrate to produce phenyl mercury chloride, then carbon tetrachloride extraction, then GLC	0.4mg L^{-1}	[799]
Complex cyanides	–	–	Electron capture	Complexes broken down with ultraviolet radiation, reacted with bromine water to produce cyanogen bromide, then GLC	100μg L^{-1}	[800]
Fluoride	–	–	Emission line of fluorine at 685.6nm	–	pg amounts	[801]
Nitrate	–	–	Electron capture	Nitrate denitrified with pseudo-amonias chlorophils to nitrogen, then GLC	0.2ng L^{-1}	[802]
Phosphate	–	–	–	Phosphate converted to phosphate with sodium borohydride at 400°C, then GLC of phosphine	–	[803]
Sulphide	Helium stripping of sulphide from water	–	Flame photometric	–	0.2p mL^{-1}	[804]

Source: Own files

diameter, operated at 50°C. Hydrogen is used as carrier gas. Chloride contents determined by this method were $5.0 \pm 1.8\mu g$ L $^{-1}$ in potable water.

Nota *et al.* [809] have devised a simple and sensitive method for the determination of amounts of bromide down to 0.05mg L $^{-1}$ in water. The procedure is in two stages. Cyanogen bromide is formed by the reaction between bromide, chlorine and cyanide,

$$Br^- + 2Cl_2 + 2CN^- \rightarrow BrCN + ClCN + 3Cl^-$$

Cyanogen bromide is then separated by gas chromatography and selectively detected with an electron capture detector.

It should be noted that the internal standard (nitromethane) must always be introduced after the addition of cyanide in order to prevent rapid darkening of the solutions, which causes non-reproducible results. No interferences are caused by oxidising or reducing substances or by mercury or cadmium at concentrations below 200mg L $^{-1}$. Mercury and cadmium are, among the metals, the strongest complex-forming agents with bromide.

Even small amounts of aromatic compounds are likely to produce some interferences owing to their tendency to bind bromide, which is formed by reaction with chlorine water.

15.4.4.2 Nitrate

A method has been described [810,811] for determining aqueous nitrates by conversion to nitrobenzene followed by electron capture gas chromatography. The procedure can also be used to determine aqueous nitrites and gaseous oxides of nitrogen if they are first converted to nitrates. The procedure was evaluated by analysing drinking water for nitrate over a period of one month. The method is sensitive and capable of measuring typical environmental levels of nitrogen compounds. The detection limit for nitrobenzene is about 10 $^{-12}$g.

For the determination of nitrate, use a 1 dram vial with a polyethylene stopper (Kimble No. 60975–L) as a reaction vessel. Introduce a 0.20ml aliquot of aqueous sample into the vial, followed by 1.0ml of thiophen free benzene. Catalyse the reaction by addition of 1.0ml of concentrated sulphuric acid. Shake the vial for 10min. Remove the benzene layer immediately from the reaction vial with a Pasteur pipette, place it in a separate vial and analyse by gas chromatography with electron capture detection for the nitrobenzene concentration generated. Treat standard solutions of potassium nitrate in the same manner to generate a standard calibration plot relating nitrobenzene concentration to peak height. If higher precision is desired (approximately 4% relative standard deviation), add 2,5-dimethylnitrobenzene to the benzene prior to reaction.

15.4.5 Waste waters

15.4.5.1 Cyanide

Funazo *et al.* [812] have described a method for the determination of cyanide in water in which the cyanide ion is converted into benzonitrile by reaction with aniline, sodium nitrite and cupric sulphate. The benzonitrile is extracted into chloroform and determined by gas chromatography with a flame ionisation detector. The detection limit for potassium cyanide is 3mg L^{-1}. Lead, zinc and sulphide ion interfere at 100mg L^{-1} but not at 10mg L^{-1}.

$$C_6H_5NH_2 + NaNO_2 \rightarrow C_6H_5N^+ = N + NaOH + OH^-$$

$$C_6H_4N = N \xrightarrow{\overset{+}{Cu^{2+}}} C_6H_5CN + N_2$$

Funazo *et al.* [794] have also described a more sensitive gas chromatographic procedure capable of determining down to 3μg L^{-1} total cyanide in waste waters. The method is based on the derivatisation of cyanide to benzonitrile, which is extracted with benzene and determined by flame thermionic gas chromatography. In the derivatisation reaction, aqueous cyanide reacts with aniline and sodium nitrite in the presence of copper(II) sulphate and forms benzonitrile.

Funazo *et al.* [794] tested this method in the presence of several ions normally found in environmental samples (Table 15.24). The peak area of benzonitrile derivativised from the standard cyanide solution (0.5mg L^{-1}) was arbitrarily assigned a value of 100.

None of the ions except thiocyanate, cyanate and sulphide interfered at a concentration of 500mg L^{-1}.

Nota and Improta [813] determined cyanide in coke oven waste water by gas chromatography. The method is based on treatment of the sample with bromine and direct selective determination of the cyanogen bromide by gas solid chromatography using a BrCN$^-$ selective electron capture detector. No preliminary treatment of the sample to remove interferences is necessary in this method, and in this sense it has distinct advantages over many of the earlier procedures. Bromine also oxidises thiocyanate to cyanogen bromide. Previous treatment of the sample with aqueous formaldehyde destroys thiocyanate and prevents its interference.

Nota *et al.* [814] have also described a gas chromatographic headspace method based on different principles for the determination of 0.01–100mg L^{-1} of cyanide and thiocyanate in coke oven waters and waste effluents. This method involves firstly transforming the cyanides, or the thiocyanates, into hydrogen cyanide by acidification, then removing hydrogen cyanide from the aqueous sample by the head space technique and finally separation of hydrogen cyanide by gas solid chromatography and selective detection with a nitrogen phosphorus detector. This

Table 15.24 Results of interference study CN⁻ concentration: 0.5µg/mL

Ion	Concentration	Added as	Peak area[a]
		(mg L⁻¹)	
None			100.0 ± 2.0
F⁻	500	NaF	99.2 ± 1.7
Cl⁻	500	NaCl	99.4 ± 0.2
Br⁻	500	NaBr	100.2 ± 1.8
I⁻	500	KI	98.6 ± 1.3
SO_4^{2-}	500	Na_2SO_4	100.4 ± 3.2
NO_3^-	500	$NaNO_3$	100.8 ± 2.5
HCO_3^-	500	$KHCO_3$	99.9 ± 1.9
$H_2PO_4^-$	500	KH_2PO_4	99.4 ± 0.8
SCN⁻	500	NaSCN	86.0 ± 2.1
SCN⁻	100	NaSCN	99.0 ± 3.0
CNO⁻	500	KCNO	104.1 ± 0.7
CNO⁻	100	KCNO	102.8 ± 1.7
S^{2-}	0.1	$Na_2S.9H_2O$	90.9 ± 4.6
S^{2-}	0.05	$Na_2S.9H_2O$	100.2 ± 2.3
NH_4^+	500	$(NH_4)_2SO_4$	101.8 ± 1.7

[a]Mean ± S.D. of five replicate analyses

Source: Reproduced with permission from Elsevier Science, UK [794]

procedure for the determination of cyanide is based on three stages: first, transformation into hydrogen cyanide by acidification; second, removal of hydrogen cyanide by the headspace technique, and third, gas chromatographic separation of hydrogen cyanide and selective detection with a nitrogen phosphorus detector. A similar procedure is adopted for the determination of thiocyanate, the only difference being the quantitative transformation of thiocyanate into hydrogen cyanide according to the reactions:

$$SCN^- + Br_2 + 4H_2O \rightarrow BrCN + SO_4^{2-} + 7Br^- + 8H^-$$
$$BrCN + Red \rightarrow HCN + Br^- + Ox$$

(where Red = SO_2 or I⁻ and Ox = SO_4^{2-} or I_3^-). If cyanide is present prior to the oxidation step, it must be transformed into unreactive cyanohydrin by an excess of formaldehyde or removed by boiling the solution previously acidified to pH 2.

The presence of iron(II), iron(III) and copper(II) in the sample decreases the response for hydrogen cyanide. The greatest effect is caused by copper (II). Reducing agents do not interfere in the analysis of cyanide but oxidising substances have to be reduced prior to heating of the sample. Oxidants and reducing agents do not interfere in the determination of thiocyanate.

15.4.5.2 Polysulphide

Borchardt and Easty [815] developed a gas chromatographic method for determining polysulphide in kraft pulping liquors. The polysulphide is decomposed to elemental sulphur in buffer solution. The elemental sulphur is derivatised with triphenyl phosphine. The resulting triphenylphosphine sulphur is determined by flame ionisation–gas chromatography.

15.4.5.3 Sulphide

Hawke *et al.* [816] described a method for the determination of sulphides in effluents using gas chromatography with a flame photometric detector. Samples were acidified and the hydrogen sulphide brought to solution vapour equilibrium. Following analysis by gas chromatography the concentration of hydrogen sulphide was determined by interpolation on a calibration graph. The precision, accuracy, bias and effect of potentially interfering substances were determined. The samples could be processed in 1h by this method.

Knoery and Cutter [817] have described a method for the determination of sulphide and carbonyl sulphide in non saline waters using specialised collection procedures followed by gas chromatography using flame photometric detector. The following species were determined: H_2S_{aq}, HS^-, S^{2-} and particulate metal sulphides. The sample was stripped with helium and the evolved gases analysed by gas chromatography. Down to 0.2pmol L^{-1} of total dissolved sulphide could be determined.

15.4.5.4 Thiocyanate

Nota *et al.* [814] have described a gas chromatographic headspace analysis technique for the determination of 0.01–100mg L^{-1} thiocyanate (and cyanide) in coke oven effluents.

15.4.6 Mineral waters

15.4.6.1 Iodide

Wu *et al.* [818] have described a method for the derivatisation of iodide into pentafluorobenzyl iodide using pentafluorobenzyl bromide. The derivative was analysed at μg levels by gas chromatography with electron capture detection. The effects of solvent, water content, base or acid concentration, amount of pentafluorobenzylbromide, reaction time and reaction temperature were examined. Interferences by common anions were minimal. The method was applicable to iodide determination in spring water.

References

1 Gouw, T.H. *Analytical Chemistry*, **42**, 1394 (1970).
2 Kawahara, F.K. Laboratory Guide for the Identification of Petroleum Products. 1014 Broadway, Cincinnati, Ohio, US. Department of the Interior, Federal Water Administration, Analytical Quality Control Laboratory, 41pp. (1969).
3 Jeltes, R. and Veldink, R. *Journal of Chromatography*, **27**, 242 (1967).
4 Lively, L. Bngng. *Bull. Ext. Serv. Perdue University*, **118**, 657 (1965).
5 Trent River Authority, private communication, Trent River Authority, Meadow Lane, Nottingham.
6 Ramsdale, S.J. and Wilkinson, R.E. *Journal of the Institute of Petroleum (London)*, **54**, 326 (1968).
7 Johnson, W. and Kawahara, F.K. Proc. 11th Conf. Great Lakes Res., p. 550 (1968).
8 Ellerker, R. *Water Pollution Control*, **67**, 542 (1968).
9 Dewitt Johnson, W. and Fuller, F.D. Proc. 13th Conf. Great Lakes Res., p. 128 (1970).
10 Blumer, M. *Marine Biology*, **5**, 195 (1970).
11 Benyon, L.R. *Methods of the Analysis of Oil in Water and Soil*. The Hague, Striching, CONCAWE. 40 pp. (1968).
12 Duckworth, D.F. Aspects of Petroleum Pollutant Analysis, in *Water Pollution in Oil* (Ed. P. Hepplee). Institute of Petroleum, London, p. 165 (1971).
13 Le Pera, M.E. *Identification and Characterization of Petroleum Fuels using Temperature-Programmed Gas–Liquid Chromatography*, Springfield, V.A. National Technical Information Service. Rep. No. AD 646–382 (1966).
14 Jeltes, R. *Analytical Abstracts*, **15**, 3633 (1968).
15 McAuliff, C. *Chemical Geology*, **4**, 255 (1969).
16 Adlard, E.R. and Matthews, P.H.D. *Natural Physical Sciences*, **233**, 83 (1971).
17 Lur'e, YuYu, Panova, V.A. and Nickolaeva, Z.V. Gidrokhim Water, 55, 108 (1971). Ref. Zhur. Khim. 199D (16) Abstract No. 16G258.
18 Cole, R.D. *Nature (London)*, **233**, 546 (1971).
19 Jeltes, R. and Van Tonkelaar, W.A.M. *Water Research*, **6**, 271 (1972).
20 Jeltes, R. and Van Tonkelaar, W.A.M. *H₂O*, **5**, 288 (1972).
21 Kawahara, F.K. *Journal of Chromatographic Science*, **10**, 629 (1972).
22 Lysyj, I. and Newton, P.R. *Analytical Chemistry*, **44**, 2385 (1972).
23 Lysyj, I. and Newton, P.R. *Analytical Abstracts*, **22**, 459 (1972).
24 Garra, M.E. and Muth, J. *Science and Technology*, **8**, 249 (1974).
25 McAucliffe, C. *Journal of Physical Chemistry*, **70**, 1267 (1966).
26 Desbaumes, E. and Imhoff, C. *Water Research*, **6**, 885 (1972).
27 Bridie, A.L., Bos, J. and Herzberg, S.J. *J. St. Petrol.*, **59**, 263 (1973).
28 Wasik, S.P. and Tsang, W.W. *Analytical Chemistry*, **42**, 1649 (1970).
29 Wasik, S.P. and Tsang, W. *Analytical Abstracts*, **21**, 1854 (1971).
30 Wasik, S.P. Private communication.
31 Mel'kanovitskaya, G.C. Gidrokhim. Mater., 53, 153 (1972). Ref. Zhur. Khim. 199D (1972). Abstract No. 129265.
32 Chatot, G., Jequir, W., Jay, M., Fontages, R. and Obaton, P. *Journal of Chromatography*, **45**, 415 (1969).
33 Harrison, R.M., Perry, R. and Wellings, R.A. *Water Research*, **10**, 207 (1976).
34 Caddy, D.E. and Meek, D.M. Technical Report TR36 Water Research Centre, Stevenage Laboratory, Elder Way, Stevenage, Herts. (1976).
35 Acheson, M.A., Harrison, R.M., Perry, R. and Welling, R.A. *Water Research*, **10**, 207 (1976).

36 Basu, D.K. and Saxena, J. *Journal of Environmental Science and Technology*, **12**, 791 (1978).
37 Favretto, L., Stancher, B. and Tunis, R. *Analyst (London)*, **103**, 955 (1978).
38 Stephanou, A. *Chemosphere*, **13**, 43 (1984).
39 Semenchenko, L.L. and Kaplin, V.T. *Zhur. Anal. Khim.*, **23**, 1257 (1968).
40 Eichelberger, J.W., Dressman, R.C. and Longbottom, A. *Journal of Environmental Science and Technology*, **4**, 57 (1970).
41 Murray, D.A. *Journal of Fisheries Research Board, Canada*, **32**, 292 (1975).
42 Yorkshire Water Authority, private communication.
43 Vonznakova, A. and Popl, M.J. *Journal of Chromatographic Science*, **17**, 682 (1979).
44 Vonzakova, Z., Popl, M.J. and Barker, M. *Journal of Chromatographic Science*, **15**, 123 (1978).
45 Goldberg, M.C., De Long, L. and Sinclair, M. *Analytical Chemistry*, **45**, 89 (1973).
46 Goldberg, M.C. and Weiner, E.R. *Analytical Chemistry*, **115**, 373 (1980).
47 Department of the Environment/National Water Council Standing Committee of Analysts. HMSO, London. Methods for the examination of waters and associated materials. Phenols in waters and effluents by gas–liquid chromatography, 4-aminoantipyrine of 3-methyl-2-benzothiazolinone hydrazone. 39 pp. (RP 22B: CENV) (1981).
48 Janda, V. and Krijt, K. *Journal of Chromatography*, **283**, 309 (1984).
49 Bethge, P.O. and Lindstroem, K. *Analyst (London)*, **99**, 137 (1974).
50 Ochiai, M. *Journal of Chromatography*, **194**, 224 (1980).
51 Murray, A.J. and Riley, J.P. *Analytica Chimica Acta*, **65**, 261 (1973).
52 Murray, A.J. and Riley, J.P. *Nature (London)*, **242**, 37 (1973).
53 Novak, J., Zluticky, J., Kubulka, V. and Mostecky, J. *Journal of Chromatography*, **76**, 45 (1973).
54 Kleopfer, R.D. and Fairless, B.T. *Environmental Science and Technology*, **6**, 1036 (1972).
55 Jensen, S., Jernelov, A., Large, R. and Palmork, K.H. Food and Agriculture Organisation of the United Nations. Report No. FIR: MP 170/E–88 November (1970).
56 Dietz, F. and Traud, J. *Vom Water*, **41**, 137 (1973).
57 Deetman, A.A., Demeulemeester, P., Garcia, M. *et al. Analytica Chimica Acta*, **82**, 1 (1976).
58 Guam, C.S. and Wong, M.K. *Journal of Chromatography*, **72**, 283 (1972).
59 Hagenmaier, H., Werner, G. and Jager, W.Z. *Wasser Abwasser Forsch*, **15**, 195 (1982).
60 Alberti, S. and Jonke. F. *Zeitschrift für Wasser und Abwasser*, **8**, 140 (1975).
61 Burgasser, A.J. and Calaruotolo, J.F. *Analytical Chemistry*, **49**, 1508 (1977).
62 Narang, R.S. and Bush, B. *Analytical Chemistry*, **52**, 2076 (1980).
63 Rudling, L. *Water Research*, **4**, 533 (1970).
64 Renberg, L. *Analytical Chemistry*, **46**, 459 (1974).
65 Chriswell, C. and Cheng, R. *Analytical Chemistry*, **47**, 1325 (1975).
66 Bauer, U. *Gas- u. Wasserfach. Wasser Abwasser*, **113**, 58 (1972).
67 Berg, O.W., Diolady, P.L. and Rees, G.A. *Bulletin of Environmental Contamination and Toxicology*, **7**, 338 (1972).
68 Beezhold, F.L. and Stout, V.F. *Bulletin of Environmental Contamination and Toxicology*, **10**, 10 (1973).
69 *Chromopak News*. June 1979 No. 20. Chromopak Ltd., PO Box 3, 4330 AA Middelburg, The Netherlands (1979).
70 Schulte, E. and Acker, L. Z. *Analyt. Chemie*, **268**, 260 (1974).

71 Hermanson, H.P., Helrich, K. and Carey, W.F. *Analytical Letters (London)*, **1**, 941 (1968).
72 Metera, J., Rabhl, V. and Mostecky, J. *Water Research*, **10**, 137 (1976).
73 Onuska, F.I. *Water Research*, **7**, 835 (1973).
74 Kimoto, W.I., Dooley, C.J., Canoe, J. and Fiddler, W. *Water Research*, **14**, 869 (1980).
75 Nikaido, M.M., Raymond, D.D., Francis, A.J. and Alexander, M. *Water Research*, **11**, 1085 (1977).
76 Tompkins, B.A. and Griest, W.H. *Analytical Chemistry*, **68**, 2533 (1996).
77 Tompkins, B.A., Griest, W.H. and Higgins, C.E. *Analytical Chemistry*, **67**, 4387 (1995).
78 Mills, A.I. and Alexander, M.J. *Environmental Quality*, **5**, 437 (1976).
79 Chau, A. and Fox, M.E. *Journal of Chromatographic Science*, **9**, 271 (1971).
80 Warren, C.B. and Malek, E. *Journal of Chromatography*, **64**, 219 (1972).
81 Williams, P.J., Benoit, F., Muzeka, K. and O'Grady, R. *Journal of Chromatography*, **130**, 423 (1977).
82 Aue, W.D., Hastings, C.R., Gerhard, K.O. *et al. Journal of Chromatography*, **72**, 259 (1972).
83 Reichert, J.K. and Linckens, A.H.M. *Environmental Technology Letters*, **1**, 42 (1980).
84 Gardiner, J. *Analyst (London)*, **102**, 120 (1977).
85 Gardiner, J. Water Research Centre, Stevenage Laboratory, Herts, UK, Technical Memorandum TM101 (1975).
86 Kari, F.G. Dissertation. ETH. No. 10698. Dübendorfer, Switzerland (1994).
87 Schürch, S. and Dubendorfer, G. *Mitt. Geb. Lebensmittelunters*, **80**, 324 (1989).
88 Simo, R., Grimalt, J.O. and Albaiges, J. *Analytical Chemistry*, **68**, 1493 (1996).
89 Andreae, M.O. *Analytical Chemistry*, **52**, 150 (1980).
90 Braman, R.S., Ammons, J.M. and Bricker, J.L. *Analytical Chemistry*, **50**, 992 (1978).
91 Murray, D.A.J. *Journal of Fisheries Research Board, Canada*, **32**, 457 (1975).
92 Gorbech, S., Harring, R., Knauf, W. and Werner, H.J. *Bulletin of Environmental Contamination and Toxicology*, **6**, 40 (1971).
93 Chau, A.S.Y. and Terry, K.J. *Ass. Off. Anal. Chemists*, **55**, 1228 (1972).
94 Chau, A.S.Y. and Terry, K. *Analytical Abstracts*, **24**, 2513 (1973).
95 Chau, A.S.Y. *J. Ass. Off. Anal. Chemists*, **55**, 1232 (1972).
96 *Chromopak News*, No. 20, June (1979).
97 American Public Health Association. Standard Methods for the Examination of Water and Waste Water, 15th ed., Method 509 A,P 293.
98 Method S73. Supplement to the 15th ed. of Standard Methods for the Examination of Water and Waste Water. Selected Analytical Methods Approved and Cited by US Environmental Protection Agency, American Public Health Association, American Water Works Association, Water Pollution Control Federation.
99 Shevchuk, I.A., Dubchenco, Y.G. and Nordenova, Y.S. *Soviet Journal of Water Chemistry and Technology*, **7**, 73 (1985).
100 Kahanovitch, Y. and Lahav, N. *Environmental Science and Technology*, **8**, 762 (1974).
101 Suzuki, M., Yamato, Y. and Wanatabe, T. *Environmental Science and Technology*, **11**, 1109 (1977).
102 Taylor, R., Bogacka, T. and Krasnicki, K. *Chemia Analit.*, **19**, 73 (1974).
103 Weil, L. and Ernst, K.E. *Gas- u. Wass. Fach.*, **112**, 184 (1971).
104 Simal, J., Crous Vidal, J., Maria-Charro Arias, A. *et al. An. Bromat. (Spain)*, **23**, 1 (1971).

105 Sackmauereva, M., Pal'usova, O. and Szokolay, A. *Water Research*, **11**, 551 (1977).
106 Johnson, L.C. *Bulletin of Environmental Contamination and Toxicology*, **5**, 542, (1970).
107 Konrad, J.G., Pionke, H.B. and Chesters, G. *Analyst (London)*, **94**, 490 (1969).
108 Pionke, H.B. *Analytical Abstracts*, **17**, 2442 (1969).
109 Pionke, H.B., Konrad, J.G., Chesters, J.G. and Armstrong, D.E. *Analyst (London)*, **93**, 363 (1968).
110 Kadoum, A.M. *Bulletin of Environmental Contamination and Toxicology*, **3**, 65 (1968).
111 Millar, J.D., Thomas, R.E. and Schattenberg, H.J. *Analytical Chemistry*, **53**, 214 (1981).
112 Brodtmann, N.V. *Bulletin of Environmental Contamination and Toxicology*, **15**, 33 (1976).
113 Thompson, J.F., Reid, S.J. and Kantor, E.J. *Arch. Environ. Contam. Toxicol.*, **6**, 143 (1977).
114 Lauren, G. *Bulletin of Environmental Contamination and Toxicology*, **5**, 542 (1970).
115 Ballinger, D.G. *Methods for Organic Pesticides in Water and Waste Water*, US Govt. Printing Office, Washington DC, 58 pp. (1971).
116 Methods for Organic Pesticides in Water and Waste Water 1972–759–31/2113. Region 5–11 US Govt. Printing Office, Washington DC. (1971).
117 Stachel, B., Bactjer, K., Cetinskaya, M. *et al. Analytical Chemistry*, **53**, 1469 (1981).
118 Malaiyandi, M., Jenkins, E., Lee, P. and Bowron, M.J. *Environmental Science and Health*, **A20**, 219 (1985).
119 Lamar, W.L., Goerlitz, D.F. and Law, L.M. *Identification and Measurement of Chlorinated Organic Pesticides in Water by Electron Capture Gas Chromatography*, US Govt. Survey Water Supply Paper 1817–B, 12 pp. (1965).
120 Brodtmann, N.V. *Journal of the American Water Works Association*, **67**, 558 (1975).
121 Breidenbach, A.W. *The Identification and Measurement of Chlorinated/Hydrocarbon Pesticides in Surface Water*, 1968–0–315–842, US Govt. Printing Office, Washington DC.
122 Kahn, L. and Wayman, C.H. *Analytical Chemistry*, **36**, 1340 (1964).
123 Lamar, W.L., Goerlitz, D.F. and Law, L.M. *Determination of Organic Pesticides in the Environment*. American Chemical Society, Advances in Chemistry, Series 60, p. 187 (1966).
124 US Federal Water Pollution Control Administration. FWPCA *Method for Chlorinated Hydrocarbon Pesticides in Water and Wastewater*, Cincinnati, Federal Water Pollution Control Administration, 29 pp. (1969).
125 Sackmauereva, M., Pal'usova, O. and Hluchan, E. *Vod. Hospod.*, **10**, 267 (1972).
126 Szokolay, A., Uhnak, J. and Madaric, A.A. *Chem. Zvesti*, **25**, 453 (1971).
127 Janak, J., Sackmauereva, M., Szokolay, A. and Madaric, A. *Chem., Zsvesti*, **27**, 128 (1973).
128 Janak, J., Sackmauereva, M., Szokolay, A. and Pal'usova, O. *Journal of Chromatography*, **91**, 545 (1974).
129 Aspila, K.I., Carron, J.M. and Chau, A.S.Y. *Journal of the Association of Analytical Chemists*, **60**, 1097 (1977).
130 *Analytical Methods Manual*, Inland Waters Directorate, Water Quality Branch, Ottawa, Ontario, Canada (1974).
131 Miles, J.R.W., Tu, C.M. and Harris, O.R. *Journal Econ. Entomol.*, **62**, 1334 (1969).
132 Chau, A.S.Y., Rosen, J.D. and Cochrane, W.P. *Bulletin of Environmental Contamination and Toxicology*, **6**, 225 (1971).
133 Chau, A.S.Y. *Journal Am. Oil Color Chem. Ass.*, **57**, 585 (1974).

134 Thompson, J.F. (ed.) *Analysis of Pesticide Residues in Human and Environmental Samples*, US Environmental Protection Agency, Research Triangle Park, NC, sec, 10A 12/2/74, p. 9 (1974).
135 Eichelberger, J.W. and Lichtenberg, J.L. *Environmental Science and Technology*, **5**, 541 (1971).
136 Sherma, J. and Shafik, T.M. *Archives of Environmental Contamination and Toxicology*, **3**, 55 (1975).
137 Kadoum, A.M. *Bulletin of Environmental Contamination and Toxicology*, **3**, 247 (1968).
138 Askew, J., Ruzicka, H.J. and Wheals, B.B. *Analyst (London)*, **94**, 275 (1969).
139 Herzel, F. *Arch. Hyg. Bakteriol.*, **154**, 18 (1970).
140 Ahling, B. and Jenson, S. *Analytical Chemistry*, **42**, 1483 (1970).
141 Department of the Environment, London. Private communication.
142 Zitko, V. and Choi, P. PCB and Other Industrial Halogenated Hydrocarbons in the Environment. Fish. Res. Board Can. Technical Report No. 272, Biological Station, St Andrews, NB. (1971).
143 Oller, W.L. and Cramer, M.F. *Journal of Chromatographic Science*, **B13**, 296 (1975).
144 Edwards, R. *Chem. Ind. (London)*, 1340 (1970).
145 Fishbein, L. *Journal of Chromatography*, **58**, 345 (1972).
146 Schulte, E., Their, H.P. and Acker, L. *Deut. Lebensm.–Rundsch.*, **72**, 229 (1976).
147 Goke, G. *Deut. Lebensm.–Rundsch.*, **71**, 309 (1975).
148 Dolan, J.W. and Hall, R.C. *Analytical Chemistry*, **45**, 2198 (1973).
149 Musty, P.R. and Nickless, G. *Journal of Chromatography*, **89**, 185 (1974).
150 Elder, G. *Journal of Chromatography*, **121**, 269 (1976).
151 Södergren, A. *Journal of Chromatography*, **160**, 271 (1978).
152 Bacaloni, A., Goretti, G., Lagano, A. and Petronio, B.M. *Journal of Chromatography*, **175**, 169 (1979).
153 Leoni, V. *Journal of Chromatography*, **62**, 63 (1971).
154 McKone, C.E., Byast, T.M. and Hance, R.J. *Analyst (London)*, **97**, 653 (1972).
155 Purkayastha, R. and Cochrane, W.P. *Journal of Agriculture and Food Chemistry*, **21**, 93 (1973).
156 Ramsteiner, K., Hoermann, W.D. and Eberle, D.O. *Journal of the Association Official Analytical Chemists*, **57**, 192 (1974).
157 Hormann, W.D., Tournayre, J.C. and Egli, H. *Pesticide Monitoring*, **13**, 128 (1979).
158 Method S69. Method for triazine pesticide in water and waste water. Methods for benzidine, chlorinated organic compounds, pentachlorophenol and pesticides in water and waste water (Interim, Pending Issuance of Methods for Organic Analysis of Water and Wastes), US Environmental Protection Agency, Environmental Monitoring and Support Laboratory (EMSL) September (1978).
159 Steinheimer, T.R. and Brooks, M.G. *International Journal of Environmental Analytical Chemists*, **17**, 97 (1984).
160 Jahda, V. and Marha, K. *Journal of Chromatography*, **329**, 186 (1985).
161 Lee, H.B. and Stokker, Y.D. *Journal of the Association of Official Analytical Chemists*, **69**, 568 (1986).
162 De Loach, H.K. and Hemphill, D.D. *Journal of the Association of Official Analytical Chemists*, **52**, 33 (1969).
163 De Loach, H.K. and Hemphill, D.D. *Journal of the Association of Official Analytical Chemists*, **53**, 1129 (1970).
164 Brazhinkov, V.V., Poshememsky Sadodynskii, I.K. and Chernjakin, J. *Journal of Chromatography*, **175**, 221 (1979).

165 Kawahara, F.K. *Journal of Water Pollution Control Federation*, **39**, 572 (1967).
166 Dale, W.E. and Miles, J.W. *Journal Agric. Food Chem.*, **17**, 60 (1969).
167 Shafik, M.T. *Bulletin of Environmental Contamination and Toxicology*, **3**, 309 (1968).
168 Miller, C.W. and Funes, A. *Journal of Chromatography*, **59**, 161 (1971).
169 Rice, J.R. and Dishberger, H.J. *Journal of Agriculture and Food Chemistry*, **16**, 867 (1968).
170 Laws, E.P. and Webley, D.J. *Analyst (London)*, **86**, 249 (1961).
171 Bates, J.A.R. *Analyst (London)*, **90**, 453 (1965).
172 Daughton, C.G., Crosby, D.G., Carnos, R.L. and Hseih, D.P.H. *Journal of Agricultural and Food Chemistry*, **24**, 236 (1976).
173 Daughton, C.G., Cook, A.M. and Alexander, M. *Analytical Chemistry*, **51**, 1949 (1979).
174 Mallet, V.N., Brun, G.L., MacDonald, R.N. and Berkane, K. *Journal of Chromatography*, **160**, 81 (1978).
175 Verwej, A., Regenhardt, C.E.A. and Boter, H.L. *Chemosphere*, **8**, 115 (1979).
176 Crosby, D.G. and Bowers, O. *Journal of Agriculture and Food Chemistry*, **16**, 839 (1968).
177 McKone, C.E. and Hance, R.J. *Journal of Chromatography*, **36**, 234 (1968).
178 Khan, S.U., Greenhalgh, R. and Cochrane, W.P. *Bulletin of Environmental Contamination and Toxicology*, **13**, 602 (1975).
179 Kirkland, J.J. *Analytical Chemistry*, **34**, 428 (1962).
180 Lolke, H. *Pest. Science*, **5**, 749 (1974).
181 Freistad, H. *Journal of the Association of Official Analytical Chemists*, **57**, 221 (1974).
182 McKone, C.E. and Hance, R.J. *Bulletin of Environmental Contamination and Toxicology*, **4**, 31 (1960).
183 McKone, C.E. and Hance, R.J. *Analytical Abstracts*, **17**, 3849 (1969).
184 Rosales, J.Z. *Analytical Chemistry*, **256**, 194 (1971).
185 Cohen, I.C. and Wheals, B.B. *Journal of Chromatography*, **43**, 233 (1969).
186 Kongovi, R.R. and Grochowski, R. *American Laboratory*, **30**, 30, (1981).
187 Lee, H.B., Stokker, D. and Chau, A.S.Y. *Journal of the Association of Official Analytical Chemists*, **69**, 557 (1986).
188 Devine, J.N. and Zweig, G. *Journal of the Association of Official Analytical Chemists*, **52**, 187 (1969).
189 Larose, R.H. and Chau, A.S.Y. *Journal of the Association of Official Analytical Chemists*, **56**, 1183 (1973).
190 Agemian, H. and Chau, A.S.Y. *Journal of the Association of Official Analytical Chemists*, **59**, 732 (1976).
191 Miles, C.J. and Moye, H.A. *Analytical Chemistry*, **60**, 220 (1988).
192 Agemian, H. and Chau, A.S.Y. *Journal of the Association of Official Analytical Chemists*, **60**, 1070 (1977).
193 Soderquist, C.J. and Crosby, D.C. *Bulletin of Environmental Contamination and Toxicology*, **8**, 363 (1972).
194 Payne, S.L. *Journal of Agriculture and Food Chemistry*, **22**, 79 (1974).
195 Cannard, A.T. and Criddle, W.J. *Analyst (London)*, **100**, 848 (1975).
196 Abbott, S.D., Hall, R.C. and Giam, G.S. *Journal of Chromatography*, **45**, 317 (1969).
197 Hall, R.C., Giam, G.S. and Merkle, M.G. *Analytical Chemistry*, **43**, 423 (1970).
198 Norris, L.A. and Montgomery, M.L. *Bulletin of Environmental Contamination and Toxicology*, **13**, 1 (1975).
199 Getzendaner, M.E. *Journal of the Association of Official Analytical Chemists*, **52**, 824 (1969).
200 Frank, P.A. and Demint, R.J. *Environmental Science and Technology*, **3**, 69 (1969).

201 Van der Poll, J.M. and de Vos, R.H. *Journal of Chromatography*, **187**, 244 (1980).
202 Vos, L., Bridie, A.L. and Henzberg, S. *H₂O*, **10**, 277 (1977).
203 Balkin, F. and Hable, M.A. *Bulletin of Environmental Contamination and Toxicology*, **40**, 244 (1988).
204 Roberts, A.J. and Thomas, T.C. *Environmental Toxicology Chem.*, **5**, 3 (1986).
205 Majid, A. *ADSTRA Journal of Research*, **2**, 241 (1986).
206 Fitzpatrick, M.G. and Tan, S.S. *Chemistry, New Zealand*, **57**, 22 (1993).
207 Guo, S. Sepu. 5, 132 (1987), Novrocik, J. Novrocikova, M. Czeck CS 235803 B1. 6pp. (1986).
208 Teply, J. and Dressler, M. *Journal of Chromatography*, **191**, 221 (1980).
209 Senn, R.B. and Johnson, M.S. *Ground Water Monitoring Review*, **7**, 58 (1987).
210 Yang, Z. and Cal, Z. *Fenxi Huaxue*, **14**, 13 (1988).
211 Natusel, D.F.S. and Tomkins, B.A. *Analytical Chemistry*, **50**, 1429 (1978).
212 Constable, D.J.T., Smith, S.R. and Tanka, J. *Environmental Science and Technology*, **18**, 975 (1984).
213 Petrick, G., Schulz, D.E. and Duinker, J.C. *Journal of Chromatography*, **435**, 241 (1988).
214 Frycka, J. *Journal of Chromatography*, **65**, 432 (1972).
215 Hellmann, H.Z. *Analytical Chemistry*, **275**, 109 (1975).
216 Bowman, M.C. and Beroza, M. *Analytical Chemistry*, **40**, 535 (1968).
217 Burchfield, H.P., Wheeler, R.J. and Bernos, J.B. *Analytical Chemistry*, **43**, 1976 (1971).
218 Freed, D.J. and Faulkner, L.R. *Analytical Chemistry*, **44**, 1194 (1971).
219 Robinson, J.W. and Goodbread, J.P. *Analytica Chimica Acta*, **66**, 239 (1973).
220 Hoffmann, D. and Wynder, O. *Analytical Chemistry*, **32**, 295 (1960).
221 Haenni, E.O., Havard, J.W. and Joe, F.L. *Journal of the Association of Official Agricultural Chemists*, **45**, 67 (1962).
222 Department of the Environment, National Water Council Standing Technical Committee of Analysts, HMSO, London. *Methods for the Examination of Water and Associated Materials. Formaldehyde, Methanol and Related Compounds in Raw and Potable Waters* (1982), Tentative Methods (1983).
223 DiCorcia, A. and Samperi, R. *Analytical Chemistry*, **51**, 776 (1979).
224 Nevinnaya, L.V. and Kofanov, V.I. *Soviet Journal of Water Chemistry Technology*, **6**, 54 (1984).
225 Kawahara, F.K. *Analytical Chemistry*, **40**, 2073 (1968).
226 Kawahara, F.K. *Analytical Abstracts*, **17**, 1827 (1969).
227 Khomenkô, A.N. and Goncharova, I.A. Gidrochim. Mater., **48**, 77 (1968). Ref. Zhur. Khim. 199D. Abstract No. 13G 267 (13) (1969).
228 Van Huyssteen, J.J. *Water Research*, **4**, 645 (1970).
229 Richard, J.J., Chriswell, C.D. and Fritz, J.S. *Journal of Chromatography*, **199**, 143 (1980).
230 Vairavamurthy, A., Andreae, M.O. and Brooks, J.M. *Analytical Chemistry*, **58**, 2684 (1986).
231 Mebran, M.F., Galker, M., Mebran, M. and Cooper, W.J. *Journal of High Resolution Chromatography and Chromatography Communications*, **11**, 610 (1988).
232 Fatoki, O.S. and Vernon, F. *Water Research*, **23**, 123 (1989).
233 Olsen, E.S., Diehl, J.W. and Froelich, M.L. *Analytical Chemistry*, **60**, 1920 (1988).
234 Ozawa, H. and Tsukioka, T. *Analytical Chemistry*, **59**, 2914 (1987).
235 Yamada, H. and Somiya, I.J. *Ozone: Science Engineering*, **2**, 251 (1981).
236 Middleditch, B.S., Sung, N.J., Alakis, A. and Settemlire, G. *Chromatographia*, **23**, 273 (1987).
237 Evans, W.H. and Dennis, A. *Analyst (London)*, **98**, 782 (1973).

238 Josefsson, B.O. *Analytica Chimica Acta*, **52**, 65 (1970).
239 Stabel, V.H.H. *Arch. Hydrobiol.*, **80**, 216 (1977).
240 Nishikawa, H., Hayakawa, T. and Sakai, T. *Analyst (London)*, **112**, 45 (1987).
241 Cooper, R. and Wheatstone, K.C. *Water Research*, **7**, 1375 (1973).
242 Ramsted, T. and Nestrick, T.J. *Water Research*, **15**, 375 (1981).
243 Di Corcia, A., Samperi, R., Sabastini, E. and Severini, C. *Chromatographia*, **14**, 86 (1981).
244 Coleman, R.A., Edstrom, R.D., Unger, M.A. and Huggett, R. *Journal of Analytical Chemistry*, **542**, 631 (1982).
245 Rennie, P.J. *Analyst (London)*, **107**, 327 (1982).
246 Korenman, Y.I., Savchina, L.A. Kanavets, R.P. and Nevinnaya, L.L. *Zhur. Anal Khim.*, **43**, 1105 (1988).
247 Nolte, J., Mayer, H. and Paschold, B. *Fresenius Journal of Analytical Chemistry*, **325**, 20 (1988).
248 Mangani, F., Fabbri, A., Crescentini, G. and Bruner, F. *Analytical Chemistry*, **58**, 3261 (1988).
249 Korenman, Y.I., Minasyants, V.A. and Fokin, V.N. *Zhur. Anal. Khim.*, **43**, 1303 (1988).
250 Kaplin, V.T., Semenchenko, L.V. and Fesenko, N.G. *Gidrokhim Mater.*, **41**, 42 (1966).
251 Seminencko, L.V. and Kaplin, V.T. *Gidrokhim. Mater.*, **43**, 744 (1967).
252 Kaplin, V.T. and Semenchenko, L.V. *Gidrokhim. Mater.*, **46**, 182 (1968).
253 Szewczyk, J. and Desal, R. *Koks Smola Gaz.*, **13**, 355 (1968).
254 Panova, V.A. *Ochistka. Proizvod Stochnykh Vod.*, **4**, 184 (1969).
255 Eichelberger, J.W., Dressman, R.C. and Longbottom, A. *Journal of Environmental Science and Technology*, **4**, 576 (1970).
256 Kawahara, F.K. *Environmental Science and Technology*, **5**, 235 (1971).
257 Paklomova, A.D. and Berendeeva, V.L. *Ukr. Zhur.*, **40**, 1211 (1974).
258 Baird, R.B., Kuo, C.L. and Shapiro, J.S. *Yankow Arch. Environ. Contam. Toxicol.*, **2**, 165 (1974).
259 Plechova, O.A., Filippovy, Yu. S. and Artemova, I.M. *Tekhnol. Ochiski Prirod Stochnvod.*, **200**, 1 (1977).
260 Baker, R.A. *Journal of the American Water Works Association*, **58**, 751 (1966).
261 Baker, R.A. *Journal of the American Water Works Association*, **1**, 977 (1967).
262 DiCorcia, A. *Journal of Chromatography*, **80**, 69 (1973).
263 Goren-Stul, S., Kleijn, H.F.W. and Mostaert, A.E. *Analytica Chimica Acta*, **34**, 322 (1966).
264 Grant, D.W. and Vaughan, G.A. In *Gas Chrmoatography*, M. Van Swaay (ed.)., Butterworths, London, p. 305 (1962).
265 Kusy, V.J. *Journal of Chromatography*, **57**, 132 (1971).
266 Dietz, F., Traud, J. and Koppe, P. *Chromatographia*, **9**, 380 (1976).
267 Smith, D. and Lichtenberg, J. J. *ASTM Spec. Techn. Publ.*, **448**, 78 (1968).
268 Baker, R.A. *Air Water Pollution*, **10**, 591 (1966).
269 Derek, J.M. *Journal Fish Res. Bd. Canada*, **32**, 292 (1975).
270 Nagasawa, K., Uchiyama, H., Ogamo, A. and Shinozuka, T. *Journal of Chromatography*, **144**, 77 (1977).
271 Zerbe, J. *Chem. Anal.*, **22**, 575 (1977).
272 Matsumoto, G., Ishiwateri, R. and Hanya, T. *Water Research*, **11**, 693 (1977).
273 Meijers, A.P. and van der Leer, R. Ch. *Water Research*, **10**, 597 (1976).
274 Rump, O. *Water Research*, **8**, 889 (1974).
275 Potin-Gautier, M., Bonastre, J. and Grenier, P. *Environmental Technology Letters*, **1**, 464 (1980).

276 Quimby, B.R., Delaney, M.F., Uden, P.C. and Barnes, R.M. *Analytical Chemistry*, **52**, 259 (1980).
277 Kroner, R.C. *Public Works*, **111**, 81 (1980).
278 Yang, M.I., Harms, S., Luo, Y.Z. and Pauliszyn, J. *Analytical Chemistry*, **66**, 1339 (1994).
279 Zitko, V. *Journal of Chromatography*, **81**, 152 (1973).
280 Zitko, V. *Journal of the Association of Official Analytical Chemists*, **57**, 1253 (1974).
281 Friedman, D. and Lombardo, P. *Journal of the Association of Official Analytical Chemists*, **58**, 703 (1975).
282 Chiba, K. and Haraguchi, H. *Analytical Chemistry*, **55**, 1504 (1983).
283 Arbon, R.E. and Grimsrudd, E.P. *Analytical Chemistry*, **62**, 1762 (1990).
284 Renberg, L. *Analytical Chemistry*, **50**, 1836 (1978).
285 Junk, G.A., Chriswell, D.C., Chiang, R.S. *et al. Fresenius Zeitschrift für Analytische Chemie*, **282**, 331 (1976).
286 Kissinger, L.D. and Fritz, J.S. *Journal of the American Water Works Association*, **68**, 435 (1971).
287 Libby, A.J. *Analyst (London)*, **111**, 1221 (1986).
288 Bognar, J.A., Knighton, W.B. and Grimsrud, E.P. *Analytical Chemistry*, **64**, 2451 (1992).
289 Oliver, B.G. and Bothen, K.D. *Analytical Chemistry*, **52**, 2066 (1980).
290 Italia, M.P. and Uden, P.C. *Journal of Chromatography*, **449**, 326 (1988).
291 Kapila, S. and Aue, W.A. *Journal of Chromatography*, **15**, 569 (1977).
292 Ryan, D.A., Argentine, S.M. and Rice, G.W. *Analytical Chemistry*, **62**, 853 (1990).
293 Lanbereau, P.G. *Korrespondens Abwasser*, **31**, 499 (1984).
294 Kozelko, T.A., Kocherovskaya, I.V., Tolstopyatova, G.V. and Vinnitskaya, S.M. *Gig Sanit*, **3**, 44 (1988).
295 Shi, M., Zhu, X. and Hu, Z. *Shanghai Huanjing Kexue*, **7**, 21 (1988).
296 Lee, H.B., Stokker, Y.D. and Chau, A.S.Y. *Journal of the Association of Official Analytical Chemists*, **700**, 1003 (1987).
297 Huang, G. and Zhu, R. *Sepu*, **5**, 113 (1987).
298 Dano, S.D., Chambon, P., Chambon, R. and Sanou, A. *Analusis*, **14**, 538 (1988).
299 Kokhonen, I.O.O. and Knuutinen, J. *Chromatographia*, **17**, 154 (1983).
300 Schulte, E. and Acker, L. *Fresenius Zeitschrift für Analytische Chemie*, **268**, 260 (1974).
301 Farrington, D.S. and Mundy, W. *Analyst (London)*, **101**, 639 (1976).
302 Hoben, H.J., Ching, S.A., Casarett, L.J. and Young, R.A. *Bulletin of Environmental Contamination and Toxicology*, **15**, 78 (1976).
303 Ashiya, K., Otani, H. and Kajino, K. *Japan Water Works Association*, **No. 536**, 12 (1979).
304 Dunn, W.J., Stalling, D.L., Schwarz, I.R. *et al. Analytical Chemistry*, **56**, 1308 (1984).
305 Webb, R.G. and McCall, A.C. *Journal of the Association of Official Analytical Chemists*, **55**, 746 (1972).
306 Tas, A.C. and Kleipool, J.C. *Bulletin of Environmental Contamination and Toxicology*, **8**, 32 (1972).
307 Sissons, D. and Welti, D. *Journal of Chromatography*, **60**, 15 (1971).
308 Albro, P.W., Haseman, J.K., Clemmer, T. and Corbett, B.J. *Journal of Chromatography*, **136**, 147 (1977).
309 Sawyer, L.D. *Journal of the Association of Official Analytical Chemists*, **61**, 272 (1978).
310 Albro, P.W. and Fishbein, O. *Journal of Chromatography*, **69**, 273 (1972).

311 Hanai, T. and Walton, J.F. *Analytical Chemistry*, **49**, 1954 (1977).
312 Sawhrey, B.L., Frink, C.R. and Glowa, W. *Journal of Environmental Quality*, **10**, 444 (1981).
313 Duinker, J.C., Schulz, D.E. and Petrick, G. *Marine Pollution Bulletin*, **19**, 19 (1988).
314 Hussun, M.N. and Juns, P.C. *Analytical Chemistry*, **60**, 978 (1988).
315 Roblat, A., Xyra-fas, G. and Marshall, D. *Analytical Chemistry*, **60**, 982 (1988).
316 Brinkman, A. Th., Seetz, J.W.F.L. and Reymer, H.G.M. *Journal of Chromatography*, **116**, 353 (1976).
317 Brinkman, A. Th., DeKok, A., De Vries, G. and Reymer, H.G.M. *Journal of Chromatography*, **128**, 101 (1976).
318 Freudenthal, J. and Greve, P.A. *Bulletin of Environmental Contamination and Toxicology*, **10**, 108 (1973).
319 Banks, K.A. and Bills, D.D. *Journal of Chromatography*, **33**, 450 (1968).
320 Potter, W.G. LIB 1609, FDA, Minneapolis District (1969).
321 Erney, D.R. *Journal of the American Oil Colour Association*, **58**, 1202 (1975).
322 Wegman, R.C.C. and De Korte, G.A.L. *International Journal of Environmental Analytical Chemistry*, **9**, 1 (1981).
323 Nevinnaya, L.V. and Klyachko, Y.A. *Zhur. Anal. Khim.*, **41**, 2257 (1988).
324 Arkell, G.H. and Croll, B.I. The Water Research Association, Medmenham, Harlow, UK, Report TP78. *The Determination of Acrylamide in Water*, December (1970).
325 Croll, B.T. and Simpkins, G.M. *Analyst (London)*, **97**, 281 (1972).
326 Hashimoto, A. *Analyst (London)*, **97**, 281 (1972).
327 Bao, Z. and Zhao, Q. *Huanjing Huaxue*, **6**, 67 (1987).
328 Kofanoo, V.I., Savchina, L.A., Kanavets, R.B. and Nevinnaya, L.L. *Zhur. Anal. Khim.*, **43**, 1105 (1988).
329 Coulte, L.T., Hargescheinger, E.E., Pascuto, F.M. and Baker, G.B. *Journal of Chromatographic Science*, **19**, 151 (1981).
330 Cranmer, M. *Bulletin of Environmental Contamination and Toxicology*, **5**, 329 (1970).
331 Bengtsson, G. *Journal of Chromatographic Science*, **23**, 397 (1985).
332 Gardner, W.G. and Lee, G.F. *Environmental Science and Technology*, **7**, 719 (1973).
333 Benonfella, F. and Gald, A. *Trib Cebedeau*, **39**, 23 (1988).
334 Warren, J.M. and Beasley, R.K. *Analytical Chemistry*, **56**, 1963 (1984).
335 Murray, D., Povoledo, D. and Fish, J. *Journal of the Research Board, Canada*, **28**, 1043 (1971).
336 Rosencrance, A.B. and Bruejgemann, E.E. Report USAMBROL–TR 8601 20 pp. (1988).
337 Stolzberg, R.G. and Hume, D.N. *Journal of Chromatography*, **49**, 374 (1977).
338 Fulton, D.B., Sayer, B.G., Bain, A.D. and Malle, H.V. *Analytical Chemistry*, **64**, 349 (1992).
339 Bruins, A.P., Weidaft, L.O.G., Henion, J.D. and Budde, W.L. *Analytical Chemistry*, **59**, 2647 (1987).
340 Steinheimer, T.R. and Ondrus, M.G. *Analytical Chemistry*, **58**, 1839 (1986).
341 Henatsch, J.J. and Jutter, E. *Journal of Chromatography*, **445**, 97 (1988).
342 Leck, C. and Bagander, L.E. *Analytical Chemistry*, **60**, 1680 (1988).
343 Leck, C. and Bagander, L.E. *Analytical Chemistry*, **60**, 1880 (1988).
344 Longbottom, J.E., Edgell, K.W., Erb, E.J. and Lopez-Avila, V. *Journal of the Association of Official Analytical Chemists International*, **76**, 1113 (1993).
345 Seidel, B.S., Duebel, O., Faubel, W. and Ache, H.J. *Fresenius Journal of Analytical Chemistry*, **354**, 900 (1996).

346 Garlucci, G., Airoldi, L. and Farelli, R. *Journal of Chromatography*, **287**, 425 (1984).
347 Waters, J. and Garrigan, J.T. *Water Research*, **17**, 1549 (1983).
348 Gerbersmann, C., Lobinski, R. and Adams, F.C. *Analytica Chimica Acta*, **316**, 93 (1995).
349 Darskus, R., Hagenmaier, H., Von Holst, C. and Berkmann, L. *Fresenius Journal of Analytical Chemistry*, **348**, 148 (1994).
350 Kolesar, E.S. and Walser, R.M. *Analytical Chemistry*, **60**, 1737 (1988).
351 Kolesar, E.S. and Walser, R.M. *Analytical Chemistry*, **60**, 1731 (1988).
352 Grate, J.W., Klusty, M., Borger, W.R. and Snow, A.W. *Analytical Chemistry*, **62**, 1927 (1990).
353 Grate, J.W., Rose Pehrsson, S.L., Venezky, D.L. *et al. Analytical Chemistry*, **65**, 1868 (1993).
354 Bargnoux, H., Pepin, D., Chakard, J.L. *et al. Analysis*, **5**, 170 (1977).
355 Puijker, L.M., Veenedaal, G., Janssen, H.M.J. and Griepink, B. *Fresenius Zeitschrift für Analytische Chemie*, **306**, 1 (1981).
356 Erney, D.R. *Analytical Letters*, **12**, 501 (1979).
357 US Environmental Protection Agency Report No. EPA–600–4–74, 1974 T 108 (1974).
358 Novikova, E.E. *Zhur Vses. Khim. Obschch.*, **18**, 562 (1973).
359 Le Roy, M.L. and Goerlitz, D.E. *Journal of the Association of Official Analytical Chemists*, **53**, 1276 (1970).
360 Woodham, R.L. and Collier, C.W. *Journal of the Association of Official Analytical Chemists*, **54**, 1117 (1971).
361 Baird, R.B., Carmona, L.G. and Kuo, C.L. *Bulletin of Environmental Contamination and Toxicology*, **9**, 108 (1973).
362 Woodham, D.W., Loftis, C.D. and Collie, C.J. *Agricultural and Food Chemistry*, **20**, 163 (1972).
363 Balayanis, F.G. *Journal of Chromatography*, **90**, 198 (1974).
364 Lester, J.F. and Smiley, J.W. *Bulletin of Environmental Contamination and Toxicology*, **7**, 43 (1972).
365 Deubert, K.H. *Bulletin of Environmental Contamination and Toxicology*, **5**, 379 (1970).
366 Weil, L. and Quentin, K.E. *Zeitschrift für Wasser und Abwasser*, **7**, 147 (1974).
367 Thomas, O. *Analytical Chemistry*, **62**, 1607 (1990).
368 Croll, B.T. *Chemy. Ind.*, **40**, 1295 (1970).
369 Mohuke, M., Rohde, K.H., Brugmann, L. and Franz, P. *Journal of Chromatography*, **364**, 323 (1986).
370 Croll, B.T. *Analyst (London)*, **96**, 810 (1971).
371 Fuka, T., Janda, U. and Triska, J. *Vodni Hospodarstvi, Series B*, **33**, 245 (1983).
372 Devine, M.J. *Agricultural Chemistry*, **21**, 1095 (1973).
373 Hughes, R.A. and Lee, G.F. *Environmental Science and Technology*, **7**, 934 (1973).
374 Pocaro, P.J. and Shubiak, P. *Analytical Chemistry*, **44**, 1865 (1972).
375 Takami, K., Yamasaki, H., Okumura, T. and Nakamoto, M. *Bunseki Kagaku*, **37**, 538 (1988).
376 Chau, A.S.Y. and Terry, K. *Journal of the Association of Official Analytical Chemists*, **55**, 1228 (1972).
377 Bergemann, H. and Hellman, H. *Deutsch. Gewasserkundliche Mitteilungen*, **24**, 31 (1981).
378 Zoun, P.E.F., Splerenburg, T.J. and Boars, A.J. *Journal of Chromatography*, **393**, 133 (1987).
379 Bondarev, N.S., Spiridonov, Y.Y., Shestakov, U.G. and Chvertkin, B.Y. *Zhur. Anal. Chim.*, **42**, 1305 (1987).

380 Ahmed, I. *Journal of the Association of Official Analytical Chemists*, **70**, 745 (1987).
381 Deutsch, M.E., Westlake, W.E. and Gunther, F.A. *Journal of Agriculture and Food Chemistry*, **18**, 178 (1970).
382 Choudbury, T.K., Gerhardt, K.O. and Mawhinney, T.P. *Environmental Science and Technology*, **30**, 3259 (1996).
383 Li, Q., Yang, J. and Jin, Z. *Zhejiang Gongxueyuan Xuebao*, **34**, 64 (1987).
384 Hindin, E., Horteen, M.J., May, D.S., Skrinde, R.T. and Dunstan, G.H. *Journal of the American Water Works Association*, **54**, 88 (1962).
385 Skinde, R.T., Caskey, J.W. and Ellespie, C.K. *Journal of the American Water Works Association*, **54**, 1407 (1962).
386 Teasley, J.T. and Cax, W.S. *Journal of the American Water Works Association*, **55**, 8 (1963).
387 Liu, S. *Huanjing Kexue*, **8**, 75 (1987).
388 Brody, S.S. and Chaney, J.E. *Journal of Gas Chromatography*, **4**, 42 (1966).
389 Burchfield, H.P., Johnson, D.E., Rhoades, J.W. and Wheeler, R.J. *Gas Chromatography*, **3**, 28 (1965).
390 Guiffrida, L., Ives, N.F. and Bostwick, J.C. *Journal of the Association of Official Analytical Chemists*, **49**, 8 (1966).
391 Kamen, A. *Journal of Gas Chromatography*, **3**, 336 (1965).
392 Bache, C.A. and Lisk, D.J. *Analytical Chemistry*, **37**, 1477 (1965).
393 Hartmann, C.H. *Aerograph Research* Notes 1–6 (1966).
394 Kanazawa, J. and Kawahara, T. *Nippon Nogei Kagaku Kaishi*, **40**, 178 (1966).
395 Claborn, H.V., Mann, H.D. and Vehler, D.D. *Journal of the Association of Official Analytical Chemists*, **51**, 1243 (1968).
396 Burchfield, H.P., Rhoades, J.W. and Wheeler, R.J. *Journal of Agriculture and Food Chemistry*, **13**, 511 (1965).
397 Ives, N.F. and Guiffrida, L. *Journal of the Association of Official Analytical Chemists*, **50**, 1 (1967).
398 Dressler, M. and Janak, J. *Colln. Czech. Chem. Commun.*, **33**, 3970 (1968).
399 Wooley, D.E. *Analytical Chemistry*, **40**, 210 (1968).
400 Novak, A.V. and Malmstadt, H.V. *Analytical Chemistry*, **40**, 1108 (1968).
401 Bowman, M.C. and Beroza, M. *Analytical Chemistry*, **40**, 1448 (1968).
402 Kamen, A. *Journal of Chromatographic Science*, **7**, 541 (1969).
403 Skolnick, M. *Journal of Chromatographic Science*, **8**, 462 (1970).
404 Bowman, M.C., Beroza, M. and Hill, K.R. *Journal of the Association of Official Analytical Chemists*, **54**, 346 (1971).
405 Grice, H.W., Yates, M.I. and David, D.J. *Journal of Chromatographic Science*, **9**, 90 (1970).
406 Craven, D.A. *Analytical Chemistry*, **42**, 1679 (1970).
407 Svojanousky, V. and Nebola, R. *Chemickchisty*, **67**, 295 (1973).
408 Greenhalgh, R. and Cochrane, W.P. *Journal of Chromatography*, **70**, 37 (1972).
409 Heberer, T., Butz, S. and Stan, H. *International Journal of Environmental Analytical Chemistry*, **58**, 43 (1995).
410 Meagher, W.R. *Journal of Agricultural Food Chemistry*, **14**, 374 (1966).
411 McCone, C.E. and Hance, R.J. *Journal of Chromatography*, **69**, 204 (1972).
412 Yip, B. *Journal of American Oil Color Chemists Association*, **45**, 367 (1962).
413 Yip, G. *Journal of American Oil Color Chemists Association*, **47**, 1116 (1964).
414 Suzuki, S.H. and Malina, M. *Journal of American Oil Color Chemists Association*, **48**, 1164 (1965).
415 Goerlitz, D.F. and Lamar, W.L. *Determination of Phenoxy Acid Herbicides in Water by Electron Capture and Microcoulometric Gas Chromatography*, Geological

Survey Water Supply, Paper 1817–CUS Government Printing Office, Washington DC. (1967).
416 Yip, G. *Journal of American Oil Color Chemists Association*, **54**, 966 (1971).
417 Howard, S.F. and Yip, J. *Journal of American Oil Color Chemists Association*, **54**, 970 (1971).
418 Colas, A., Lerenard, A. and Rover, J. *Journal Chim. Anal.*, **54**, 7 (1972).
419 Chau, A.S.Y. and Terry, K. *Journal of the Association of Official Analytical Chemists*, **58**, 1294 (1975).
420 Johnson, L.C. *Journal of the Association of Official Analytical Chemists*, **56**, 1503 (1973).
421 Carnac, V.C. *Zhurnal Analiticheskoi Khimmi*, **30**, 2444 (1975).
422 Chau, A.S.Y. and Terry, K. *Journal of the Association of Official Analytical Chemists*, **59**, 633 (1976).
423 Stankova, O., Cap, L., Smicka, J. and Lankova, J. *Acta Univ. Patacki. Olomuc. Fac. Rerum Nat.*, **88**, 167 (1987).
424 De Kok, A., Vos, Y.J., Van Garderen, C. *et al. Journal of Chromatography*, **288**, 71 (1984).
425 Jraczyk, H.J. *Pflanz. Nachr. Bayer*, **25**, 21 (1972).
426 Jarczyk, H.J. *Plflanz. Nachr. Bayer*, **28**, 334 (1975).
427 Lowen, W.K., Bleidner, W.E., Kirkland, J.J., Pease, H.L. and Zweig, G. (eds.) *Analytical Methods for Pesticides, Plant Growth Regulators and Food Additives*, Vol. IV, Herbicides. Academic Press, New York, p. 157 (1964).
428 Faerber, H. and Schoeler, H.F. *Journal of Agriculture and Food Chemistry*, **41**, 217 (1993).
429 Cohen, I.C., Norcup, J., Kuzicka, J.H.A. and Wheals, B.B. *Journal of Chromatography*, **49**, 215 (1970).
430 Trehy, M.L., Yosf, R.A. and McCreary, J.J. *Analytical Chemistry*, **56**, 1281 (1984).
431 Zhang, W.Z., Lemley, A.T. and Spalik, J. *Journal of Chromatography*, **299**, 269 (1984).
432 Wesetlake, W.E., Monika, I. and Gunther, F.A. *Bulletin of Environmental Contamination and Toxicology*, **8**, 109 (1972).
433 Reeves, R.G. and Woodham, D.W. *Journal of Agriculture and Food Chemistry*, **22**, 76 (1974).
434 Rolls, J.W. and Cortes, A. *Journal of Gas Chromatography*, **2**, 132 (1964).
435 Holden, E.R., Jones, W.M. and Beroza, M. *Journal of Agriculture and Food Chemistry*, **17**, 56 (1969).
436 Gutenmann, W.H. and Lisk, D.J. *Journal of Agriculture and Food Chemistry*, **3**, 48 (1965).
437 Sundaram, K.M.S., Szeto, S.Y. and Hindle, R. *Journal of Chromatography*, **177**, 29 (1979).
438 Lewis, D.L. and Paris, D.F. *Journal of Agriculture and Food Chemistry*, **22**, 148 (1974).
439 Coburn, J.A., Ripley, B.D. and Chau, A.S.Y. *Journal of the Association of Official Analytical Chemists*, **59**, 188 (1976).
440 Ballesteros, E., Gallego, M. and Valcarcel, M. *Environmental Science and Technology*, **30**, 2071 (1996).
441 Mo, H., An, F., Liu, W. and Chang, X. *Huanjing Kexue*, **8**, 73 (1987).
442 Wood, G.W. *Analytical Abstracts*, **23**, 4219 (1972).
443 Louter, A.J.H., Brinkman, U.A.T. and Ghijsen, R.T.J. *Microcolumn Separation*, **5**, 303 (1993).
444 Goosens, E.C., de Jong, D., de Jong, G.J., Rinkema, F.D. and Brinkman, U.A.T. *Journal of High Resolution Chromatography*, **18**, 38 (1995).

445 Kulikova, G.S., Krishenko, U.E. and Pashkevich, K.I. *Zhur. Anal. Khim.*, **42**, 723 (1987).
446 Ripley, B.D., Hall, J.A. and Chau, A.S.Y. *Environmental Letters*, **7**, 97 (1974).
447 Grift, N. and Lockhart, W.L. *Journal of the Association of Official Analytical Chemists*, **57**, 1282 (1974).
448 NRC Associate Committee on Scientific Criteria for Environmental Quality, *Fenitrothion: The Effects of its use on Environmental Quality and its Chemistry*, NRCC No. 14104, p. 106 (1975).
449 Cassita, A. and Mallett, V.N. *Chromatographia*, **16**, 305 (1984).
450 Suffet, L.H. and Faust, S.D. *Journal of Agriculture and Food Chemistry*, **20**, 52 (1972).
451 Ishibashi, M. and Suzuki, M. *Journal of Chromatography*, **456**, 382 (1988).
452 Grob, R.L. and Kaiser, O.J. *Journal of Environmental Science and Health*, **A11**, 623 (1976).
453 Petersen, J.C., Guiochom, G., Demorate, C. and Dugang, M. *International Journal of Environmental Analytical Chemistry*, **14**, 23 (1983).
454 Rooney, T.A. and Freeman, R.R. Hewlett Packard Ltd., Avondale, USA. Technical Paper No. 69, *Analysis of organic contaminants in surface water using high resolution gas chromatography and selective detectors*, 174th American Chemical Society National Meeting, Chicago, Illinois, August 28–Sept. 2 (1977). Also in High Resolution Gas Chromatography, ed. S.P. Cram (1978).
455 Suffet, I.H. and Glazer, E.R. *Journal of Chromatographic Science*, **16**, 12 (1978).
456 Dix, F.D. and Fritz, J.S. *Journal of Chromatography*, **408**, 201 (1987).
457 Liu, S., Chichester-Constable, D.J., Huball, J., Smith, S.R. and Stuart, J.D. *Analytical Letters*, **20**, 2073 (1987).
458 Melcher, R.G. and Morabito, G.L. *Analytical Chemistry*, **62**, 2183 (1990).
459 Hankemeier, T., Louter, A.J.H., Rinkema, F.D. and Brinkman, U.A.T. *Chromatographia*, **40**, 119 (1995).
460 Miller, A. and Vallet, G. *Chemom. Intell. Lab. Syst.*, **17**, 153 (1992).
461 Mueller, S., Efer, J. and Engewald, W. *Chromatographia*, **38**, 694 (1993).
462 Morabito, P.L., McCabe, T., Hiller, J.F. and Zakett, D. *Journal of High Resolution Chromatography*, **16**, 90 (1993).
463 Okuno, I. and Higgins, W.H. *Bulletin of Environmental Contamination and Toxicology*, **18**, 428 (1977).
464 Kiigemagi, U., Burnard, J. and Terrier, L.C. *Journal of Agricultural and Food Chemistry*, **23**, 717 (1975).
465 Singley, J.E., Kirhmer, C.J. and Minra, R. US Environment Protection Agency Series EPS–600/2–74–02, US Government Printing Office, Washington DC, No. 126 (1974).
466 Kataoka, H., Rhu, S., Sakiyama, N. and Makita, M. *Journal of Chromatography*, **726**, 253 (1996).
467 Kawano, Y. and Benvenue, A. *Journal of Chromatography*, **72**, 51 (1972).
468 Van der Hoff, G.R., Pelusio, F., Brinkman, U.A.T., Baumann, R.A. and Van Zoonen, P. *Journal of Chromatography*, **719**, 59 (1996).
469 West, S.D. and Turner, L.G. *Journal of the Association of Official Analytical Chemists*, **71**, 1049 (1988).
470 Duffy, M., Driscoll, J.N., Pappas, S. and Sonford, W. *Journal of Chromatography*, **44**, 73 (1988).
471 Koei, M., Kalurand, M. and Kussik, E. *Analytica Chimica Acta*, **199**, 197 (1987).
472 Slabodnik, J., Hogenboom, A.C., Louter, A.G.H. and Brinkman, U.A.T. *Journal of Chromatography*, **730**, 353 (1996).

473 Yoshioka, M., Tsuji, M., Yamasaki, T. and Okuno, T. *Hyogo-ken. Kogal Kenkyusho Kenkyu Hakaku*, **18**, 60 (1988).
474 Matsabara, H. and Ikeda, Y. *Fufuoka-shi Eisel Shkenschoho*, **12**, 126 (1987).
475 Lopes-Avila, V., Heath, N. and Hu, A. *Journal of Chromatographic Science*, **25**, 356 (1987).
476 Pankow, J.F. *Journal of High Resolution Chromatographic Communications*, **10**, 409 (1987).
477 Pankow, J.F. and Rosen, M.E. *Environmental Science and Technology*, **22**, 398 (1988).
478 Cochran, J.W. *Journal of High Resolution Chromatography Communications*, **10**, 573 (1987).
479 Melanson, E.G., Johnson, E.K. and Poppiti, J. *Journal of Proceedings*. Water Quality Technical Conference 1986 (Advances in Water Anal. Treat.), pp. 203–204 (1987).
480 Sakodynskii, K.I., Panina, L., Reznikova, Z.A. and Kargman, V.B. *Journal of Chromatography*, **364**, 455 (1988).
481 Engst, R. and Knoll, R. *Nahrung*, **17**, 837 (1973).
482 Kawamura, K. and Kaplan, I.R. *Analytical Chemistry*, **48**, 1616 (1984).
483 Adams, I.M. *Process Biochemistry*, **2**, 33 (1967).
484 Institute of Petroleum Standardization Committee. *Journal of the Institute of Petroleum (London)*, **56**, 107 (1970).
485 Wilson, C.A., Ferreto, E.P. and Coleman, H.J. *American Chemical Society Div. Chem. Pepr.*, **60**, 613 (1975).
486 Garza, M.E. and Muth, J. *Environmental Science and Technology*, **8**, 249 (1974).
487 Blumer, M. and Sass, J.J. *Science*, **176**, 1120 (1972).
488 Boylan, D.B. and Tripp, B.W. *Nature (London)*, **230**, 44 (1971).
489 McKay, T.R. Proceedings of the Ninth Symposium on Gas Chromatography, Montreux, Switzerland, pp. 33–38 October (1972).
490 Zafiron, O.C., Myers, J. and Freestone, F. *Marine Pollution Bulletin*, **4**, 87 (1973).
491 Bulten, J.N., Morris, B.F. and Sass, J.J. Bermuda Biological Research, Special Publication No. 10. 350 pp. (1973).
492 Blumer, M. *Marine Biology*, **5**, 195 (1970).
493 Brunnock, J.V., Duckworth, D.F. and Stephens, G.G. *Journal of the Institute of Petroleum*, **54**, 310 (1966).
494 Zafiron, O.C. and Oliver, C. *Analytical Chemistry*, **45**, 952 (1973).
495 Rasmussen, D.V. *Analytical Chemistry*, **48**, 1562 (1976).
496 Girenko, D.B., Klisenko, M.A. and Dishcholka, Y.K. *Hydrobiological Journal*, **11**, 60 (1975).
497 Wilson, A.J., Forester, J. and Knight, J. US Fish Wildlife Circular, 355, 18–20 (1969). Centre for Estuaries and Research Gulf Breeze, Florida, US. (1969).
498 Wilson, A.J. *Bulletin of Environmental Contamination and Toxicology*, **15**, 515 (1976).
499 Wilson, A.J. and Forester, J. *Journal of the Association of Official Analytical Chemists*, **54**, 525 (1971).
500 Quinn, J.G. and Meyers, P.A. *Limnology and Oceanography*, **16**, 129 (1971).
501 Eklund, G., Josefsson, B.O. and Roos, C. *Journal of Chromatography*, **142**, 595 (1977).
502 Bullister, J.L. and Weiss, R.F. *Deep Sea Research, Part A*, **35**, 839 (1988).
503 Elder, D. *Marine Pollution Bulletin*, **7**, 63 (1976).
504 Savchuck, S.A., Rudenko, B.A., Brodskii, E.S. and Soifer, V.S. *Journal of Analytical Chromatography*, **50**, 1081 (1995).
505 Duane, L. and Stock, J.T. *Analytical Chemistry*, **50**, 1891 (1978).

506 Wu, T.L., Lambert, L., Hastings, D. and Banning, D. *Bulletin of Environmental Contamination and Toxicology*, **24**, 411 (1980).
507 Garrett, W.D. *Limnology and Oceanography*, **10**, 602 (1965).
508 Fam, S., Stenstrau, M.K. and Silverman, G.J. *Environmental Engineering*, **113**, 1032 (1987).
509 Wade, T.L. and Quinn, J.G. *Marine Pollution Bulletin*, **6**, 54 (1975).
510 Roe, V.D., Lacy, M.J., Stuart, J.D. and Robbins, G.A. *Analytical Chemistry*, **61**, 2584 (1989).
511 Ritter, J., Stromquist, V.K., Mayfield, H.T., Henley, M.V. and Lavine, B.K. *Microchemical Journal*, **54**, 59 (1996).
512 Saxena, J., Basu, D.K. and Kozuchowski, J. US National Technical Information Service, Springfield, Virginia. Report No. PB276635, p. 94 (1977).
513 Saxena, J., Kozuchowski, J. and Basu, D.C. *Environmental Science and Technology*, **11**, 682 (1977).
514 Bellar, U.B., Lichtenberg, J.J. and Eichelberger, J.W. *Environmental Science and Technology*, **10**, 926 (1976).
515 Nicholson, A.A. and Meresz, O. *Bulletin of Environmental Contamination and Toxicology*, **14**, 453 (1975).
516 Eklund, G., Josefson, B. and Roos, C.J. *Journal of Chromatography*, **142**, 575 (1977).
517 Eklund, G., Josefson, B. and Roos, C.J. *Journal of High Resolution Chromatography Commmunication*, **1**, 34 (1978).
518 Trussell, A.R., Umphres, M.D., Leong, L.Y.C. and Trussell, R.R. *Journal of the American Waterworks Association*, **71**, 385 (1979).
519 Fine, D.H., Rainbehler, D.P., Huffman, F. and Epstein, S.S. *Bulletin of Environmental Contamination and Toxicology*, **14**, 404 (1975).
520 Sackmauereva, M., Pal'usova, O. and Szokolay, A. *Water Research*, **11**, 551 (1977).
521 Le'Bel, G.L. and Williams D.T. *Bulletin of Environmental Contamination and Toxicology*, **24**, 397 (1980).
522 Le'Bel, G.L., Williams, D.T., Griffith, G. and Benoit, F.M. *Journal of the Association of Official Analytical Chemists*, **62**, 281 (1971).
523 Swinnerton, J.W. and Linnenbom, V.J. *Journal of Gas Chromatography*, **5**, 570 (1967).
524 McCarthy, L.V., Overton, E.B., Raschke, C.K. and Laseter, J.L. *Analytical Letters (London)*, **13**, 1417 (1980).
525 Kisinger, L.D. and Fritz, J.S. *Journal of the American Water Works Association*, **68**, 435 (1976).
526 National Survey for Halocarbons in Drinking Water. Health and Welfare, Canada, 77–EHD–9 (1977).
527 Bush, B. and Narang, R.S. *Bulletin of Environmental Contamination and Toxicology*, **13**, 436 (1977).
528 Friant, S.I. Thesis. Drexel University. University Microfilms Ltd., London. 468 PR 29162 (1972).
529 Smillie, R.D., Nicholson, A.A., Meresz, O. et al. *Organics in Ontario Drinking Waters, Part II. A Survey of Selected Water Treatment Plants*, Ontario Ministry of the Environment, April 1977.
530 Richard, J.J. and Junk, G.A. *Journal of the American Water Works Association*, **69**, 62 (1977).
531 Mieure, J.P. *Journal of the American Water Works Association*, **69**, 62 (1977).
532 Nicholson, A.A., Meresz, O. and Lemyk, B. *Analytical Chemistry*, **49**, 814 (1977).

533 Hammerstrand, K. *Chloroform in Drinking Water, Varian Instrument Application,* **10,** 22 (1976).
534 Viryasov, M.B. and Nikitin, Y.S. *Soviet Journal of Water Chemistry and Technology,* **9,** 71 (1987).
535 US Environmental Protection Agency. Federal Register No. 28, 43, 5756 (1978).
536 US Environmental Protection Agency. Control of Organic Chemical Constituents in Drinking Water. Federal Register, **44,** 5755 *et seq.* February 9. (1978).
537 Reding, R.J. *Chromatographic Science,* **25,** 338 (1987).
538 Peruzi, P. and Griffini, O. *Journal of High Resolution Chromatography,* **8,** 450 (1985).
539 Mehran, F.F., Slither, R.A. and Cooper, W.J. *Journal of Chromatographic Science,* **22,** 241 (1984).
540 Kirschen, N.A. *Varian Instrument Applications,* **15,** 2 (1981).
541 Norin, H. and Renberg, L. *Water Research,* **14,** 1397 (1980).
542 Miller, J.W. and Uden, P.C. *Environmental Science and Technology,* **17,** 150 (1983).
543 Janda, V., Marha, K. and Mitera, J. *Journal of High Resolution Chromatography, Chromatography Communications,* **11,** 54 (1988).
544 Dingynan, H. and Jianfei, T. *Analytical Chemistry,* **63,** 2078 (1991).
545 Inoke, M., Tsuchiya, M. and Matsumio, T. *Environmental Bulletin (Series B),* **7,** 129 (1984).
546 Kissinger, L.P. US National Technical Information Service, Springfield, Virginia. No. 15–T–845 161 pp. (32411) (1979).
547 Marshall, J.W. and Wampler, T.P. *American Laboratory,* **20,** 72 (1988).
548 Comba, M.E. and Kaiser, K.L.E. *International Journal of Environmental Analytical Chemistry,* **16,** 17 (1983).
549 Nicolson, B.C., Bursill, D.B. and Conche, D.J. *Journal of Chromatography,* **325,** 221 (1985).
550 Dressman, R.C., Stevens, A.A., Fair, J. and Smith, B.G. *Journal of the American Water Works Association,* **71,** 392 (1979).
551 Reichert, J.K. and Lochtman, J. *Environmental Science Letters,* **4,** 15 (1983).
552 Kirschen, N.A. *Varian Instrument Applications,* **14,** 10 (1980).
553 Method 501.1. *The Analysis of Trihalomethanes in Finished Waters by the Purge and Trap Method.* US Environmental Protection Agency, EMSL, Cincinnati, Ohio, 45268, November 6 (1979).
554 Fielding, M., McLoughlin, K. and Steel, C. Water Research Centre. Enquiry Report ER 532. August 1977. Water Research Centre, Stevenage Laboratory, Elder Way, Stevenage, Herts, UK (1977).
555 Report of the Environmental National Water Council Standing Committee of Analysts. HMSO, London. *Methods for the Examination of Waters and Associated Materials. Chloro and Bromo Trihalogenated Methanes in Water* (1981).
556 Van Rensberg, J.F.T., Van Huyssteen, J.T. and Hassett, A.J. *Water Research,* **12,** 127 (1978).
557 Benoit, G.M. and Williams, D.T. *Bulletin of Environmental Contamination and Toxicology,* **27,** 303 (1981).
558 Morgade, C., Barquet, A. and Pfaffenberger, C.D. *Bulletin of Environmental Contamination and Toxicology,* **24,** 257 (1980).
559 Rodriquez, F., Bollain, M.H., Garcia, C.H. and Cela, R. *Journal of Chromatography,* **733,** 405 (1996).
560 Patil, S.S. *Analytical Letters (London),* **21,** 1397 (1988).
561 Webb, R.G. and McCall, A.C. *Journal of Chromatographic Science,* **11,** 366 (1973).

562 Department of the Environment. *Methods for the Examination of Waters and Associated Materials. Organochlorine Insecticides and Polychlorinated Biphenyls in Waters*, 1978. Tentative method (32313). HMSO, London, 28 pp. (1979).
563 Van Loon, W.M.G.M., Bron, J.S. and de Groot, B. *Analytical Chemistry*, **65**, 1726 (1993).
564 Rosen, E. *American Laboratory*, **12**, 49 (1980).
565 O'Keefe, P., Meyer, C., Smith, R. *et al. Chemosphere*, **15**, 1127 (1986).
566 Cranmer, M.E. and Peoples, A. *Analytical Biochemistry*, **55**, 255 (1973).
567 Cranmer, M.F. and Peoples, A. *Analytical Abstracts*, **22**, 1779 (1972).
568 Van Huyssteen, J.J. *Water Research*, **4**, 645 (1970).
569 Hunter, I.R., Orgeren, V.H. and Pence, J.W. *Analytical Chemistry*, **32**, 682 (1960).
570 Murtaugh, J.J. and Bunch, R.L. *Water Pollution Control Federation*, **37**, 410 (1965).
571 Gehrke, G.W. and Larkin, W.J. *Journal of Agriculture and Food Chemistry*, **9**, 85 (1961).
572 Harivank, J. *Vodni Hsparastvi Chem. Abstr.*, **14**, 394 (1964).
573 Smith, E.M. and Dila, R. *Pharm. Belg.*, **20**, 225 (1965).
574 Emery, E.M. and Koenrner, W.E. *Analytical Chemistry*, **33**, 146 (1961).
575 Erwin, E.S., Marco, G.C. and Emery, E.M. *J. Dairy Science*, **44**, 1768 (1961).
576 Cochrane, G.C. *Journal of Chromatographic Science*, **13**, 440 (1975).
577 Baker, R.A. *Journal of Chromatography*, **4**, 18 (1960).
578 Narkis, N., Henfield-Furie, S. *Water Research*, **12**, 437 (1978).
579 American Public Health Association. WPCF and AWWA. *Standard Methods for the Determination of Water and Waste Water*, 14th edn. New York (1976).
580 Mattson, D.E. and Nygren, S. *Journal of Chromatography*, **124**, 265 (1976).
581 Jenson, S., Johnels, A.G., Olsson, M. and Otterlind, G. *Ambio, Spec. Rep.*, **1**, 71 (1972).
582 Methods for the Analysis of Water. Association of Official Analytical Chemists, Washington DC. 10th edn., p. 393 (1965).
583 Erne, K. *Acta Pharmacol. Toxicol.*, **14**, 158 (1958).
584 Bartlett, J.K. and Skoog, D.A. *Analytical Chemistry*, **26**, 1008 (1954).
585 Zimmerli, B., Sulser, H. and Marek, B. *Mitt. Geb. Lebensmittelunters. Hyg.*, **62**, 60 (1971).
586 Cochrane, W.P. and Maybury, R.B. *Journal of the Association of Official Analytical Chemists*, **56**, 1324 (1973).
587 Miller, G.A. and Wells, C.E. *Journal of the Association of Official Analytical Chemists*, **52**, 548 (1969).
588 Jensen, S., Renberg, L. and Reutergard, L. *Analytical Chemistry*, **49**, 316 (1977).
589 Pearson, J.R., Aldrich, F.D. and Stone, A.W. *Journal of Agriculture and Food Chemistry*, **15**, 938 (1967).
590 Ahnoff, M. and Josefsson, B. *Bulletin of Environmental Contamination and Toxicology*, **13**, 159 (1975).
591 Goerlitz, D.F. and Law, L.H. *Bulletin of Environmental Contamination and Toxicology*, **6**, 9 (1971).
592 Copin, A., Deten, R., Monteau, A. and Rocher, M. *Science of the Total Environment*, **22**, 179 (1982).
593 Fox, M.W. *Environmental Science and Technology*, **7**, 838 (1973).
594 Kuo, P.P.K., Chian, E.S.K. and Chang, B.T. *Environmental Science and Technology*, **11**, 1177 (1977).
595 Kuo, P.P.K., Chian, E.S.K., DeWalle, F.B. and Kim, J.H. *Analytical Chemistry*, **49**, 1023 (1977).
596 Webb, R.G., Garrison, A.W., Kieth, L.H. and McGuire, J.J. US Environmental Protection Agency Report No. EPA–R2–277 (1973).

597 Kramer, D.N., Klein, N. and Bascellce, R.A. *Analytical Chemistry*, **31**, 250 (1959).
598 Somiya, I., Yamada, H., Izumi, Y. and Odagaki, M. *Japanese Journal of Water Pollution Control*, **2**, 181 (1979).
599 Hindin, E. *Water Sewage Works*, **92**, 94 (1964).
600 Abbatichio, P., Balice, N. and Cera, O. *Environmental Technology Letters*, **4**, 179 (1983).
601 Hignit, E.C. and Azamoff, P.F. *Lite Science*, **20**, 337 (1977).
602 Rhoades, J.W., Thomas, R.E., Johnson, D.E. and Tillery, J.B. US National Information Service, Springfield, Virginia. Report No. PB81–232167. *Determination of Phthalates in Industrial and Municipal Waste Waters* (1981).
603 UK Department of the Environment, Standing Committee. *Analytical Methods. Examination of Waters and Associated Materials. Chlorobenzenes Water, Organochlorine Pesticides, PCBs, Turbid Waters, Halogenated Solvents, Related Compounds, Sewage, Sludge Waters.* 1985. 44 pp. (1986).
604 Richardson, M.L., Webb, K.S. and Gough, T.A. *Ectoxicology and Environmental Safety*, **4**, 207 (1980).
605 Rudling, L. *Water Research*, **6**, 871 (1972).
606 Rudling, L. *Water Research*, **5**, 831 (1971).
607 Aue, W.H., Hastings, C.R., Gerhardt, K.O. *et al. Journal of Chromatography*, **72**, 259 (1972).
608 McIntyre, A.E., Perry, R. and Lester, J.N. *Bulletin of Environmental Contamination and Toxicology*, **26**, 116 (1981).
609 Krueger, C.J. and Field, J.A. *Analytical Chemistry*, **67**, 3363 (1995).
610 Bailey, J.C. and Viney, N.J. Gas chromatographic investigation of odour and a sewage treatment plant. Water Research Centre, Medmenham Laboratory, Medmenham, UK. Technical Report No. TR 125, December (1979).
611 Jenkins, L.L., Gute, J.P., Krasner, S.W. and Baird, R.B. *Water Research*, **14**, 441 (1980).
612 McIntyre, A.E., Perry, R. and Lester, J.N. *Environmental Technology Letters*, **1**, 57 (1980).
613 McIntyre, A.E., Lester, J.N. and Perry, R. *Analysis of Organic Substances of Concern in Sewage Sludge.* Final report to the Department of the Environment for contracts DGR/480/66 and DGR/480/240, pp. 42–45, Imperial College, London, UK. November (1979).
614 Holden, A.V. and Marsden, K. *Journal of Chromatography*, **44**, 481 (1969).
615 Canagay, A.B. and Levine, P.L. US Environmental Protection Agency, Cincinnati, Ohio. Report EPA 600/2–79–166 (1974). 100 pp. (P22–CAR) (1979).
616 Dmitriev, M.T. and Kitrosski, N.A. Gig. Svait., 10, 69 (1969). Ref. Zhur. Khim. 19GD (7) Abstract 7G256 (1970).
617 Waggott, A. and Saunders, C.L. Water Research Centre, Medmenham, UK. Technical Report TR44. Gas liquid chromatography applied to the analysis of complex environmental samples, May. (1977).
618 Ellison, W.K. and Wallbank, T.E. *Water Pollution Control*, **73**, 656 (1974).
619 Baird, R., Selna, M., Hoskins, J. and Chappelle, D. *Water Research*, **13**, 493 (1979).
620 Roeraade, J. *Journal of Chromatography*, **330**, 263 (1985).
621 The Oil Companies Study Group of Conservation of Clean Air and Water, Europe. CONCAWE. Report 6/82. *Analysis of Trace Substances in Aqueous Effluents from Petroleum Refineries* (1982).
622 Environmental Protection Agency Report EPA 610. Environmental Protection Agency, USA (1977).

623 Vitenberg, A.G., Kuzvetsova, L.H., Butaeva, I.L. and Ishakov, M.D. *Analytical Chemistry*, **49**, 128 (1977).
624 Supplement to the 15th edn. of Standard Methods for the Examination of Water and Waste Water. Selected Analytical Methods Approved and Cited by US Environmental Protection Agency, American Public Health Association, American Water Works Association, Water Pollution Control Federation. Method S51 September (1978).
625 Supplement to the 15th edn. of Standard Methods for the Examination of Water and Waste Water. Selected Analytical Methods Approved and Cited by US Environmental Protection Agency, American Public Health Association, American Water Works Association, Water Pollution Control Federation. Method S78. Method for PCBs in water and waste water, September. (1978).
626 Goerlitz, D.F. and Law, L.M. *Method for Chlorinated Hydrocarbons in Water and Waste Water*, US Environmental Protection Agency, Interim Method. p. 7. September. (1978).
627 Code, R.D. *Journal of the Institute of Petroleum (London)*, **54**, 288 (1968).
628 Voss, R.H. and Rapsomatiotis, A. *Journal of Chromatography*, **346**, 205 (1985).
629 Larsson, M. and Roos, C. *Chromatographia*, **17**, 185 (1983).
630 Esposito, G.G. and Schaeffer, K.R.H. *Journal of American Industrial Hygiene Association*, **37**, 268 (1976).
631 American Public Health Association. *Standard Methods for the Examination of Waters and Wastewaters. Methods for Volatile Chlorinated Organic Compounds in Water and Wastewater* (1985).
632 Buisson, R.S.K., Kirk, P.W.W. and Lester, J.N. *Journal of Chromatographic Science*, **22**, 339 (1984).
633 Chau, A.S.Y. and Coburn, J.A. *Journal of the Association of Official Analytical Chemists*, **57**, 389 (1974).
634 Matthew, J. and Elzerman, A.W. *Analytical Letters (London)*, **14**, 1351 (1981).
635 McIntyre, A.E., Perry, R. and Lester, J.N. *Environmental Pollution (Series B)*, **2**, 223 (1981).
636 Hill, N.F., McIntyre, A.E., Perry, R. and Lester, J.N. *International Journal of Environmental Analytical Chemistry*, **15**, 107 (1982).
637 Lysyj, I. *Environmental Science and Technology*, **8**, 31 (1974).
638 Austern, B.M., Dobbs, R.A. and Cohen, J.M. *Environmental Science and Technology*, **9**, 658 (1975).
639 Stoly'arov, B.V. *Chromatographia*, **14**, 699 (1981).
640 Urano, K., Maeda, H., Ogura, K. and Wada, H. *Water Research*, **16**, 323 (1982).
641 Searl, T.D., King, W.H. and Brown, R.A. *Analytical Chemistry*, **42**, 954 (1970).
642 Mirzayanov, U.S. and Bugrov, Y.F. *Zavod Lab.*, **38**, 656 (1792).
643 Pilipenko, A.T., Golovka, N.V., Lantukh, G.V. and Tulyupa, F.M. *Soviet Journal of Water Chem. Technology*, **9**, 61 (1987).
644 British Coke Research Association, Chesterfield, Derby, UK. Research Report No. 55, 21 pp. (1969).
645 *Determinations of phenols in coke oven effluents with special reference to consent conditions.* British Coke Research Association Coke Research Report No. 79 June. (1973).
646 Hood, L.V.S. and Winefordner, J.D. *Analytica Chimica Acta*, **42**, 199 (1968).
647 Fogelquist, E., Josefsson, B. and Ross, C. *Journal of High Resolution Chromatography, Chromatography Communications*, **3**, 568 (1980).
648 Harrington, K.J. and Hamilton, J. *Canadian Journal of Chemistry*, **48**, 2607 (1970).
649 Karmil'chik, A. Ya., Stonkuz, V. and Korchargova, E. *Kh. Zhur. Analit. Khim.*, **26**, 1231 (1971).

650 Reubecker, O. *Environmental Science and Technology*, **19**, 1232 (1985).
651 Bork, L.S., Cooper, R.L. and Wheatsone, K.C. *Water Research*, **6**, 1161 (1971).
652 Hassler, J. and Rippa, F. Vyskummy Ustai Vodneko Hospodarstra; Veda a Vyskum Praxi. No. 50. 50 pp. (1977).
653 De Leer, E.W.B. *H₂O*, **13**, 171 (1980).
654 Nulton, C.P., Haile, C.L. and Redford, D.P. *Analytical Chemistry*, **56**, 598 (1984).
655 Paama, L., Maeorg, U. and Kokk, K. *Zhur. anal. Khim.*, **43**, 904 (1988).
656 Delfino, J.J. and Easty, N. *Analytical Chemistry*, **51**, 2235 (1979).
657 Easty, D.B. and Wabers, B.A. *Tappi*, **61**, 71 (1978).
658 Zatkovetski, V.M., Satonova, Z.B., Koren'kov, V.N., Nevskii, A.B. and Sosin, S.L. *Zavod. Lab.*, **37**, 1434 (1971).
659 Bark, L.S., Cooper, P.L. and Wheatstone, K.C. *Water Research*, **6**, 117 (1972).
660 Riggin, R.M., Cole, T.F. and Billets, C. *Analytical Chemistry*, **55**, 1862 (1983).
661 Jenkins, R.L. *Bulletin of Environmental Contamination and Toxicology*, **13**, 436 (1975).
662 Polishchuk, O.M. and Gorbonds, T.V. *Soviet Journal of Chemical Technology*, **5**, 50 (1983).
663 Richard, J.J. and Junk, G.A. *Analytical Chemistry*, **58**, 723 (1986).
664 Belkin, F., Bishop, R.W. and Sheely, G.A. *Journal of Chromatographic Science*, **23**, 532 (1985).
665 Tian, L.X. and Fu, R.N. International Jahrestag-Franunhafer-Inst. Trub-Explosivst. 1988. 17th Ana. Propellants Explos. Chem. Phys. Methods, 63/1–63/9 (1988).
666 Nevinnaya, L.V. and Klyachko, Y.A. *Soviet Journal of Water Chemistry and Technology*, **10**, 71 (1988).
667 Kirk, A.W.W., Perry, R. and Lester, J.N. *International Journal of Environmental Analytical Chemistry*, **12**, 293 (1982).
668 Scoggins, M.W. and Skurcenski, L.J. *Chromatographic Science*, **15**, 573 (1977).
669 Matisova, E., Lehotang, J., Garaj, J. and Violova, A. *Journal of Chromatography*, **237**, 164 (1982).
670 Stevens, R. *Analytical Chemistry*, **56**, 1608 (1984).
671 Olson, E.S., Diehe, W. and Miller, D.J. *Analytical Chemistry*, **55**, 1111 (1983).
672 Jaroslav, N. *Chemicky. Prumsyl.*, **20**, 575 (1970).
673 Vitenberg, A.G., Butaeva, I.L. and Dimitova, Z.St. *Chromatographia*, **8**, 693 (1975).
674 Devonald, R.H., Serenius, R.G. and McIntyre, A.D. *Pulp Paper Magazine, Canada*, **73**, 50 (1972).
675 Girenko, D.B. and Klisenko, M.A. *Gig. Sanit.*, (3), 48 (1987).
676 Barcelo, D., Maris, F.A., Geerdink, R.H., Frei, R.H. and Brinkman, U.A.T. *Journal of Chromatography*, **394**, 65 (1987).
677 Crompton, T.R. Private communication.
678 Merz, W. and Nen, H.J. *Chemosphere*, **16**, 469 (1987).
679 Schmeltz, I., Matasker, J. and Rachick, E. *Journal of Chromatography*, **268**, 21 (1983).
680 Jiao, Y. and Sun, S. Huanjing Huaxue, 6, 62 (1987).
681 Kimoto, W.L., Dooby, C.J., Canoe, J. and Fiddler, W. *Water Research*, **14**, 869 (1980).
682 Weil, L., Jandik, J. and Eicheldorfer, D. Z. *Wasser Abwasser Forschung*, **13**, 141 (1980).
683 Andreae, M.O. *Analytical Chemistry*, **48**, 820 (1977).
684 Chau, Y.K., Wong, P.T.S., Bengert, G.A. and Kramer, O. *Analytical Chemistry*, **51**, 186 (1979).

685 Chau, Y.K., Wong, P.T.S., Saitoh, H. *Journal of Chromatographic Science*, **162**, 14 (1976).
686 Segar, D.A. *Analytical Letters (London)*, **7**, 89 (1974).
687 Chau, Y.K., Wong, P.T.S. and Kramer, O. *Analytica Chimica Acta*, **146**, 211 (1983).
688 Chau, Y.K., Wong, P.T.S. and Goulden, P.D. *Analytica Chimica Acta*, **85**, 421 (1976).
689 Chau, Y.K., Wong, P.T.S., Bengeut, G.A. and Kramer, O. *Analytical Chemistry*, **186**, 51 (1979).
690 Chau, Y.K., Wong, P.T.S. and Goulden, P.D. *Analytica Chimica Acta*, **421**, 85 (1976).
691 Aue, W.A., Miller, B. and Xun Yen Sen *Analytical Chemistry*, **62**, 2453 (1990).
692 Nishi, S. and Horimoto, H. *Japan Analyst*, **17**, 1247 (1968).
693 Nishi, S. and Horimoto, H. *Japan Analyst*, **19**, 1646 (1970).
694 Dressman, R.C. *Journal of Chromatographic Science*, **10**, 472 (1972).
695 Ealy, J., Schulz, W.D. and Dean, D.A. *Analytica Chimica Acta*, **64**, 235 (1974).
696 Zarnegar, P. and Mushak, P. *Analytica Chimica Acta*, **69**, 389 (1974).
697 Cappon, C.J. and Crispin-Smith, V. *Analytical Chemistry*, **49**, 365 (1977).
698 Bowles, K.C. and Apte, C.A. *Analytical Chemistry*, **70**, 395 (1998).
699 Hodge, V.F., Seidel, S.L. and Goldberg, E.D. *Analytical Chemistry*, **51**, 1256 (1979).
700 Mahone, L.G., Garner, B.J., Buch, R.R. et al. *Environmental, Toxicological Chemistry*, **2**, 307 (1983).
701 Soderquist, C.J. and Crosby, D.C. *Analytical Chemistry*, **50**, 1435 (1978).
702 Braman, P.S. and Tompkins, M.A. *Analytical Chemistry*, **51**, 12 (1979).
703 Chau, Y.K., Wong, P.T.S. and Bengert, G.A. *Analytical Chemistry*, **54**, 246 (1982).
704 Agemian, H. and Chau, V. *Analytica Chimica Acta*, **101**, 193 (1978).
705 Kolthoff, T.M. and Belcher, R. In *Volumetric Analysis*. Vol. 3. pp. 511–513. Interscience Publishers, New York (1957).
706 Stringer, D.E. and Attrep, M. *Analytical Chemistry*, **51**, 731 (1979).
707 Braman, R.S. and Foreback, C.C. *Science*, **182**, 1247 (1973).
708 Braman, R.S., Johnson, D.L., Craig, C. et al. *Analytical Chemistry*, **49**, 612 (1977).
709 Gifford, P.R. and Bruckenstein, S. *Analytical Chemistry*, **52**, 1028 (1980).
710 Talmi, Y. and Bostick, D.T. *Analytical Chemistry*, **47**, 2145 (1975).
711 Crecelius, E.A. *Analytical Chemistry*, **50**, 826 (1978).
712 Shalk, A.U. and Tallman, D.E. *Analytica Chimica Acta*, **98**, 251 (1978).
713 Von Endt, D.W., Kearny, P.C. and Kaufman, D.D. *Journal of Agriculture and Food Chemistry*, **16**, 17 (1968).
714 Portman, J.E. and Riley, J.P. *Analytica Chimica Acta*, **31**, 509 (1964).
715 Casvalho, M.B. and Hercules, D.M. *Analytical Chemistry*, **50**, 2030 (1978).
716 Lodmell, J.D. PhD Thesis, University of Tennessee, Knoxville, Tenn. (1973).
717 Johnson, L.D., Gerhart, K.O. and Aue, W.A. *Science of the Total Environment*, **1**, 108 (1972).
718 Fickett, A.W., Doughtrey, E.H. and Mishak, P. *Analytica Chimica Acta*, **79**, 93 (1975).
719 Soderquist, C.J., Crosby, D.G. and Bowers, J.B. *Analytical Chemistry*, **46**, 155 (1974).
720 Parris, G.E., Blair, W.R. and Brinckmann, F.E. *Analytical Chemistry*, **49**, 378 (1977).
721 Bergmann, K. and Neidart, B. *Fresenius Journal of Analytical Chemistry*, **356**, 57 (1996).
722 Estes, S.A., Uden, P.C. and Barnes, R.M. *Analytical Chemistry*, **53**, 1336 (1981).
723 Rapsaniankis, S., Donard, O.F.Y. and Weber, J.H. *Analytical Chemistry*, **58**, 35 (1986).

724 Chakraborti, D., De Jongh, W.R.A., Van Mol, W.E., Von Leuvenbergen, R.J.A. and Adams, F.C. *Analytical Chemistry*, **56**, 2692 (1984).
725 Forsythe, D.D. and Marshall, W.D. *Analytical Chemistry*, **55**, 2132 (1983).
726 Potter, H.R., Jarview, A.W.P. and Markell, R.N. *Water Research*, **76**, 123 (1977).
727 Emteborg, H., Sinemus, H.W., Radzuik, B., Baxter, D.C. and Frech, W. *Spectrochemica Acta, Part B*, **51B**, 829 (1996).
728 Longbottom, J.E. *Analytical Chemistry*, **44**, 1111 (1972).
729 Filippelli, M. *Applied Organometallic Chemistry*, **8**, 687 (1994).
730 Mushak, P., Tibetts, F.E., Zarneger, P. and Fisher, G.B. *Journal of Chromatography*, **87**, 215 (1973).
731 Westhoo, G. *Acta. Chem. Sci. Anal.*, **21**, 1790 (1967).
732 Westoo, G. *Acta. Chem. Sci. Anal.*, **22**, 2277 (1968).
733 Lee, Y.H. *International Journal of Environmental Analytical Chemistry*, **29**, 263 (1987).
734 Lobinski, R., Dirk, W.M.R., Ceulemans, M. and Adams, F.C. *Analytical Chemistry*, **64**, 159 (1992).
735 Muller, M.D. *Analytical Chemistry*, **59**, 617 (1987).
736 Neubert, G. and Wirth, H.Q. *Fresenius Zeitschrift für Analytische Chemie*, **273**, 19 (1975).
737 Neubert, G. and Andreas, H. *Fresenius Zeitschrift für Analytische Chemie*, **280**, 31 (1976).
738 Hattori, Y, Kabayashi, A., Takemito, S. et al. Journal of Chromatography, **315**, 341 (1984).
739 Cai, Y. and Bayona, J.M. *Journal of Chromatographic Science*, **33**, 89 (1995).
740 Carber-Pinasseau, C., Lespes, G. and Autric, M. *Applied Organometallic Chemistry*, **10**, 505 (1996).
741 Stab, J.A., Cofino, W.P., Van Haltum, B. and Brinkmann, U.A.T. *Fresenius Journal of Analytical Chemistry*, **347**, 247 (1993).
742 Nelson, H.D. Doctoral Dissertation. University of Utrecht, 29 (1967).
743 Beckert, W.F. *Journal of High Resolution Chromatography*, **16**, 106 (1993).
744 Lui, Y., Lopez-Avila, V., Alcaraz, M. and Beckert, W.F. *Analytical Chemistry*, **66**, 3788 (1994).
745 Mienema, L.A., Burger, W.T., Versluis-Dehaau, G. and Gevers, E.C. *Environmental Science and Technology*, **12**, 288 (1978).
746 Unger, M.A., MacIntyre, W.G., Greaves, J. and Hergett, R.J. *Chemosphere*, **15**, 461 (1986).
747 Lobinski, R., Bontrom, C.F., Candelone, J.P. et al. *Analytical Chemistry*, **65**, 2510 (1993).
748 Braman, P.S. and Tompkins, M.A. *Analytical Chemistry*, **51**, 12 (1979).
749 Chiba, K., Wong, P.T.S., Tanabe, K. et al. *Analytical Chemistry*, **55**, 1 (1983).
750 Brinckmann, F.E. Trace metals in seawater. In: Proceedings of a NATO Advanced Research Institute on Trace Metals in Seawater, 30/3–30/4/81, Sicily, Italy (eds. C.S. Wong et al.) Plenum Press, New York (1981).
751 Aue, W.A. and Flinn, C.S. *Journal of Chromatography*, **142**, 145 (1977).
752 Jackson, J.A., Blair, W.R., Brinckmann, F.E. and Iverson, W.P. *Environmental Science and Technology*, **16**, 110 (1982).
753 Evans, W.H., Jackson, F.I. and Dellar, D. *Analyst (London)*, **104**, 16 (1979).
754 Takahashi, K. *Nippon Kagaku Kaishi*, (9), 1591 (1988).
755 Valkirs, A.D., Seligman, P.F., Olsen, G.J., Brinckman, F.E. *Marine Pollution Bulletin*, **17**, 320 (1986).
756 Valkirs, A.O., Seligman, P.F., Stang, P.M. *Marine Pollution Bulletin*, **17**, 319 (1986).
757 Estes, S.A., Uden, P.C. and Barnes, R.M. *Analytical Chemistry*, **54**, 2402 (1982).

758 Chau, Y.K., Bengert, G.A. and Dunn, J.L. *Analytical Chemistry*, **56**, 271 (1984).
759 Maguire, R.J. and Huneault, H. *Journal of Chromatography*, **209**, 458 (1981).
760 Kusaka, Y., Tguji, H., Fujimoto, Y. and Ishida, K. *Journal of Radioanalytical Chemistry*, **71**, 7 (1982).
761 Andreae, M.O. and Froelich, P.M. *Analytical Chemistry*, **53**, 287 (1981).
762 Vien, S.H. and Fry, R.C. *Analytical Chemistry*, **60**, 465 (1988).
763 Cutter, L.S., Cutter, G.A., Maria, L.C. and McGlove, S.D. *Analytical Chemistry*, **63**, 1138 (1991).
764 Measures, C.I. and Edmond, J.M. *Analytical Chemistry*, **58**, 2065 (1986).
765 Tao, H., Mlyazaki, A. and Bansho, K. *Analytical Science*, **4**, 299 (1988).
766 Shimoishi, Y. and Toei, K. *Analytica Chimica Acta*, **100**, 65 (1978).
767 Flinn, C.G. and Aue, W.A. *Journal of Chromatography*, **153**, 49 (1978).
768 Johansson, K., Oernemark, U. and Olin, A. *Analytica Chimica Acta*, **274**, 129 (1993).
769 De La Calle Guntinas, M.B., Lobinski, R. and Adams, F.C. *Journal of Analytical Atomic Spectroscopy*, **10**, 11 (1995).
770 Nakashima, S. and Toei, K. *Talanta*, **15**, 1475 (1968).
771 Monteil, A. *Analysis*, **9**, 112 (1981).
772 Talmi, Y. and Andren, A.W. *Analytical Chemistry*, **46**, 2122 (1974).
773 Nakamuro, N., Sayato, Y., Tonomura, M. and Ose, Y. *Eisei Kagaku*, **18**, 237 (1972).
774 Gosink, T.A. and Reynolds, D.J. *Journal of Marine Science Communications*, **1**, 101 (1975).
775 Shimoishi, Y. *Analytica Chimica Acta*, **64**, 465 (1973).
776 Young, J. and Christian, G.D. *Analytica Chimica Acta*, **65**, 127 (1973).
777 Measures, C.I. and Burton, J.D. *Nature (London)*, **273**, 293 (1978).
778 Measures, C.I. and Burton, D. *Analytica Chimica Acta*, **120**, 177 (1980).
779 Measures, C.I. and Burton, J.D. *Earth Plan. Sci. Letters*, **46**, 385 (1980).
780 Measures, C.I., McDuff, R.E. and Edmond, J.M. *Earth Plan. Sci. Letters*, **49**, 102 (1981).
781 Lee, M.L. and Burrell, D.C. *Analytica Chimica Acta*, **66**, 245 (1973).
782 Mugo, R.K. and Orions, K.I. *Analytica Chimica Acta*, **271**, 1 (1992).
783 Schaller, H. and Neeb, R. *Fresenius Zeitschrift für Analytische Chemie*, **327**, 170 (1987).
784 Mena, M.L., McLeod, C.W., Jones, P. *et al. Fresenius Journal of Analytical Chemistry*, **351**, 456 (1995).
785 Jones, R. and Nickless, G. *Journal of Chromatography*, **76**, 285 (1973).
786 Jones, R. and Nickless, G. *Journal of Chromatography*, **89**, 201 (1974).
787 Gallus, S.M. and Henmann, K.G. *Journal of Analytical Atomic Spectroscopy*, **11**, 887 (1996).
788 Galluss, M. Private communication.
789 Kuo, C.J., Lin, I.H., Shih, J.S. and Yeh, J.C. *Journal of Chromatographic Science*, **20**, 455 (1982).
790 Mack, R.S. and Grimsrud, E.P. Private communication.
791 Funazo, K., Kuseno, K., Tanaka, T. and Shono, T. *Analyst (London)*, **107**, 182 (1982).
792 Funazo, K., Tanaka, M. and Shono, T. *Analytical Chemistry*, **52**, 1222 (1980).
793 Uchida, H., Shimoishi, Y. and Toei, K. *Environmental Science and Technology*, **14**, 541 (1980).
794 Funazo, K., Kusano, K., Wu, H.L. Tanaka, M. and Shono, T. *Journal of Chromatography*, **245**, 93 (1982).
795 Narasaki, P., Histaomi, K. and Matsura, T. *Eisel Kagaku*, **33**, 158 (1987).

796 Ando, M. and Sayato, V. *Water Research*, **17**, 1823 (1983).
797 Nota, G., Vernassi, G., Acampara, A. and Sannolo, N. *Journal of Chromatography*, **173**, 228 (1979).
798 Maros, L., Kaldy, M. and Iqaz, S. *Analytical Chemistry*, **61**, 733 (1989).
799 Belcher, R., Major, J.R., Rodriguez-Vasquez, J.A., Stephen, W.I. and Uden, P.C. *Analytica Chimica Acta*, **57**, 73 (1971).
800 Nota, G., Poliemhari, R. and Imperota, C. *Journal of Chromatography*, **123**, 411 (1976).
801 Campanella, L. and Sprilli, R. *Metodi Analtica per le Acque*, **No. 1** (1981).
802 Christensen, S. and Tiedje, J.M. *Applied and Environmental Microbiology*, **54**, 1409 (1988).
803 Hashimoto, S., Fujimara, K. and Fuwa, K. *Limnology and Oceanography*, **32**, 729 (1987).
804 Radford-Kuoery, J. and Cutter, G.A. *Analytical Chemistry*, **65**, 976 (1993).
805 Faiqle, W. and Klochow, D. *Fresenius Zeitschrift für Analytische Chemie*, **306**, 190 (1981).
806 Cutter, G.A. and Oatts, T.J. *Analytical Chemistry*, **59**, 717 (1987).
807 Grandet, M., Weil, L. and Quentin, K.E. *Zeitschrift für Wasser und Abwasser*, **16**, 66 (1983).
808 Bachmann, K. and Matusca, P. *Fresenius Zeitschrift für Analytische Chemie*, **315**, 243 (1983).
809 Nota, G., Vernassi, G., Acampora, A. and Sonnolo, N. *Journal of Chromatography*, **173**, 228 (1979).
810 Ross, W.D., Buttler, G.W., Tuff, Y.D.G., Rehg, W.R. and Winiger, M.T. *Journal of Chromatography*, **112**, 719 (1975).
811 Tesche, J.W., Rehg, W.R. and Sievers, R.E. *Journal of Chromatography*, **126**, 743 (1976).
812 Funazo, K., Kusano, K., Tanaka, M. and Shomo, T. *Analytical Letters (London)*, **13**, 751 (1980).
813 Nota, G. and Improta, C. *Water Research*, **13**, 177 (1979).
814 Nota, G., Miraglia, U.R. and Acampora, A. *Journal of Chromatography*, **207**, 47 (1981).
815 Borchardt, L.T. and Easty, D.B. *Journal of Chromatography*, **299**, 471 (1984).
816 Hawke, D.J., Lloyd, A., Martinson, D.M., Slater, P.B. and Excell, C. *Analyst (London)*, **110**, 269 (1985).
817 Knoery, J. and Cutter, G.H. *Analytical Chemistry*, **65**, 976 (1993).
818 Wu, H.L., Lin, S.J., Funazo, K., Tanaka, M. and Shono, T. *Fresenius Zeitschrift für Analytische Chemie*, **322**, 409 (1985).
819 Suzuki, M., Yamoto, Y. and Wanatabe, T. *Environmental Science and Technology*, **11**, 1109 (1977).
820 Henderson, J.E., Peyton, G.R. and Glaze, W.H. Convenient liquid–liquid extraction method for the determination of halomethanes in water at the parts per billion level. In *Analysis and Identification of Organic Pollutants in Water*. Ed. L.H. Keith. Ann Arbor Science, Ann Arbor, Michigan, 105–195 (1976).

Chapter 16

Gas chromatography–mass spectrometry

16.1 Organic compounds

16.1.1 Non saline waters

16.1.1.1 Alkylbenzene sulphonates

Fig. 16.1 shows a gas chromatogram of alkylbenzene sulphonate methyl esters in a river water sample [5]. The pattern of the gas chromatogram is analogous to that of the linear alkybenzene sulphonate standard. The assignment of the peaks was performed on the basis of the retention times and mass spectrum, and the individual components of alkylbenzene sulphonate were determined by mass fragmentography. For overlapped peaks on the gas chromatogram, more than two mass spectra were recorded. The total amounts of alkylbenzene sulphonate determined by mass fragmentography were in good accord with those determined by gas chromatography. Desulphonation gas chromatography has been applied [1–4] to the analysis of partially degraded linear alkylbenzene sulphonate mixtures.

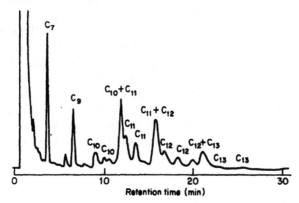

Fig. 16.1 Gas chromatogram of ABS as methyl esters in river water. Column temperature, 230°C
Source: Reproduced by permission from Elsevier Science, UK [5]

16.1.1.2 Other applications

Other applications of gas chromatography–mass spectrometry include (Table 16.1) the following: polyaromatic hydrocarbons, esters, phenols, aromatic acids, chlorophenols, glycols, hydroxybenzenes, 1:4 dioxane, propiolactone, vinyl chloride, polychlorobiphenyls, chlorinated solvents, chlorolignosulphonic acids, chloroanisoles, chlorodioxins, chlorobenzofurans, volatile organics, 2,2,*bis*-4-(2-hydroxyethoxy)-3,–5 dichromopropane, aliphatic and alicyclic amines, pyridine bases, nitriloacetic acid, ethylene diamine tetraacetic acid, aldehydes, Methoxychlor, Kepone, Heptachlor, Trifluralin, s-triazines, Acetochlor, Carbofuran, melamine, acidic herbicides, benzothiazole, napthols, chloroethyl phosphates, Dicamba, 4-(chloromethyl sulphonyl) bromobenzene, pendimethalin, hydroxylcarbamates, Benzthiocarb, Oxidiazone, dialkyl and aryl phosphates, tri-(2-chloroethyl)phosphate, isopropyl (ethylphosphono-fluoridate), diisopropylfluorophosphate, geosmin, nonyl phenyl acetic acid, octyl phenols, polyethoxylate, Tricyclazone, 3-chloro-4-(dichloro-methyl)-5-hydroxy-2-(5H)-furanone, squalane and total organic nitrogen and carbon.

16.1.2 Aqueous precipitation

16.1.2.1 Miscellaneous organic compounds

Watabe *et al.* [99] identified trace amounts of phthalate esters, polyaromatic hydrocarbons, higher fatty acids and their esters in rain water by means of a gas chromatography mass spectrometer–computer system.

Pankow *et al.* [100] give details of a sampler developed for collection of rain water. The sampler is controlled electronically, provides *in situ* filtration of the sample, and carries out preconcentration of non polar organic compounds by means of cartridges containing the sorbent Tenax GF. It is possible to incorporate cartridges of ion exchange resin for preliminary concentration of organic acids. Samples of rain water collected by this equipment from several rainfall events at two sites in Oregon were analysed by gas chromatography and mass spectrometry and 27 organic compounds were identified. The results were used in conjunction with available Henry's law constants to estimate the local atmospheric levels of these compounds at the sampling sites.

16.1.3 Seawater

16.1.3.1 Petroleum hydrocarbons

Albaiges and Albrecht [107] propose that a series of petroleum hydrocarbons of geochemical significance (biological markers) such as C_{20}–C_{40}

Table 16.1 Gas chromatography–mass spectrometry of organic compounds in non saline waters

Organic	Extraction from water	GLC column conditions	Detector	Comments	LD	Ref
Polyaromatic compounds	–	–	–	GC–MS methods not affected by method bias as are methods using optical detection	–	[6]
Dimethyl phthalate, dibutyl phthalate, di(2-ethyl-hexyl)phthalate, benzoyl butyl phthalate	Solvent extraction	–	–	Clean up on deactivated Florasil	0.5ng	[7]
Dibutyl phthalate, bis(2-2 butoxyethoxy) methane, bis(2-ethyl-hexyl adipate) dioctylphthalate diisodecyl phthalate isomers, trichlorobenzane, biphenyl benzoate, butyl benzoate	–	–	–	Comparison of GC–MS, high resolution mass spectrometry and high pressure liquid chromatography	0.001 mg L^{-1}	[8–10]
Phthalates	–	–	–	GC–MS MS with chemical ionisation	–	[11]
Silylated alkyl phenols, alkyl phenyl ethoxylate	–	–	–	GC–MS	–	[12]
Phenols, aromatic acids, eg pentachlorophenol, bisphenol, phthalic acid, trimesic acid, p-hydroxy-benzoic acid, vanilic acid, syringic acid, p-coumaric acid, ferulic acid	Ethyl acetate	–	–	GC, mass fragmentography	–	[13]

Table 16.1 continued

Organic	Extraction from water	GLC column conditions	Detector	Comments	LD	Ref
Carboxylic acids, glycols, dihydroxybenzene	–	–	–	GC–MS. In situ derivativisation with n-hexyl chloroformate	–	[14]
Carboxylic acids	–	–	–	Combined GC–fast atom bombardment, calcium tri-ethonolamine used as matrix	–	[15]
1,4 dioxane	–	–	Flame ionisation	Purge and trap GC–MS or adsorbtion/desorption of 1,4 dioxane then GC with flame ionisation detector	ppb	[16]
Propiolactone	–	–	–	GC–MS. Propiolactone bromated to 3-bromopropionate then bromoacetylated to p-bromo-phenacyl 3-bromopropionate	0.25ppm	[17]
Vinyl chloride	–	–	–	Mass fragmentography then GC–MS recording m/e 62 and 64 peaks. Quadrupole m/s equipped with multiple ion detection	sub μg L^{-1}	[18]
Polychlorobiphenyls	–	–	–	Gas chromatography–mass spectrometry	–	[19–22, 228]
Tetra to nona poly-chlorobiphenyls	–	–	–	Study of exchange reaction of chlorine by oxygen with PCB anions as a method for detection of PCB cogeners in a GC–MS/MS system	–	[23]
Polychlorobiphenyls	–	–	–	Thermal desorption GC–MS	–	[24]

Table 16.1 continued

Organic	Extraction from water	GLC column conditions	Detector	Comments	LD	Ref
Polychlorobiphenyls	–	Capillary	Electron capture	Study of application of negative ion chemical ionisation MS, selected ion monitoring and pulsed positive ion/negative ion chemical ionisation MS	–	[25]
Polychlorobiphenyls	–	–	Electrolytic conductivity, electron capture	Gas chromatography–mass spectrometry	–	[26]
Polychloroterphenyls	Hexane	3% OV–17 on Varoport 30, at 280–325°C (m/e 436) and 200–285°C (for m/e 470)	–	Hexane extract cleaned on alumina or Florasil, then GC–MS in mass fragmentography mode at m/e 436 and m/e 470	0.1ppm	[27]
Chlorinated solvents	–	–	–	Combination of purge and trap analysis with GC–MS	ppt	[28]
Chloroligno sulphonic acids	–	–	–	Combination of pyrolysis GC and MS with single ion monitoring	0.1µg L⁻¹	[29]
Chloroanisoles, chloro-methylanisoles	Methylene dichloride	–	Electron capture	Methylene dichloride extracts clean upon Florasil column	0.002–0.02 µg L⁻¹	[30]
2,3,7,8 tetrachlorodi-benzo-p-dioxin	Alkali extraction, clean up and fractionation on active carbon	High resolution GC	Electron capture	Mass spectrometric identification of individual dioxins	–	[31]

444

Table 16.1 continued

Organic	Extraction from water	GLC column conditions	Detector	Comments	LD	Ref
2,3,7,8 tetrachlorodibenzo-p-dioxin	–	–	–	Mass spectrometric detection	–	[32]
Polychlorodibenzo-p-dioxins, tetra, octa, chloro and polychlorobenzofurans	–	–	–	High resolution MS coupled with ion monitoring	–	[33]
15 polychlorodibenzo-p-dioxins	–	–	–	Negative ion MS	–	[34]
Polychlorodibenzo-p-dioxins, polychlorodibenzofurans, polychlorobiphenyls	–	–	–	Study of molar responses of these substances using a mass spectrometric detector	–	[35]
Polychlorodibenzo-p-dioxins, polychloro-dibenzofurans	–	–	–	Mass spectrometry of these compounds in a quadrupole ion trap, comparison with single frequency modulation and multi-frequency resonant excitation modes	–	[36]
Volatile organic compounds incl. aromatics haloforms and chloro-aromatics	–	–	–	Membrane MS	–	[37]

Table 16.1 continued

Organic	Extraction from water	GLC column conditions	Detector	Comments	LD	Ref
Volatile chlorocompounds	–	–	–	Purge and trap GC–MS	–	[38]
624 purgeable organic compounds eg vinyl chloride, chloro naphthalene, anthracene	–	–	–	Spray and trap method using thermal desorption GC–MS method for field use	–	[39]
2,2 bis,4-(2-hydroxy ethoxy)-3,5, dibromophenyl propane	Methylene dichloride	–	–	Methylene dichloride extract silated with N,N(trimethyl) silyl trifluoroacetamide, then GC–MS	–	[40]
Aliphatic amines	Hexane	–	–	Amines converted into imines using pentafluorobenzaldehyde, then hexane extracted then high resolution GC–MS using multiple ion monitoring	10µg L^{-1}	[41]
Aliphatic and alicyclic amines	–	–	–	Amines derivativised to tri-chloroethyl carbamates then GC–MS	–	[42]
Ethylene, dibromide, 1,2 dibromo-3-chloropropane	Hexane	–	–	Negative ion chemical ionisation GC–MS using isobutane as reagent gas	0.1µg L^{-1}	[43]
Pyridine bases	–	Capillary column coated with DB wax	–	GC–MS, recovery 90%	0.01–1ng	[44]
Nitriloacetic acid	–	Capillary column	–	Extract acetylated then converted to trimethyl ester, then GC–MS	–	[45]

Table 16.1 continued

Organic	Extraction from water	GLC column conditions	Detector	Comments	LD	Ref
Ethylene diamine tetraacetic acid, DTPA	-	-	Nitrogen–phosphorus detector	These compounds esterified	10ppb	[46]
Nitriloacetic acid	-	-		These compounds converted to methyl esters, then GC–MS	sub ppb	[47]
Non volatile organic compounds, ethylene diamine tetraacetic acid, carboxy alkyl phenoxy ethoxy carboxylates, di-carboxylic acids, aldehydes	-	-		These compounds propylated with propanol–formic acid–acetyl chloride, then GC–electron impact MC or GC chemical ionisation–MS and GC–MS/MS	-	[48]
Ethylene ferric diamine tetraacetic acid	-	3% QF-ion GasChrom Q at 110–260°C	Flame ionisation	Converted to N-trifluoryl-n-butyl ester, photolysed, esters identified from fragmentation pattern double beam mass spectrometer	-	[49]
Methoxychlor	Hexane	-	Electron capture	-	sub ppb	[50]
Kepone (chlorodecone)	-	-		GC–MS on kepone and its photoproducts	-	[51]
Heptachlor, Trifluralin, S-triazines	-	-		GC–MS with selected ion recording	1ppb	[52]
Acetochlor	-	-		GC–MS	sub ppb	[53]
Carbofuran, Atrazine	-	-		GC–MS	0.1 μg L^{-1}	[54]
Atrazine, Lindane, Diazinon, Pentachlorophenol	-	-		Isotope dilution GC–MS	0.1–1μg L^{-1}	[55]

Table 16.1 continued

Organic	Extraction from water	GLC column conditions	Detector	Comments	LD	Ref
Atrazine	Solid phase extraction	–	–	GC–MS interferences removed on Florasil	sub ppt	[56]
Atrazine, Cyonazine, Metochlor, pendi-methalin, Simazine	–	–	–	GC–MS	–	[57]
Melamine (2,4,6 tri-amino-2,3,5 triazine)	–	–	–	Melamine derivitivised with N,O-bis(trimethyl silyl) trifluoro-acetamide then GC–MS	sub μg L^{-1}	[58]
Atrazine	–	–	–	Comparison of magnetic particle immunoassay with GC–MS	–	[59]
Triazine herbicides	On-line solid phase extraction	–	–	On-line solid phase extraction, then GC–MS	ppb	[60]
Acidic herbicides	–	–	–	Herbicides converted to penta-fluoro benzyl bromide derivatives, then GC–MS	sub ppb	[61]
Benzothiazole, naphtholols, 3-chloroethyl-phosphate, 3-methyl-3-octanol, Diazinon, Chlorfenvinphas	–	–	–	GC–MS	–	[62]
Dicamba 2,4D	–	–	–	Isotope dilution GC–MS, recoveries 84%	low μg L^{-1}	[63]
4-(chloromethylsulph-benyl)bromo benzene	–	–	–	GC–MS	–	[64]
Pendimethalin (N-methyl propyl) 3,4, dimethyl-2,6, dinitrobenzene	Benzene	–	–	GC–MS	1μg L^{-1}	[65]

Table 16.1 continued

Organic	Extraction from water	GLC column conditions	Detector	Comments	LD	Ref
Hydroxycarbamates	–	–	–	In situ derivitivisation with n-hexyl chloroformate then GC–MS with chemical ionisation	10ppb	[66]
Benzthiocarb(S-(4-chloro-benzyl), N,N-diethylthiol-carbamate) oxidiazon (2-tert-butyl-4-(2,4, di-chloro-5-isopropoxy-phenyl, 2-1,3,4-oxadiazolin-5-one), CNP (2,4,6 tri-chlorophenyl-4'-nitro phenyl ether)	–	–	–	GC–MS	–	[67,68]
Dialkyl phosphates	–	–	–	Extractive pentafluoro benzylation, then GC–MS	–	[69]
Alkyl and aryl phosphate	Dichloromethane	–	Flame photometric	Florasil column, then GC–MS	–	[70]
Tri(2-chloroethyl) phosphate, oxidiazone, phthalic and esters	–	–	–	GC–MS	–	[71]
Volatile organic compounds	–	Vocol(diphenyl dimethyl–cross-linked polysiloxone and 95% methyl-silicone and 5% phenylsilicone	–	Comparison of two columns, vocol column test, trap GC–MS method	–	[72]

Table 16.1 continued

Organic	Extraction from water	GLC column conditions	Detector	Comments	LD	Ref
Volatile organic compounds	–	–	–	Construction of passing samples with sorbent tube that fits into sample chamber fitted with a diffusional membrane permeable to organic vapours. Organics desorbed from sorbent and analysed by GC–MS	–	[73]
266 pollutants	–	–	–	GC–MS (ion trap)	–	[74]
Volatile organic compounds	–	–	–	Spray extraction into gas phase of volatile organics then GC–MS	10–30µg L^{-1}	[75]
Environmental organic contaminants	–	–	–	Multidimensional GC with infra-red and MS detection	–	[76]
Volatile organic compounds	–	–	–	Negative ion chemical ionisation GC–MS	1ng	[77]
Volatile organic compounds	–	–	–	Purge and trap GC–MS	–	[78]
Polar volatile organic compounds	–	–	–	Trap GC–MS	–	[79]

Table 16.1 continued

Organic	Extraction from water	GLC column conditions	Detector	Comments	LD	Ref
27 compounds including methyl dichloroethanoate, 2,3-dichloro-2-methyl-butane, diethylenylbenzene, dodecane ethenylethyl-benzene, 2-ethoxypropane, 3-ethyl-4-methylfuran-2,5-dione, methyl and ethyl-naphthalenes, hexane, hex-1-ene, 1-methyl-1 H-indene, 2-methylpentane, 1,1,1-tri-chloropropan-2-one, 1,1,3-trichloropropan-2-one and undecane	–	–	–	Mass fragmentography	–	[80]
Miscellaneous organic compounds	–	–	–	GC–MS	$0.1\,\mu g\,L^{-1}$	[81–84]
21 volatile compounds	–	–	–	GC–MS	–	[85]
100 compounds	Gas purging	–	–	GC–MS	–	[86]
78 compounds	–	–	–	GC–MS	–	[87]
Isopropyl ethyl phos-phorofluoridate, diiso-propylfluoro phosphate	–	–	–	GC–MS	–	[88]
Geosmin, 2-methyiso-borneol	–	–	–	Computer controlled GC–mass fragmentography	$10\,ng\,L^{-1}$	[89]

Table 16.1 continued

Organic	Extraction from water	GLC column conditions	Detector	Comments	LD	Ref
(Monyl phenyl) acetic acid, octylphenol polyoxyethylate, C_2-C_4 diols, fatty acids	–	Capillary column GC	–	Capillary column GC–MS	–	[90–94]
Tricydazone	Methylene dichloride	Chromosorb W	Flame thionic	GC–MS	1ng	[95]
3-chloro-4-(dichloro-methyl)-5-hydroxy-2 (5H)-furanone	–	–	–	Compound converted to methyl derivative then GC–MS	ppt	[96]
Squalane	Ethyl acetate	–	–	GC–MS	ng	[97]
Total organic nitrogen and carbon	–	–	–	Pyrolysis GC–MS	–	[98]

Source: Own files

acyclic isoprenoids and C_{27} steranes and triterpenes are used as passive tags for the characterisation of oils in the marine environment. They use mass fragmentography of samples to make evident these series of components without resorting to complex enrichment treatments. They point out that computerised gas chromatography–mass spectrometry permits multiple fingerprinting from the same gas chromatographic run. Hence rapid and effective comparisons between samples and long-term storage of the results for future examination can be carried out.

Generally, the most apparent is the n-paraffin distribution that has proved to be useful in differentiating the main types of pollutant samples (crude oils, fuel oils and tank washings [102] or even types of crude oils [103,104], although in this case the method involves the quantification of the previously isolated n-paraffins, therefore lengthening the analysis time.

All the above fingerprints exhibit a different usefulness for characterising oils. Their variation between crudes and their resistance to the sea-weathering process are not enough, in many cases, for providing the unequivocal identification of the pollutant. The n-paraffins can, apparently, be removed by biodegradation as well as the lower acyclic isoprenoids at respectively slower rates [104]. The flame photometric detector chromatogram is less sensitive to modification by bacterial metabolism but can also be affected by evaporation, in spite of its higher retention range, as will be shown later. However, the last part of the flame ionisation detector chromatogram appears to be highly promising in overcoming these limitations. In fact, this part corresponds to a hydrocarbon fraction that boils at over 400°C, so it cannot easily be evaporated under environmental marine conditions. Moreover, it contains a wealth of compounds of geochemical significance, namely isoparaffins and polycyclic alkanes of isoprenoid, sterane and triterpane structure [105,106] as a result of a complete reduction of precursor isoprenyl alcohols, sterols and triterpanes, respectively. Therefore, their occurrence and final distribution in crude oils will be related to their particular genetic history, that is to the original sedimentary organic matter and to the processes undergone during its geochemical cycle. In consequence, it was assumed by Albaigés and Albrecht [107] that these factors can provide unique hydrocarbon compositions for each crude oil, by which the unambiguous identification of the samples can be brought about. Besides their geochemical stability, these compounds do also remain unaltered after biodegradation [108], being, in this respect, valuable passive tags for characterising marine pollutants.

The problem is that such components are present frequently at trace levels, as part of very complex mixtures and can only be recognised after long and tedious enrichment treatments that are not practical from the standpoint of the routine or monitoring analysis.

Mass fragmentography provides a satisfactory tool for obtaining specific fingerprints for classes and homologous series compounds, resolved by gas chromatography. In addition, computerised gas chromatography–mass spectrometry allows multiple fingerprinting from the same chromatogram, that is especially important for a quick survey of any compound class in a scanty sample and permits storing the information for further processing or correlation studies. However, to carry out the analysis successfully a precise knowledge of the nature and the gas chromatographic and mass spectrometric behaviour of such compounds is needed. Albaigés *et al.* have done considerable work on this topic [109–112].

Brown and Huffman [113] reported an investigation of the concentration and composition of non volatile hydrocarbons in Atlantic Ocean and nearby waters. Seawater samples were taken at depths of 1 and 10m and the non volatile hydrocarbons were identified by mass spectrometric techniques. The results show that the non volatile hydrocarbons in Atlantic and nearby waters contained aromatics at lower concentrations than would be expected if the source of the hydrocarbons were crude oil or petroleum refinery products. Hydrocarbons appeared to persist in the water to varying degrees, with the most persistent being the cycloparaffins, then isoparaffins, and finally the aromatics.

Walker *et al.* [114] examined several methods and solvents for use in the extraction of petroleum hydrocarbons from estuarine water and sediments, during an *in situ* study of petroleum degradation in sea water. The use of hexane, benzene and chloroform as solvents is discussed and compared, and quantitative and qualitative differences were determined by analysis using low-resolution computerised mass spectrometry. Using these data, and data obtained following the total recovery of petroleum hydrocarbons, it is concluded that benzene or benzene–methanol azeotrope are the most effective solvents.

Smith [115] classified large sets of hydrocarbon oil spectra data by computer into 'correlation sets' for individual classes of compounds. The correlation sets were then used for determining the class to which an unknown compound belongs from its mass spectral parameters. A correlation set is constructed by use of ion series summation, in which a low-resolution mass spectrum is expressed as a set of numbers representing the contribution to the total ionisation of each of 14 ion series. The technique is particularly valuable in the examination of results from coupled gas chromatography–mass spectrometry of complex organic mixtures.

16.1.3.2 Phenols

Boyd [116] used aqueous acetylation followed by gas chromatography–mass spectrometry to determine ppt concentrations of phenols in seawater. Ppq detection limits were achieved by a large sample volume extraction set-up.

16.1.4 Potable water

16.1.4.1 Nitrosamines

Richardson *et al.* [117] have applied gas chromatography with a chemiluminescence detection system to the determination of microgram levels of nitrosamines (N-nitrosodimethylamine, N-nitrosodiethylamine, N-nitrosomorpholine and N-nitrosodiethanolamine, N-nitrosopyrrolidine, N-nitrosopiperidine and N-nitroso-5-methyl-1,3-oxazolidine) in potable water supplies. Nitrosamines may be removed from aqueous media by solvent extraction and subsequently concentrated by evaporation of the solvent, in order to detect levels as low as $0.01\mu g\ L^{-1}$.

For the estimation of volatile dialkyl nitrosamines and N-nitrosopiperidine, N-nitrosopyrrolidine and N-nitrosomorpholine, 10N sulphuric acid was added to the sample which was then extracted with redistilled dichloromethane. 1.5m sodium hydroxide was added to the combined extract. After separation, the organic layer was dried over sodium sulphate and evaporated to 2.5mL at 46°C on a water bath. Hexane was added, and evaporation continued to about 250μL. Aliquots of 5μL were analysed for volatile nitrosamines using gas chromatography.

In the chemiluminescence procedure, which detects all nitrosamines amenable to gas chromatography, effluent from a chromatograph passes into a catalytic chamber, whereupon the nitrosamine is fragmented to give rise to nitric oxide. This reacts with ozone and results in a chemiluminescent emission in the near infrared, which is detected with a photomultiplier tube. Interferences are minimised by placing a cold trap between the catalyst and ozone chamber, and by incorporating an optical filter in front of the photomultiplier.

The presence and amounts of nitrosamines were confirmed in some of the samples using the gas chromatograph coupled to an AEI MS902 mass spectrometer. The nitrosamines were detected by parent–ion monitoring using peak matching, in the manner described by Gough and Webb [118]. The detection limit was $0.1\mu g\ L^{-1}$ for each of the nitrosamines. While measurement of the nitric oxide fragment by mass spectrometry is applicable to all nitrosamines, it results in a significantly poorer detection limit. Further, the mass spectrometer will respond to any compound giving rise to NO^+, including C^- and N-nitroso compounds, nitro compounds and nitramines. The upper limits of N-nitrosodimethylamine, N-nitrosodiethylamine, N-nitrosomorpholine and N-nitrosodiethanolamine detected were 0.2, 2.0 100 and $60\mu g\ L^{-1}$ respectively.

16.1.4.2 Polyaromatic hydrocarbons

Benoit *et al.* [119] have investigated the use of macroreticular resins, particularly Amberlite XAD–2 resin, in the preconcentration of Ottawa tap water samples prior to the determination of 50 different PAHs by gas chromatography–mass spectrometry.

Water samples were prepared as follows: sampling cartridges, containing 15g Amberlite XAD–2 macroreticular resin that had previously been cleaned by the method of McNeil et al. [120] were rinsed with 250mL acetone and washed with at least 1L of purified water. The cartridges were attached to a potable water supply and the flow of water was controlled at ca. 70mL min^{-1}. When 300mL of water had been passed through the cartridge, the cartridge was disconnected from the tap and as much water as possible was removed from the cartridge by careful draining followed by the application of vacuum from a water aspirator. The XAD–2 resin was eluted with 300mL of 15:85v/v acetone:hexane solution at a flow rate of ca. 5mL min^{-1} (all solvents were of 'distilled in glass' quality and were redistilled in an all-glass system). The organic layer was dried by passage through a drying column containing anhydrous sodium sulphate over a glass–wool plug. Both the sodium sulphate and the glass–wool plug were cleaned by successive washings with methylene chloride, acetone and hexane prior to use. The dried solution was concentrated to a volume of ca. 3mL, using a rotary evaporator, then quantitatively transferred with acetone to a graduated vial and was further concentrated, using a gentle stream of dry nitrogen gas, to a final volume of 1mL.

To analyse the solvent extracts a 10μL aliquot of the concentrated extract was injected into a Finnigan 4000 gas chromatograph–mass spectrometer coupled to a 6110 data system. A 3% OV–17 provided the best separation of the detectable PAHs. Quantitative estimations of the detectable O–PAHs in Ottawa drinking water were obtained by comparison of the areas of the two characteristic ion peaks (Tables 16.2 and 16.3) in the mass chromatograms of the reference standard and the field sample, respectively. No corrections were made for incomplete recovery. Of the 50 PAHs in the standard used by Benoit et al. [119], 38 are detected in at least one of the drinking water samples tested.

Grimmer et al. [121] used capillary gas chromatography for the fingerprint analysis of PAHs in potable water samples. Following extraction with cyclohexane (after addition of an internal standard) and clean-up using Sephadex LH 20, the extract is subjected to gas chromatography with a flame ionisation detector. The ratio of the peak heights to that of the internal standard is used to calculate the amounts of PAHs present in the extract. The limit of detection is about 0.1ng L^{-1}, less than 1L of drinking water being required for the analysis.

16.1.4.3 Multiorganic mixtures

A tremendous amount of material has been published on this subject, which is beyond the reach of this book to cover in detail. The work

Table 16.2 Polycyclic aromatic hydrocarbons detected in Ottawa drinking water

Compound	Ions monitored		Rel. ret. time[a]	Concentration (ng L^{-1})
Napthalene	128	102	1.00	6.8
2-Methylnaphthalene	142	141	1.59	2.4
1-Methylnaphthalene	142	141	1.75	1.0
Azulene	128	102	1.90	n.d.
2-Ethylnaphthalene	156	141	2.26	>0.70
2,6-Dimethylnaphthalene	156	141	2.32	
Biphenyl	154	153	2.30	0.70
1,3-Dimethylnaphthalene	156	141	2.51	1.9
2-Vinylnaphthalene	154	153	2.68	n.d.
2,3-Dimethylnaphthalene	156	141	2.69	>0.68
1,4-Dimethylnaphthalene	156	141	2.69	
3-Phenyltoluene	168	167	2.74	0.20
Diphenylmethane	168	91	2.88	1.4
4-Phenyltoluene	168	167	2.94	0.20
Acenaphthylene	152	151	3.00	0.05
Acenaphthene	154	153	3.25	0.20
Bibenzyl	182	91	3.41	1.9
1,1-Diphenylethylene	180	179	3.48	>7.4
cis-Stilbene	180	179	3.59	
2,2-Diphenylpropane	196	181	3.62	n.d.
2,3,5-Trimethylnaphthalene	170	155	3.71	0.65
3,3'-Dimethylbiphenyl	182	167	4.03	0.31
Fluorene	166	165	4.15	0.15
4,4'-Dimethylbiphenyl	182	167	4.18	0.57
4-Vinylbiphenyl	180	178	4.56	n.d.
Diphenylacetylene	178	89	4.90	0.05
9,10-Dihydroanthracene	180	179	5.03	0.66
trans-Stilbene	180	179	5.20	
9,10-Dihydrophenanthrene	180	179	5.29	>0.47
10,11-Dihydro-5H-dibenzo-(a,d)cycloheptane	194	179	6.03	0.40
Phenanthrene	178	89	6.08	
Anthracene	178	89	6.14	>0.52
1-Phenylnaphthalene	204	203	6.77	n.d.
1-Methylphenanthrene	192	191	7.06	n.d.
2-Methylanthracene	192	191	7.25	0.51
9-Methylanthracene	192	191	7.59	n.d.
9-Vinylanthracene	204	203	7.82	n.d.
Triphenylmethane	244	167	8.04	n.d.
Fluoroanthene	202	101	8.41	0.55
Pyrene	202	101	8.90	0.53
9,10-Dimethylanthracene	206	191	8.99	0.19
Triphenylethylene	256	178	9.25	0.08
p-Terphenyl	230	115	9.44	n.d.
1,2-Benzofluorene	216	108	9.64	n.d.
2,3-Benzfluorene	216	108	9.73	n.d.

Table 16.2 continued

Compound	Ions monitored		Rel. ret. time[a]	Concentration (ng L[-1])
Benzylbiphenyl	244	167	9.75	n.d.
1,1'-Binaphthyl	254	126	10.95	n.d.
Triphenylene	228	114	11.5	
Benz(a)anthracene	28	115	11.6	>8.1
Chysene	228	114	11.8	

Note: [a]Retention times are relative to the retention time of naphthalene (3.81 min)

Source: Reproduced by permission from Gordon and Breach, Netherlands [119]

Table 16.3 Oxygenated polycyclic aromatic hydrocarbons detected in Ottawa drinking water

Compound	Ions monitored		Rel. ret. time[a]	Concentration (ng L[-1])
Xanthene	182	181	4.83	0.20
9-Fluorenone	180	152	5.93	0.90
Perinaphthenone	180	152	7.70	0.28
Anthrone	194	165	7.90	1.4
Anthraquinone	208	180	8.11	2.4
Naphthalene	128	102	1.00	

Note: [a]Retention times are relative to the retention time of naphthalene (3.81 min)

Source: Reproduced by permission from Gordon and Breach, Netherlands [119]

concerned solely with the application of gas chromatography combined with mass spectrometry to the identification and determination of mixtures of organic substances in potable water will be discussed.

In Table 16.4 some of the more important papers on the subject are listed which, it is hoped, will assist the reader in tracking down any information he or she requires.

Perhaps some idea of the power of the gas chromatography–mass spectrometry approach to the analysis of organic micropollutants can be gauged by the work of Coleman et al. [134] on mutagenic extracts of potable water. Samples of the water (1818L) were concentrated to a small volume using reverse osmosis and lyophilisation. The concentrates were then fractionated by sequential extraction with petroleum ether, diethyl

Table 16.4 Published work on the application of gas chromatography–mass spectrometry to multiorganic analysis of potable water

Paper title	Ref.
Identification of volatile organic contaminants in Washington DC municipal water	[122]
Organic micropollutants in air and water. Sampling gas chromatographic–mass spectrometric analysis and computer identification	[123]
Recent advances in the identification and analysis of organic pollutants in water	[124]
Guidelines for qualitative and quantitative screening of organic pollutants in water supplies	[125]
Quantitative analysis of volatile organic compounds by GC–MS	[126]
Trace organics in water	[127]
An overview of the analysis of trace organics in water	[128]
GC–MS analysis of volatile organic compounds in water	[129]
Organic pollution profiles in drinking water	[130]
Some applications of gas chromatography–mass spectrometry in the water industry	[131]
GC–MS identification of trace organics in Philadelphia drinking waters during a 2-year period	[132]
Quantitative aspects of the determination of organic micropollutants in raw and potable water	[133]
Identification of organic compounds in a mutagenic extract of a surface drinking water by a computerised gas chromatography–mass spectrometry system	[134]
identification and determination of trace organic substances in tap water by computerised gas chromatography–mass spectrometry and mass fragmentography	[135]
Determination of organic contaminants by the Grob closed-loop stripping technique	[136]
Analysis of organic micropollutants in water	[137]
Trace organic contaminants in water – problems of analysis and legal requirements	[138]
Investigation of a comprehensive approach for trace analysis of dissolved organic substances in water	[139]
Sampling techniques for analysis of trace organics	[140]
Application of gas chromatography–mass spectrometry and field desorption mass spectrometry to the identification of organic compounds leached from epoxy resin and lined water mains	[141]
Organic contaminants in the aquatic environment IV. Analytical techniques	[142]
Combination of continuous liquid–liquid extraction and coupled GC–MS for analysis of pesticides in water	[143]
Precision and accuracy of concurrent multicomponent multiclass analysis of drinking water extracts by GC–MS	[144]
A study of the trace organics profile of raw and potable water systems	[145]

Source: Own files

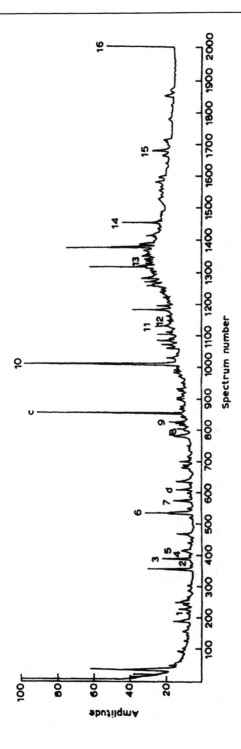

Fig. 16.2 Total ion chromatogram of the unpartitioned diethyl ether extract (CE) chromatographed on SP 1000.

Numbers identify the following GC peaks: (1) methylpyridine isomer; (2) benzyl chloride; (3) 4-methyl-2-pentanol; (4) ethyl chloroacetate; (5) *N,N*-dimethylformamide; (6) cyclohexanol; (7) cyclohexanone; (8) isophorone; (9) 2-(2-ethoxyethoxy)ethanol; (10) 1-(2-butoxyethoxy)ethanol; (11) benzothiazole; (12) N-acetylmorpholine; (13) tetramethylene sulphone; (14) diethyl phthalate; (15) disobutyl phthalate; (16) dibutyl phthalate

Source: Reproduced by permission from the American Chemical Society [134]

Table 16.5 Gas chromatography–mass spectrometry of organic compounds in potable waters

Organic	Extraction from water	GLC column conditions	Detector	Comments	LD	Ref
Aliphatic hydrocarbons	Inert gas bubbling	10% Carbowax 20M on Chromosorb WAW, 60–200°C	Mass spectrometer	GC–MS limited to organics boiling below 100°C	0.1–1 µg L^{-1}	[146]
Trihaloforms, bromodi-chloromethane, dibromo-chloromethane, bromoform carbons tetrachloride	—	—	—	Direct aqueous injection–mass fragmentography–GC–MS water removed from sample on diglycerol stationary phase precolumn	<1 µg L^{-1}	[147–149]
Trihaloforms	Gas phase stripping and desorption on to porous polymer	Glass capillary column	—	GC–MS	—	[150]
Trihaloforms	—	—	—	GC–MS	5 µg L^{-1}	[151, 152]
Halogenated aliphatic compounds	—	—	—	GC–MS	—	[153]
1,2 dibromopropane, 1,2 dibromoethane	—	—	—	GC–MS	—	[154]
Chloroform	—	—	—	Photo-assisted degradation study, GC–MS	—	[155]
Chlorinated humic acid	—	—	—	GC–MS	—	[156, 157]
Vinyl chloride	—	—	—	Mass fragmentography followed by GC–MS, recon. in m/e 62 and 64 peaks, (quadrupole MS with multiple ion detector)	—	[158]

Table 16.5 continued

Organic	Extraction from water	GLC column conditions	Detector	Comments	LD	Ref
Chlorinated aromatic compounds and poly-aromatic compounds, organophosphates	–	–	–	GC–MS	low ng	[144]
Chlorophenols	–	Electron capture and mass spectrometry	–	In situ derivativisation via acetylation or formation of pentafluorobenzyl derivatives, selected ion GC–MS more than 80% recovery	–	[159]
Chlorolignosulphonates	–	–	–	Pyrolysis–GC–MS	$10\mu g\ L^{-1}$	[29]
Aliphatic primary amines	Hexane	–	–	Amines converted to imines using pentafluorobenzaldehyde, hexane extraction, then GC–MS	$10\mu g\ L^{-1}$	[41]
Geosmin, 2-isopropyl, 3-methyox pyrazine, 2-iso-butyl-3-methoxy-pyrazine, 2-methyl isoborneol, 2,3, 6 trichloroanisole	–	–	–		$0.8\mu g\ L^{-1}$	[160]
Geosmin, 2-methyl-isoborneol	Adsorption/ extraction Dichloro-methane	–	–	GC–MS	–	[161]
Geosmin		15% Deoplex–400 on Chromosorb WNAW and 25% polyethylene glycol 21M on Chrom-osorb W	Flame ionisation and mass spectrometry	Sample passed through charcoal, charcoal extracted with dichloromethane	–	[162]

Source: Own files

ether and acetone. Most of the mutagenic activity was found in the diethyl ether extract and this was subjected to the gas chromatography–mass spectrometry technique. Fig. 16.2 shows the total ion chromatogram obtained following separation on an SP 1000 gas chromatographic column. This figure indicates several organic substances that were positively identified in this extract.

16.1.4.4 Other applications

Gas chromatography–mass spectrometry has also been applied to the determination of other types of organic compounds in potable water (Table 16.5) including: aliphatic hydrocarbons, chlorolignosulphonates, haloforms, halogenated aliphatic hydrocarbons, 1,2 dibromopropane, chlorinated aromatic hydrocarbons, halogenated phenols, aliphatic amines, geosmin and other odour-producing components of potable water.

16.1.5 Sewage effluents

16.1.5.1 Chlorinated insecticides and polychlorobiphenyls

Erikson and Pellizzari [163] analysed municipal sewage samples in the USA by a gas chromatography–mass spectrometry–computer technique for chlorinated insecticides and PCBS. The samples (300g) were extracted at pH11 six times with a total of 350mL chloroform to remove neutral and basic compounds. The extract was dried with sodium sulphate, vacuum filtered, and concentrated to 2mL using Kuderna–Danish apparatus. In cases where the sample background interfered significantly, an aliquot of the sample was chromatographed on a 1.0 × 30cm silica gel column (Snyder and Reinert [164]. PCBs and related compounds were eluted with 50mL hexane; pesticides and other compounds were eluted with 50mL toluene.

Acidic components of the samples were treated as diazomethane and dimethyl sulphate [165–167]. Analysis of all samples for PCBs was accomplished using a Finnigan 3300 quadruple gas chromatograph–mass spectrometer with a PDP/12 computer. The 180cm × 2mm i.d. glass column, packed with 2% OV–101 on Chromosorb W, was held at 120°C for 3min, programmed to 230°C at 12°C min^{-1} and held isothermally until all peaks had eluted. Helium flow was 30mL min^{-1}. The ionisation voltage was nominally 70eV and multiplier voltages were between 1.8 and 2.2kV. Full scan spectra were obtained from m/e 100–500. Samples were analysed under the following temperature conditions: up to 150°C for 3min, programmed to 230°C at 80min^{-1} isothermally until all peaks had eluted. PCBs were quantitated by gas chromatograph–mass spectrometry–computer using the selected ion–monitoring mode to provide maximum sensitivity and precision. This technique has been used in similar work on polychlorinated naphthalenes. Ten ions were selected for monitoring: one from the parent cluster for each of the chlorinated biphenyls ($C_{12}H_9Cl$ to $C_{12}Cl_{10}$). PCBs were quantitated using anthracene as external standard and a previously determined relative molar response (anthracene parent ion mass 178; 27ng mL $^{-1}$). Anthracene does not interfere with PCB determination, nor do PCBs or their fragment ions interfere with the determination of anthracene.

The retention time results for 34 chlorinated compounds found in sewage are given in Table 16.6.

16.1.5.2 Other applications

Other organic compounds that have been determined in sewage effluents by gas chromatography–mass spectrometry include the following (Table 16.7): nanylphenoxycarboxylic acids, chlorinated hydrocarbons, polychlorobiphenyls, organochlorine insecticides, hexachlorophene, pentachlorophenol, N-(phenylsulphonyl) sarcosine, dimethyl di, tri and tetrasulphides, Abrazine, Nirex, and volatile organic compounds.

16.1.6 Waste effluents

16.1.6.1 Insecticides

Gas chromatography has been used [183] to determine the following at organophosphorus insecticides at the microgram per litre level in water and waste water samples: Azinphos–methyl, Demeton–O, Demeton–S, Diazinon, Disulfoton, Malathion, Parathion–methyl and Parathion–ethyl. This method is claimed to offer several analytical alternatives, dependent on the analyst's assessment of the nature and extent of interferences and the complexity of the pesticide mixtures found. Specifically, the procedure uses a mixture of 15% v/v methylene chloride in hexane to extract organophosphorus insecticides from the aqueous sample. The method provides, through use of column chromatography and liquid–liquid partition methods for the elimination of non-pesticide interference and the pre-separation of pesticide mixtures. Identification is made by selective gas chromatographic separation and may be corroborated through the use of two or more unlike columns. Detection and measurement are best accomplished by flame photometric gas chromatography using a phosphorus-specific filter. The electron capture detector, though non-specific, may also be used for those compounds to which it responds. Confirmation of the identity of the compounds should be made by gas chromatography–mass spectrometry when a new or

Table 16.6 Summary of chlorinated compounds found in sewage sludge

Compound[a]	Retention time (min)[b]
Dichlorobenzene	0.5
Mol. wt. = 194, Cl₁	0.8
Mol. wt. = 222, Cl₁	0.9
Trichlorobenzene	1.0
Chloroaniline (tent.)	1.2
Dichloroaniline	2.1–4.7 (1.1)
Tetrachlorobenzene	2.2 (0.7)
Mol. wt. = 187, Cl₂	2.3–2.7 (0.6)
Mol. wt. = 171, Cl₂	3.0
Mol. wt. = 240, Cl₄[c]	3.0, 4.3 (1.1)
Trichloroaniline[d]	3.2, 4.4
Dichloronaphthalene	3.7
Trichlorophenol	3.7
Mol. wt. = 302, C₁	4.6
Mol. wt. = 210, Cl₃	5.1
Chlorobiphenyl	6.2
Dichlorobiphenol[e]	6.3–8.2
Trichlorobiphenyl[e]	7.5–9.6
Mol. wt. = 192, Cl₁	7.6
Mol. wt. = 288, Cl₁	8.6–11.1 (5.8)
Tetrachloronaphthalene	8.6
Mol. wt. = 218, Cl₁ (tent.)	9.1
Mol. wt. = 256, Cl₁	9.1
Dichlorobenzophenone	9.2 (7.2)
Mol. wt. - 269, Cl₁	9.9
Mol. wt. = 256, Cl₂	10.1
Pentachlorobiphenyl[e]	10.2–11.1
Mol. wt. = 288, Cl₃	10.5
Mol. wt. = 280, Cl₁	10.7
Mol. wt. - 241, Cl₁	10.8
Mol. wt. = 285, Cl₁	12.6
DDE[e]	12.7, 13.2 (8.6)
Mol. wt. = 356, Cl₂	12.8
Mol. wt. = 397, Cl₁	15.0

Notes:
[a] Unidentified compounds are listed with the apparent molecular weight and number of chlorines. If the identification of a compound is tentative, it is denoted by (tent.).
[b] Retention times are listed for the chromatographic temperature conditions, 12°C for 3min, then 12°min⁻¹ to 230°C, then hold. Values in parentheses are for chromatographic temperature conditions, 150°C for 3min, then 8°min⁻¹ to 230°C, then hold.
[c] Two separate isomers observed in some samples.
[d] Differences in retention times possibly indicate different isomers.
[e] Several isomers observed.

Source: Reproduced by permission from Gordon and Breach, Netherlands [163]

Table 16.7 Gas chromatography–mass spectrometry of organic compounds in sewage effluents

Organic	Extraction from water	GLC column conditions	Detector	Comments	LD	Ref
Nonyl(phenoxy) carboxylic acids	–	High resolution GC	–	GC–MS and high performance liquid chromatography	1 µg L^{-1}	[168]
Chlorinated hydrocarbons	–	–	–	GC–MS producing total ion chromatogram	–	[169, 170]
Polychlorobiphenyls and chlorinated insecticides	–	–	–	GC–MS, mass fragmentography	–	[171, 172]
Pentachlorophenol, hexachlorophene	–	–	–	GC–MS	–	[173]
N-(phenyl-sulphonyl) sarcosine	–	–	–	GC–MS	–	[174]
Dimethyl di-, tri- and tetra-sulphides	–	–	–	GC–MS	–	[175]
Atrazine residues	Dichloro-methane	–	Electrolytic conductivity	82–98% recovery of Atrazine	<0.01 µg L^{-1}	[176, 177]
Nirex	–	–	–	Computer controlled GC–MS, no interference from other chlorinated insecticides in sample	–	[178, 179]
Volatile organic compounds	–	–	–	Direct water sample into GC–MS	1–50mg L^{-1}	[151, 180]
54 semi-volatile organic compounds (benzidines, phenols, neutrals)	Methylene chloride or chloroform	High resolution capillary column	–	GC–MS	0.02µg L^{-1}	[181, 182]

Source: Own files

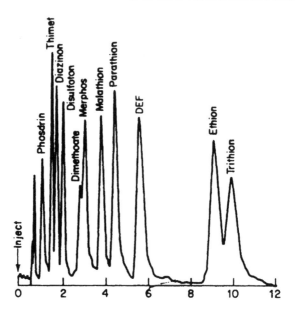

Fig. 16.3 Column packing 1.5% OV–17 + 1.95% QF–1. Carrier gas: nitrogen at 70ml min $^{-1}$. Column temperature, 215°C. Detector, flame photometric (phosphorus).
Source: Reproduced by permission from the Environmental Protection Agency, Cincinnati, USA [183]

undefined sample type is being analysed and the concentration is adequate for such determination.

Compounds such as organochlorine insecticides, PCBS and phthalate esters interfere with the analysis of organophosphorus insecticides by electron capture gas chromatography. When encountered, these interferences are overcome by the use of the phosphorus-specific flame photometric detector. Elemental sulphur will interfere with the determination of organophosphorus insecticides by flame photometric and electron capture gas chromatography. The elimination of elemental sulphur as an interference is discussed in detail. A typical gas chromatogram obtained by this procedure is shown in Fig. 16.3. Retention data for various gas chromatographic column packings are reproduced in Table 16.8.

16.1.6.2 Other applications

A particularly noteworthy example of the application of this technique to the examination of industrial and municipal waste waters is that of Burlingham [184]. This worker reports the findings of experiments to determine the potential of high resolution gas chromatography–high

Table 16.8 Retention times of some organophosphorus insecticides relative to parathion

Liquid phase[a] Column temperature (°C) Nitrogen carrier flow Pesticide	1.5% OV–17 + 1.95% QF–1[b] 215 70mL min⁻¹ RR	6% QF–12 + 4% SE–30 215 70mL min⁻¹ RR	5% OV–210 200 60mL min⁻¹ RR	7% OV–1 200 60mL min⁻¹ RR
Demeton[c]	0.46	0.26	0.20	0.74
		0.43	0.38	
Diazinon	0.40	0.38	0.25	0.59
Disulfoton	0.46	0.45	0.31	0.62
Malathion	0.86	0.78	0.73	0.92
Parathion–methyl	0.82	0.80	0.81	0.79
Parathion–ethyl	1.00	1.00	1.00	1.00
Azinphos–methyl	6.65	4.15	4.44	4.68
Parathion (min. absolute)	4.5	6.6	5.7	3.1

Notes:
[a] All columns glass, 180cm × 4mm i.d., solid support Gas–Chrom Q, 100–120mesh
[b] May substitute OV–210 for QF–1
[c] Anomalous, multipeak response often encountered

Source: Reproduced by permission from the Environmental Protection Agency, Cincinnati, USA [183]

468

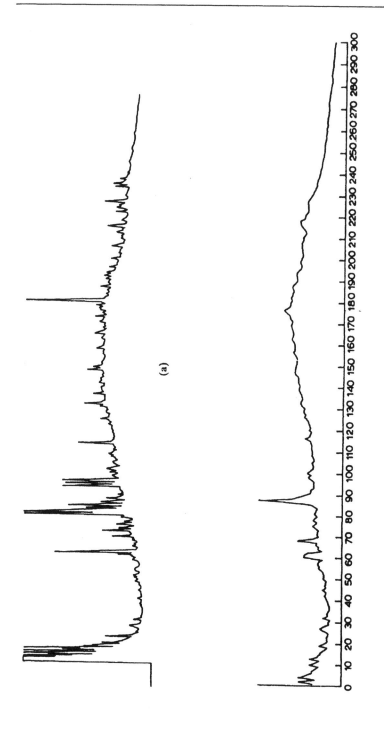

(a)

Fig. 16.4 (a) Flame ionisation detector (FID) capillary gas chromatogram of secondary effluents, 6% fraction; (b) total ionisation chromatogram
Source: Ectoxicology and Environmental Safety [184]

resolution mass spectrometry for assessing trace levels of organic compounds in waste waters. The technique was applied to effluent from a Southern California waste water treatment plant and to primary effluent from a petroleum refinery, and it is concluded that it can provide useful data on the organic constituents of waste water samples, even when they are present in extremely complex mixtures.

Complete high resolution mass spectra were recorded during elution of components from a glass capillary gas chromatographic column connected to a flame ionisation detector; data were subsequently examined with an on-line real-time software system for fast cyclic scanning high resolution mass spectrometry. Searches for specific compounds were accomplished through the generation of accurate mass spectrograms specific to particular elemental compositions, ie elemental composition chromatography.

Fig. 16.4 shows the capillary gas chromatograms obtained for a fraction. Fig. 16.4(b) shows the total ionisation chromatograms for the fraction obtained from the high resolution mass spectral data set. In general, the correlation between the total ionisation chromatogram profile and the flame ionisation detector profile is low due to both the differing relative detector responses and the consideration of scan cycle time (9.6s) with respect to chromatographic peak elution time (~20s).

The availability of accurate mass measurements on all the peaks in each mass spectrum provides a very accurate and highly specific method for locating compounds of interest in such mixtures. This simply involves searching the data set for particular accurate masses (specific elemental compositions) versus scan number (chromatographic retention time) ie elemental composition chromatograms.

Gas chromatography–mass spectrometry has been applied to the determination of other types of organic compounds in waste waters including the following (Table 16.9): hydrocarbons, phenols, organic halogen compounds, polychlorobiphenyls, chlorophenols, chlorinated tyrosine, nitrosamines, alkyl benzene sulphonates, N,N dialkyldithiocarbamates, volatile organic compounds.

16.1.7 Trade effluents

16.1.7.1 Miscellaneous applications

Various types of organic compounds have been determined in trade effluents by gas chromatography–mass spectrometry including the following (Table 16.10): hydrocarbons, mercaptons, phenols, polyaromatic hydrocarbons, vinyl chloride, chlorinated alkyl naphthalenes, chlorophenols, nitrosamines, Triclophon, organophosphorus, pesticides, organosulphur compounds and mixtures of organic compounds.

Table 16.9 Gas chromatography–mass spectrometry of organic compounds in waste waters

Organic	Extraction from water	GLC column conditions	Detector	Comments	LD	Ref
Aliphatic and aromatic hydrocarbons	Solid phase micro-extraction or methylene chloride extraction	–	Flame ionisation	–	1–30ppt	[185]
Hydrocarbons	–	–	–	Automated GC–MS	–	[186]
3,5 xylenol	Diethyl ether	–	–	Ether methylated with diazo-methane, then GC–MS	–	[187]
4-nonyl phenol	–	–	–	Phenol derivatised to derivative, cleaned up on silica gel then GC–MS	100ppt	[188]
Semi- or non-volatile chlorinated organic compounds, hexachloro-butadiene, hexachloro-cyclopentadiene, octa, chlorocyclopentadiene, hexachlorobenzene	Extractions perform-ed on homogeniser using hexane, hexane –benzene or hexane –toluene	–	–	Purge and trap method, recoveries 88–125%	0.1ppb	[189]
Polychlorobiphenyls	–	–	–	GC–MS	–	[190]
o-chloro-2-(2.4 di-chlorophenoxy)phenol	Solvent	–	–	Purification of solvent extract on silica gel derivitivisation with diazomethane then GC–MS	0.2ppt	[191]
Chlorinated typrosine	–	–	–	GC–MS	–	[192]
Volatile nitrosamines, dimethyl, diethyl and dipropyl nitrosamines	–	Capillary column	–	GC–MS	–	[193]

Table 16.9 continued

Organic	Extraction from water	GLC column conditions	Detector	Comments	LD	Ref
Linear alkylbenzene, sulphonates, dialkyl-tetraline sulphonates, and biodegradation intermediates	–	Electron capture	–	Derivativisation then electron capture GC–MS	–	[194]
S-alkyl derivatives of N,N dialkyldithio-carbamates	–	10% Apiezon L on Varoport 80 at 250°C	Flame ionisation	GC–MS on GC column effluent	–	[195]
US Environmental Protection Agency priority pollutants	–	–	–	Pollutants trapped on charcoal, gas purging at 49°C, then GC–MS	–	[196, 197]
Miscellaneous organics	–	–	–	Automated mass spectrum matching in GC–MS	–	[186]
Acid, base and neutral organics	–	Capillary column	–	Combination of robotics with capillary GC–MS	–	[198]

Source: Own files

Table 16.10 Gas chromatography–mass spectrometry of organic compounds in trade effluents

Organic	Extraction from water	GLC column conditions	Detector	Comments	LD	Ref
Hydrocarbons, Mercaptans, phenols	—	—	—	GC–MS	—	[199]
Vinyl chloride	Carbon tetrachloride	—	Electrolytic conductivity, mass spectrometry	GC–MS, recovery of vinyl chloride 77–100%	100µg L^{-1}	[200]
Chlorinated alkyl naphthalenes	—	Capillary column	—	GC–MS	—	[201]
Chlorophenols	—	—	—	GC–MS	—	[202, 203]
Neutral chlorinated organic compounds	—	—	—	Direct injection–GC–MS	10ppt	[204]
Nitroaromatic compounds, 2,4 and 2,6 dinitrotoluene, 1,3 dinitrobenzene	—	—	—	GC–MS	—	[205]
Organophosphorus pesticides	—	—	Nitrogen phosphorus and mass spectrometric detectors	GC–MS	—	[206]
Benzothiazole, 2-mercaptobenzothiazole, 2-(4-morpholinyl) benzothiazole	—	3% SP1000 methyl silicone fluid) on Supelcoport at 70–300°C	Flame ionisation	—	0.03–0.06 µg L^{-1}	[207]
2-(4-morpholinyl) benzothiazole	6:4 benzene: methanol	Capillary column	Sulphur selective	GC–MS	0.2µg L^{-1}	[208]

Table 16.10 continued

Organic	Extraction from water	GLC column conditions	Detector	Comments	LD	Ref
570 organic compounds	Methylene dichloride	Capillary column	–	GC–MS	–	[209]
Miscellaneous type stabilisers etc	–	–	–	GC–MS	–	[207]
Miscellaneous organic compounds	Freeze concentration or isolation on carbon	–	Flame ionisation	GC–MS	–	[210]

Source: Own files

16.2 Organometallic compounds

16.2.1 Non saline waters

16.2.1.1 Organoarsenic compounds

Lussi–Schlatter and Brandenberger [211] have reported a method for inorganic arsenic and phenylarsenic compounds based upon gas chromatography with mass specific detection after hydride generation with headspace sampling. However, methylarsenic species were not examined.

Odanaka *et al.* [212] have reported that the combination of gas chromatography with multiple ion detection system and a hydride generation heptane cold trap technique is useful for the quantitative determination of arsine, monomethyl-, dimethyl- and trimethylarsenic compounds and this approach is applicable to the analysis of environmental and biological samples.

In this method, arsine and methylarsines produced by sodium borohydride reduction are collected in *n*-heptane (–80°C) and then determined. The limit of detection for a 50mL sample was 0.2–0.4µg L $^{-1}$ of arsenic. Relative standard deviations ranged from 2% to 5% for distilled water replicates spiked at the 10µg L $^{-1}$ level. Recoveries of all four arsenic species from river water ranged from 85% to 100%.

Dimethyl arsenic acid yields predominantly dimethylarsine, while methylarsonic acid yields predominantly methylarsine and trimethylarsine oxide yields predominantly trimethylarsine.

The following ions were characteristic and intense ions in the mass spectra of arsines, arsine m/z 78 M$^+$, 76 (M – 2)$^+$, methylarsine; m/z 92 M$^+$, 90 (M – 2)$^+$, 76 ((M – CH$_3$) – 1)$^+$, dimethylarsine m/z 106 M$^+$, 90 ((M – CH$_3$) – 1)$^+$, trimethylarsine, m/z 120 M$^+$, 105 ((M – CH$_3$) – 1)$^+$, 103 ((M – CH$_3$)$^+$ – 2)$_+$. To achieve simultaneous measurement and to assess the specificity of the analysis, for instance the m/z 76, 78, 89 and 90 were monitored for arsine, methylarsine, and dimethylarsine and/or m/z 90, 103, 105 and 106 for alkylarsines as methylarsine, dimethylarsine and trimethylarsine. However, simultaneous determination of all four arsenicals could not be done at one injection because of a limited range of detectable mass spectra in the system used.

16.2.1.2 Organotin compounds

Meinema *et al.* [213] have described a sensitive and interference–free method for the simultaneous determination of tri-, di- and mono-butyltin species in the aqueous systems at tin concentrations of 0.01–5µg L $^{-1}$. The species are concentrated from hydrobromic acid solutions into an organic solvent by extraction with 0.05% tropolene in benzene in the presence of a metal coordinating ligand. The butyltin species in the organic extract are transformed into butylmethyltin compounds by reaction with a Grignard reagent and analysed by a gas chromatography–mass spectrometry method. The inorganic tin(IV) species in the organic extract are butylated to tetrabutyltin, which is detected by the same technique.

Maguire and Huneault [214] developed a gas chromatographic method with flame photometric detection for the determination of bis(tri-*n*-butyltin) oxide and some of its possible dialkylation products in potable and non saline waters. This method involves extraction of *bis*(tri-*n*-butyltin) oxide, Bu_2Sn^{2+}, $BuSn^{3+}$ and Sn^{4+} from the water sample with 1% tropolone in benzene, derivatisation with a pentyl Grigand reagent to form the various Bu_n pentyl (4_{-n}) Sn species followed by analysis of these by flame photometric gas chromatography–mass spectrometry in amounts down to 25ng. The pentyl derivatives are all sufficiently non volatile compared with benzene that none are lost in solvent 'stripping', yet they are volatile enough to be analysed by gas chromatography.

Mueller [215] determined down to 1µg L^{-1} of tributyltin compounds in river and lake waters using capillary column gas chromatography–mass spectrometry. The tributyltin compounds were converted to tributyl-methyltin derivatives prior to gas chromatography.

Matthias *et al.* [216] have described a comprehensive method for the determination of aquatic butyltin and butylmethyltin species at ultra-trace levels using simultaneous sodium borohydride hydridisation, extraction with gas chromatography–flame photometric detection and gas chromatography–mass spectrometric detection. The detection limits for a 100mL sample were 7ng L^{-1} of tin for tetrabutyltin and tributyltin, 3ng L^{-1} of tin for dibutyltin and 22ng L^{-1} tin for monobutyltin. For 800mL samples detection limits were 1–2ng L^{-1} tin for tri- and tetrabutyltin and below 1ng L^{-1} tin for dibutyltin. The technique was applied to the detection of biodegration products of tributyltin in non saline waters.

Unger *et al.* [217] determined butyltins in non saline water by gas chromatography with flame photometric detection and confirmation by mass spectrometry. The sample was extracted with tropalone in *n*-hexane and organotin compounds derivatised with *n*-hexyl magnesium bromide to form tetraalkyltins. The *n*-hexyl derivatives of methyltin and butyltin species were easily separated and quantified relative to an internal standard (triphenyltin chloride) which was not found to be present in environmental samples and did not interfere.

Greaves and Unger [218] have used chemical ionisation mass spectrometry with positive ion chemical ionisation to qualitatively (full scan) and quantitatively (SRMS) determine tributyl tin, dibutyltin and mobutyltin in environmental waters. Detection limits for the method, which included extraction of the water samples with hexane 10.2% tropolone and derivativisation of the organotin compounds with hexyl magnesium bromide (to form hexabutyltins) were <2ng L $^{-1}$.

Colby *et al.* [219] used laser ionisation gas chromatography–mass spectrometry to determine tetraethyltin in non saline water with a detection limit of 2.5µg L $^{-1}$ as Et$_4$Sn or 1.5fg absolute as Et$_4$Sn.

Plzak *et al.* [220] extracted organotin compounds into hexane and derivitivised by methylation with a Grignard reagent or reduced to tin hydride with sodium borohydride prior to determination by gas chromatography–mass spectrometry.

16.2.2 Seawater

16.2.2.1 Organoarsenic compounds

Talmi and Bostick [221] extracted methylarsenic compounds from seawater with cold toluene, then analysed the extract by gas chromatography using a mass spectrometric detector. Down to 0.25mg L $^{-1}$ of organoarsenic compunds were detectable.

16.2.3 Potable water

16.2.3.1 Organolead compounds

Gorecki and Pawliszyn [222] have described a new gas chromatographic method for the determination of tetraethyllead and ionic lead in water by solid-phase microextraction gas chromatography. Tetraethyllead is extracted from the headspace over the sample. Inorganic lead is first derivatised with sodium tetraethylborate to form tetraethyllead, which is extracted in the same way as pure tetraethyllead samples. The analytical procedure was optimised with respect to pH, amount of derivatising reagent added, stirring conditions and extraction time. The detection limit obtained for tetraethyllead was found to be 100ppt when using a flame ionisation detector and 5ppt when using an ion trap mass spectrometer. The detection limit for Pb^{2+}, limited by the non-zero blank, was found to be 200ppt. Linear calibration curves were obtained for both analytes when a flame ionisation detector was used for detection. For lead they spanned over four orders of magnitude. The ion trap mass spectrometry offered excellent sensitivity and selectivity, but the calibration curves were non-linear when the m/z = 295 ion was used for quantitation. The method has been verified on spiked tap water samples. An excellent agreement was found between the results obtained for standard solutions prepared using NANO pure water and spiked tap water samples.

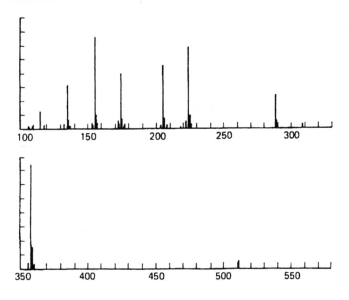

Fig. 16.5 Mass spectrum of Cr(tfa)₃
Source: Own files

16.3 Cations

16.3.1 Seawater

16.3.1.1 Chromium

Isotope dilution gas chromatography–mass spectrometry has been used for the determination of µg L^{-1} levels of total chromium in seawater [223–225]. The samples were reduced to produce chromium(III) and then extracted and concentrated as tris (1,1,1-trifluoro-2,4-pentanediono) chromium(III) (Cr(tfa)₃) into hexane. The Cr(tfa)₂⁺ mass fragments were monitored into a selected ion monitoring (SIM) mode.

Isotope dilution techniques are attractive because they do not require quantitative recovery of the analyte. One must, however, be able to monitor specific isotopes which is possible by using mass spectrometry.

In this method, chromium is extracted and preconcentrated from seawater with trifluoroacetylacetone (H(tfa) which complexes with trivalent but not hexavalent chromium. Chromium reacts with trifluoroacetylacetone in a 1:3 ratio to form an octahedral complex, Cr(tfa)₃. The isotopic abundance of its most abundant mass fragment, Cr(tfa)₂+ was monitored by a quadrupole mass spectrometer.

A mass spectrum of Cr(tfa)₃ is shown in Fig. 16.5. The isotopic distribution of the Cr(tfa)₂+ fragment (m/e 358 and 359 here) is evident.

Table 16.11 Natural abundance of Cr(tfa)

m/e	% calculated	% measured
356	3.8	3.8
357	0.5	0.6
358	75.2	74.9
359	17.0	16.6
360	3.5	4.0

Source: Own files

Table 16.12 Abundance of Cr(tfa)$_2^+$ for the chromium–53 spike

m/e	% calculated	% measured
358	3.1	3.3
359	86.5	86.7
360	9.9	9.1
361	0.5	0.9

Source: Own files

Table 16.13 Mean (±SD) chromium concentration in seawater ($\mu g\ L^{-1}$) (n≥3)

ID–GC/MS	ID–SSMS	GFAAS
0.177 ± 0.009	0.17 ± 0.03	0.19 ± 0.03
0.19 ± 0.01*	0.18 ± 0.01	ND

* Seawater reference material NASS–1
ND = not determined

Source: Own files

This is readily calculable if the individual elemental abundances are known. Assuming the isotopic abundance of 12–carbon and 13–carbon to be 98.89% and 1.11 and 50–, 52–, 53– and 54–chromium to be 4.31, 83.76, 9.55 and 2.38% respectively, and neglecting any isotopic abundances less than 1%, one can obtain a set of calculated abundances for the Cr(tfa)$_2^+$ ion. These and the measured isotopic abundances (by SIM) are listed inTable 16.11. The agreement between the two sets is excellent. The same calculation can be made for the 53–chromium spike solution by using isotopic abundances given by the supplier: 52–chromium 3.44%, 53–chromium 96.4% and 54–chromium 0.18%. Table 16.12 lists the calculated and the measured isotopic abundances for the spike solution.

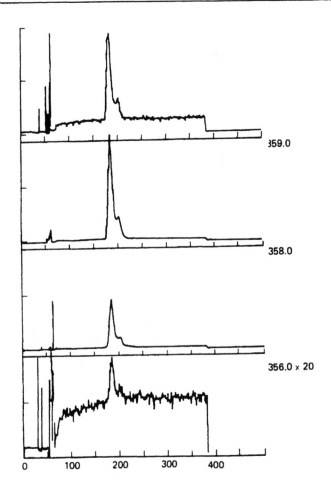

Fig. 16.6 Chromatograms of m/e 356, 358, 359 and 360 of a spiked seawater sample.
Multiplication factor
Source: Own files

A series of typical chromatograms of a spiked seawater sample is shown in Fig. 16.6. The geometric isomers of chromium trifluoroacetyl-acetone are not fully resolved.

Table 16.13 shows results of two seawater sample analyses. Agreement with data obtained by isotope dilution spark source mass spectrometry [226] and graphite furnace [227] was excellent.

The effect of calcium interference is somewhat different. At its concentration in seawater, 0.010M calcium ion had no effect upon

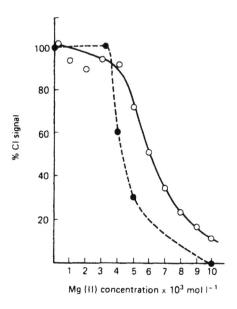

Fig. 16.7 Mg" interference of Cl analysis for Cr. —O—O— = in the presence of 0.3 M Br⁻; —•—• = in the absence of Br⁻ Cr" = 6 × 10⁻⁸ M. EDTA = 2.5 × 10⁻³ M
Source: Own files

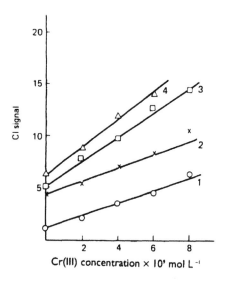

Fig. 16.8 Cl analysis for Cr in natural seawater. Curves 2–4: standard addition curves with 2, 3 and 4ml seawater added, respectively. Br⁻ = 0.30 M. EDTA = 2.5 × 10⁻³M
Source: Own files

chemiluminescence analysis of a 6×10^{-8}M chromium(III) solution in the absence of bromide ion. The chemiluminescence signal dropped to zero, however, if the calcium ion concentration was increased to 0.013M. In the presence of 0.3M bromide ion, no interference was observed for analysis of 6×10^{-8}M chromium(III) when the calcium concentration was less than or equal to 0.002M. The chemiluminescence signal increased linearly with increasing calcium ion concentration when the calcium concentration exceeded 0.002M.

The combined effect of cation interference for both magnesium(II) and calcium(II) is almost identical with the solid curve in Fig. 16.7 indicating that the magnesium ion interference is the dominant one. Fig. 16.8 shows calibration curves obtained upon spiking a seawater sample with chromium(III).

References

1 Swisher, R.D. *Journal of the Water Pollution Control Federation*, **35**, 887 (1963).
2 Huddleston, R.L. and Allred, R.C. *Dev. Ind. Microbiol.*, **4**, 24 (1963).
3 Swisher, R.D. *Journal of the Water Pollution Control Federation*, **35**, 1557 (1963).
4 Leidner, H., Gloor, R. and Wuhrmann, O. *Tenside Deterg.*, **13**, 122 (1976).
5 Hon Nami, H. and Hanya, T. *Journal of Chromatography*, **161**, 205 (1978).
6 Sim, P.G., Boyd, R.K., Gershey, R.M. *et al. Journal of Biomedical Environmental Mass Spectrometry*, **14**, 375 (1987).
7 Thuren, A. *Bulletin of Environmental Contamination and Toxicology*, **36**, 33 (1986).
8 Hites, R.A. *Journal of Chromatographic Science*, **11**, 570 (1973).
9 Hites, R.A. and Biemann, O. *Science N.Y.*, **158**, 178 (1972).
10 Hites, R.A. *Environmental Health Perspectives*, **3**, 17 (1973).
11 Brumby, W.C., Shafter, E.M. and Tillander, P.E. *Journal of the Association of Official Analytical Chemists International*, **77**, 1230 (1994).
12 Stephanou, E. Committee of European Communities. Rep EUR, EUR 10388. Organic Micropollutants in the Aqueous Environment, pp.155–161 (1988).
13 Matsumoto, G., Ishiwatari, R. and Hanya, T. *Water Research*, **11**, 693 (1977).
14 Minero, C., Vincent, M., Lago, S. and Pelizetti, E. *Fresenius Journal of Analytical Chemistry*, **350**, 403 (1994).
15 Gallegos, E.J. *Journal of Chromatographic Science*, **25**, 296 (1987).
16 Epstein, P.S., Mauer, T., Wagner, M., Chase, S. and Giles, B. *Analytical Chemistry*, **59**, 1987 (1987).
17 Matsunaga, K., Okamoto, Y., Imanaka, M. *et al. Okayama-ken Kankyo Hogen Senta Nepno*, **9**, 170 (1985).
18 Fujii, T. *Analytical Chemistry*, **49**, 1985 (1977).
19 Karlruher, B.A., Hormann, W.D. and Ramsfeiner, K.A. *Analytical Chemistry*, **47**, 2453 (1975).
20 Ahnoff, M. and Josefsson, B. *Analytical Letters (London)*, **6**, 1036 (1973).
21 Voyksner, R.D., Hass, J.R. and Sovocoal, G.W. *Analytical Chemistry*, **55**, 744 (1983).
22 Stalling, D.L. and Huckins, J.N. *Journal of the Association of Official Analytical Chemists*, **54**, 801 (1972).
23 Lopshire, L.F., Wolson, J.T. and Enke, C.G. *Toxicology and Industrial Health*, **12**, 375 (1996).

24 Abraham, B.M., Liu, T.Y. and Robbat, A. *Hazard Waste and Hazard Material*, **10**, 461 (1993).
25 Pellazari, E.E., Moseley, M.A. and Cooper, S.D. *Journal of Chromatography*, **334**, 277 (1985).
26 Webb, R.G. and McCall, A.C. *Journal of Chromatographic Science*, **11**, 366 (1973).
27 Freudenthal, J. and Greve, P.A. *Bulletin of Environmental Contamination and Toxicology*, **10**, 108 (1973).
28 Buszka, P.M., Rose, D.L., Ozuna, G.B. and Groschen, G.E. *Analytical Chemistry*, **67**, 3659 (1995).
29 Van Loon, W.H.G.M., Bron, J.S. and de Groot, B. *Analytical Chemistry*, **65**, 1726 (1993).
30 Lee, H.P. *Journal of the Association of Official Analytical Chemists*, **71**, 803 (1988).
31 Tanabe, S., Kannan, N., Wakimoto, T. and Tatsukawa, R. *International Journal of Environmental Analytical Chemistry*, **29**, 199 (1987).
32 Yasuhara, A. and Itoh, M.M. *Environmental Science and Technology*, **21**, 971 (1987).
33 Taguchi, V.V., Reiner, E.J., Wong, D.T., Meresz, D. and Hassas, B. *Analytical Chemistry*, **60**, 1429 (1988).
34 Laramee, J.A., Arbogast, B.C. and Dienzer, M.L. *Analytical Chemistry*, **60**, 1937 (1988).
35 Schimmel, H., Schmid, B., Backer, R. and Ballschmitter, K. *Analytical Chemistry*, **65**, 640 (1993).
36 Plomley, J.B., March, R.E. and Mercer, R.S. *Analytical Chemistry*, **68**, 2345 (1996).
37 LePack, M.A., Tan, J.C. and Enke, C.G. *Analytical Chemistry*, **62**, 1625 (1990).
38 Barber, L.B., Thruman, E.M., Takahashi, Y. and Noriega, M.C. *Groundwater*, **30**, 836 (1992).
39 Matz, G. and Kesners, P. *Analytical Chemistry*, **65**, 2366 (1993).
40 Okamoto, T. and Shirane, Y. *Hiroshima-ken Kankyo Senta Kenkyu-Hokoku*, **7**, 75 (1985).
41 Avery, M.J. and Junk, G.A. *Analytical Chemistry*, **57**, 790 (1985).
42 Pietsch, J., Hampel, S., Schmidt, W., Branch, H.J. and Worch, E. *Fresenius Zeitschrift für Analytische Chemie*, **355**, 164 (1996).
43 Koida, K., Tokumari, Y., Saimatsu, J. *et al. Hiroshima-Shi Eisel Kenkyusho Nenpo*, **(5)**, 34 (1988).
44 Tsukioka, T. and Murakami, T. *Journal of Chromatography*, **396**, 319 (1987).
45 Zoccolino, L. and Ronchetti, M. *Ann. Chim. (Rome)*, **77**, 735 (1987).
46 Sillanpaa, M., Sorvari, J. and Schvonen, M.L. *Chromatographia*, **42**, 578 (1996).
47 Nishikava, Y. and Okumura, T. *Chromatography*, **690**, 109 (1995).
48 Ding, W.H., Fijita, Y., Aeschimann, R. and Reinhard, M. *Fresenius Zeitschrift für Analytische Chemie*, **354**, 48 (1996).
49 Lockhart, H.B. and Blakeley, R.W. *Environmental Science and Technology*, **9**, 1035 (1975).
50 Fredeen, F.J.H., Saha, J.G. and Baba, M.J. *Pesticide Monitoring Journal*, **8**, 241 (1975).
51 Harless, R.L., Harris, D.E., Sovocoal, W. *et al. Biochemical Mass Spectrometry*, **5**, 232 (1978).
52 Bagnati, R., Benfenati, E., Davoll, E. and Fenelli, R. *Chemosphere*, **17**, 59 (1988).
53 Lindley, C.E., Stewart, J.T. and Sandstrom, M.W. *Journal of the Association of Official Analytical Chemists*, **79**, 962 (1996).
54 Deleu, R. and Copin, A. *Bulletin of Research Agronomy Gembloux*, **22**, 121 (1987).

55 Lopez-Avila, V., Hirata, P., Kraska, S., Flanage, H. and Taylor, J.H. *Analytical Chemistry*, **57**, 2797 (1985).
56 Zangwei Lai Sodagopa, O., Ramanujam, N.M., Giblin, D. and Gross, M.L. *Analytical Chemistry*, **65**, 21 (1993).
57 Gruessner, B. and Watzin, M. *Environmental Science and Technology*, **29**, 2806 (1995).
58 Kadokami, K. and Shinohara, R. *Bunseki Kagaku*, **35**, 875 (1988).
59 Gruessner, B., Schambaugh, N.C. and Watzin, M.C. *Environmental Science and Technology*, **29**, 251 (1995).
60 Vreuls, J.J., Bultermann, A.J., Ghijsen, R.T. and Brinkman, U.A.T. *Analyst (London)*, **117**, 1701 (1992).
61 Vink, M. and Van der Poll, J.M. *Journal of Chromatography*, **733**, 361 (1996).
62 Burkhard, L.D., Durham, E.J. and Lucasewycz, M.T. *Analytical Chemistry*, **63**, 277 (1991).
63 Lopez-Avila, V., Hirta, V., Kraska, S. and Taylor, J.H. *Journal of Agriculture and Food Chemistry*, **34**, 530 (1986).
64 Shiraishi, N. and Otsuki, A. *Water Research*, **21**, 843 (1987).
65 Copin, A. and Deleu, R. *Analytica Chimica Acta*, **208**, 331 (1988).
66 Vincenti, M., Minero, C., Lago, S. and Ravida, C. *Journal of High Resolution Chromatography*, **18**, 359 (1995).
67 Yamato, Y., Suzuki, M. and Wanatabe, T. *Biochemical Mass Spectrometry*, **6**, 205 (1979).
68 Yamato, Y., Suzuki, M. and Wanatabe, T. *Journal of the Association of Official Analytical Chemists*, **61**, 1135 (1978).
69 Mika, A., Tsuchihashi, H., Ueda, K. and Yamashita, M. *Journal of Chromatography, A*, **718**, 383 (1995).
70 Ishikawa, S., Taketomi, M. and Shinokara, R. *Water Research*, **19**, 119 (1985).
71 Kenmotsu, K., Okamoto, Yl., Ogino, Y., Matsunaga, K. and Ishida, T. *Okayama-ken Kankyo Hoken Senta Nenpo*, **9**, 160 (1985).
72 Lopez-Avila, V., Wood, R., Flanagan, M. and Scott, R. *Journal of Chromatographic Science*, **25**, 286 (1987).
73 Karp, K.E. *Water*, **31**, 735 (1993).
74 Kodakami, K., Sato, K., Hanada, Y. *et al. Analytical Science*, **11**, 375 (1995).
75 Baykut, G. and Volgt, A. *Analytical Chemistry*, **64**, 677 (1992).
76 Krock, K.A. and Wilkins, C.L. *Journal of Chromatography*, **726**, 167 (1996).
77 Koida, K., Fukuda, Y., Tokumori, Y. *et al. Hiroshima-shi Eisel. Kankyusho Nenpo*, (**4**), 40 (1985).
78 Michael, L.C., Pellizzari, E.D. and Wiseman, R.W. *Environmental Science and Technology*, **22**, 565 (1980).
79 Shoemaker, J.A., Bellar, T.A., Eichelberger, J.W. and Budde, W.L. *Journal of Chromatographic Science*, **31**, 279 (1993).
80 Koga, M., Shinohara, R., Kod, A. *et al. Japanese Journal of Water Pollution Res.*, **1**, 23 (1978).
81 Heller, S.R., McGuire, J.M. and Budde, W.L. *Environmental Science and Technology*, **9**, 210 (1975).
82 Yasuhara, A., Shirashi, H., Tsuji, M. and Okuno, T. *Environmental Science and Technology*, **15**, 570 (1981).
83 Guillemin, C.L., Martinez, F. and Thialt, S. *Journal of Chromatographic Science*, **17**, 677 (1979).
84 Scott, S.P., Sutherland, N. and Vincent, R.J. *Analytical Proceedings (London)*, **21**, 179 (1984).

85 Pereira, W.H. and Hughes, B.A. *Journal of the American Water Works Association*, **72**, 220 (1980).
86 Kozloski, R.P. and Sawnhey, B.L. *Bulletin of Environmental Contamination and Toxicology*, **29**, 1 (1982).
87 Scott, D.R. *Analytica Chimica Acta*, **211**, 11 (1988).
88 Tripathi, D.N., Kaushuk, M.P. and Bhattacharya, A. *Journal of Canadian Society of Forensic Science*, **20**, 151 (1987).
89 Yasuhara, A. and Fuwa, F. *Journal of Chromatography*, **172**, 453 (1979).
90 Abel, M., Conrad, and Giger, W. *Environmental Science and Technology*, **21**, 708 (1987).
91 Stephanou, E., Reinhard, M. and Ball, H.A. *Biomedical Environmental Mass Spectrometry*, **15**, 275 (1988).
92 Hayakawa, S., Okuma, K., Arakai, K., Nilnomi, J. and Kangmaru, T. *Misken Kanakyo Senta Kenkyu Hokoku*, (**7**), 51 (1987).
93 Yasuhara, A. *Agricultural Biological Chemistry*, **51**, 2259 (1987).
94 Fujimoto, N. and Ashitani, K. *Osaka-shi Suldokyoku Komubu Sulshita Shikenko Chosa Hokoku Marabini Shiken Selsek*, (**36**), 92 (1988).
95 Tsukloka, T. *Analyst (London)*, **113**, 193 (1988).
96 Zou, H., Xu, X., Zhang, J. and Zhu, Z. *Chemosphere*, **30**, 2219 (1995).
97 Matsumoto, G. and Hanya, T. *Journal of Chromatography*, **194**, 199 (1980).
98 Nakajima, K. *Water Research*, **20**, 233 (1986).
99 Wanatabe, K., Nakanishi, T., Shishido, J. *et al*. *Eisel Kagaku*, **34**, 25 (1988).
100 Pankow, J.F., Isabelle, L.M. and Asher, W.E. *Environmental Science and Technology*, **18**, 310 (1984).
101 J. Albrecht, private communication.
102 Ramsdale, S.J. and Wilkinson, R.E. *Journal of the Institute of Petroleum (London)*, **54**, 326 (1968).
103 Bunnock, J.V., Duckworth, D.F. and Stephens, G.G. *Inst. Petrol.*, **54**, 310 (1968).
104 Bailey, N.J.L., Jobson, A.M. and Rogers, M.A. *Chemical Geology*, **11**, 203 (1973).
105 Henderson, W., Wollrab, V. and Eglington, G. *Advances in Organic Geochemistry* 1968 (eds. P.A. Schenk and I. Havenaar), p. 181. Pergamon, Oxford (1969).
106 Gallegos, E.J. *Analytical Chemistry*, **43**, 1151 (1971).
107 Albaigés, J. and Albrecht, P. *International Analytical Chemistry*, **6**, 171 (1979).
108 Rubinstein, I., Strausz, O.P., Spychkerelle, C., Crawford, R.J. and Westlake, D.W.S. *Geochem. Cosmochim. Acta*, **41**, 1341 (1977).
109 Albaigés, J., Borbon, J. and Salgre, P. *Tetrahedron Lett.*, 595 (1978).
110 Rubinstein, I. and Albrecht, P. *Journal of the Chemical Society Chem. Commun.*, 957 (1975).
111 Van Dorsselaar, A., Albrecht, P. and Ourisson, G. *Bulletin of the Chemical Society of France*, **1**, 65 (1977).
112 Kimball, B.J., Maxwell, J.R., Philip, R.D. *et al*. *Geochimica Cosmochimica Acta*, **38**, 1165 (1974).
113 Brown, R.A. and Huffman, H.L. *Science*, **101**, 847 (1976).
114 Walker, J.D., Colwell, R.P., Hamming, M.C. and Ford, H.I. *Bulletin of Environmental Contamination and Toxicology*, **13**, 245 (1975).
115 Smith, D.H. *Analytical Chemistry*, **44**, 536 (1972).
116 Boyd, T.J. *Journal of Chromatography*, **662**, 281 (1994).
117 Richardson, M.L., Webb, K.S. and Gough, T.A. *Ecotoxicology and Environmental Safety*, **4**, 207 (1980).
118 Gough, T.A. and Webb, K.S. *Journal of Chromatography*, **79**, 57 (1973).

119 Benoit, F., Label, G.L. and Williams, D.T. *International Journal of Analytical Chemistry*, **6**, 277 (1979).
120 McNeil, E.E., Olson, R., Miles, W.F. and Rajalalee, R.J.M. *Journal of Chromatography*, **132**, 277 (1977).
121 Grimmer, G., Dettbarn, G. and Schneider, D.W. *Wasser und Abwasser Forshung*, **14**, 100 (1981).
122 Saunders, R.A., Blackly, C.H., Kovacina, T.A. *et al. Water Research*, **9**, 1143 (1975).
123 Versino, B., Knoppel, H., Grant, M. *et al. Journal of Chromatography*, **122**, 373 (1976).
124 Keith, L.H. *Life Science*, **19**, 1631 (1976).
125 Suffet, I.H. and Radziul, J.V. *Journal of the American Water Works Association*, **68**, 520 (1976).
126 Lingg, R.D., Melton, R.D., Kopfler, F.C., Coleman, W.E. and Mitchell, D.E.J. *Journal of the American Water Works Association*, **69**, 605 (1977).
127 Donaldson, W.T. *Environmental Science and Technology*, **11**, 348 (1977).
128 Trussell, A.R. and Umphres, M.D. *Journal of the American Water Works Association*, **70**, 595 (1978).
129 Biggs, D.P. *American Laboratory*, **10**, 81 (1978).
130 Van Rensburg, J.F.J. *Water Report*, **No. 10**, 4 (1979).
131 Leahy, J.S. and Purvis, M. *Journal of the Institute of Water Engineers Scient.*, **33**, 311 (1979).
132 Suffet, I.H., Brenner, L. and Cairo, P.R. *Water Research*, **14**, 853 (1980).
133 James, H.A., Fielding, M., Gibson, T.M. and Steel, C.P. In: *Advances in Mass Spectrometry* (ed. A. Quayle), Heydon, London, p. 1429 (1980).
134 Coleman, W.E., Melton, R.G., Kopfler, F.C. *et al. Environmental Science and Technology*, **14**, 576 (1980).
135 Shinohara, R., Kido, A., Eto, S. *et al. Water Research*, **15**, 535 (1981).
136 Coleman, W.E., Melton, R.G., Slater, R.W. *et al. Journal of the American Water Works Association*, **73**, 119 (1981).
137 Bjorsath, A. and Angeletti, G. (eds). *Proceedings of the 2nd European Symposium*, Killarney, Ireland. 17–19 November 1981, Commission of the European Communities, D. Reidel Publishing Co., Dordrecht, Netherlands (1982).
138 Quentin, K.E. *Fresenius Zeitschrift für Analytische Chemie*, **311**, 326 (1982).
139 Roland, L., Reuter, J.H. and Chian, E.S.K. *Journal of Chromatography*, **279**, 373 (1983).
140 Bruchet, A., Cognet, L. and Mallevialle, J. *Rev. Franc. Sci. L'eau*, **2**, 297 (1983).
141 Watts, C.D., James, H.A., Gibson, T.M. and Steel, C.P. *Environmental Technology Letters*, **4**, 59 (1983).
142 McIntyre, A.E. and Lester, J.N. *Science of the Total Environment*, **27**, 201 (1983).
143 Bruchet, A., Cognat, L. and Mallevaille, J. *Water Research*, **18**, 1401 (1984).
144 Benoit, F.M. and Lebel, G.L. *Bulletin of Environmental Contamination and Toxicology*, **37**, 685 (1986).
145 Pinchin, M.J. *Journal of the Institute of Water Engineering Scient.*, **40**, 81 (1986).
146 Novak, J., Zlutick, J., Kubelka, V.M. and Mosteck, J. *Journal of Chromatography*, **76**, 45 (1973).
147 Fujii, T. *Journal of Chromatography*, **139**, 297 (1977).
148 Fujii, T. *Analytica Chimica Acta*, **92**, 117 (1977).
149 Fujii, T. *Bulletin of the Chemical Society of Japan*, **50**, 2911 (1977).
150 Bertsch, W., Andersson, E. and Holzer, G. *Journal of Chromatography*, **112**, 701 (1975).
151 Harris, L.E., Budde, W.L. and Eichelberger, J.W. *Analyst (London)*, **46**, 1912 (1974).

152 Nicolson, A.A. and Meresz, O. *Bulletin of Environmental Contamination and Toxicology*, **14**, 453 (1975).
153 Dowty, B.K., Carlisle, D.R. and Laseter, J.L. *Environmental Science and Technology*, **9**, 762 (1975).
154 Vogel, T.M. and Reinhard, M. *Environmental Science and Technology*, **20**, 992 (1986).
155 Pruden, A.L. and Ollis, D.F. *Environmental Science and Technology*, **17**, 628 (1983).
156 Coleman, W.E., Munch, J.W., Kayler, W.H. *et al. Environmental Science and Technology*, **18**, 674 (1984).
157 Christman, R.F., Norwood, D.L., Webb, M.R. and Robenrath, M.J. US National Technical Information Service, Springfield, Virginia. Report No. PB81161962 Chlorination of Aquatic Humic Substances (1981).
158 Fujii, T. *Analytical Chemistry*, **49**, 1985 (1977).
159 Sithole, B.B., Williams, D.T., Lastaria, C. and Robertson, J.L. *Journal of the Association of Official Analytical Chemists*, **69**, 466 (1986).
160 Hwang, C.J., Krasner, S.W., McQuire, M.J., Mahlan, M.S. and Dale, M.S. *Environmental Science and Technology*, **15**, 535 (1984).
161 Otsuhara, M. and Suwa, M.J. *Waste Water*, **19**, 31 (1977).
162 Kikuchi, T., Minimura, T., Masada, Y. and Inoue, T. *Chem. Pharm Bulletin (Tokyo)*, **21**, 1847 (1973).
163 Erikson, M.D. and Pellizzari, E.D. *Bulletin of Environmental Contamination and Toxicology*, **22**, 688 (1979).
164 Snyder, D. and Reinert, R. *Bulletin of Environmental Contamination and Toxicology*, **6**, 385 (1971).
165 Keith, L. *Analysis of Organic Compounds in Two Kraft Mill Waste Waters*, EPA–600/4–75–005 (1975).
166 Keith, L. *Environmental Science and Technology*, **10**, 555 (1976).
167 Keith, L. *Identification and Analysis of Organic Pollutants in Water*, ed. L.H. Keith, Ann Arbor Science, Ann Arbor, MI, Ch. 36 (1976).
168 Abel, M., Conrad, T. and Giger, W. *Environmental Science and Technology*, **21**, 693 (1987).
169 Von Duzzeln, J., Lahl, U., Stachel, B. and Thiemann, W.Z. *Wasser Abwasser Forschung*, **15**, 272 (1982).
170 Henderson, J.E. and Glze, W.H. *Water Research*, **16**, 211 (1982).
171 Garcia-Gutierrez, A., McIntyre, A.E., Lester, J.N. and Peary, R. *Environmental Technology Letters*, **4**, 129 (1983).
172 Skinner, R.F., Fins, W.F. and Banelli, E.J. *Finnigen Spectra*, **3**, No. 1 (1973).
173 Buhler, D.R., Rasmussun, M.E. and Nakaue, H.S. *Environmental Science and Technology*, **7**, 929 (1973).
174 Heberer, T. and Stan, H. *Fresenius Zeitschrift für Analytische Chemie*, **3**, 639 (1994).
175 Konig, W.A., Ludwig, K., Rinker, M., Salting, K.H. and Gunther, W. *Journal of High Resolution Chromatography, Chromatography Communications*, **3**, 415 (1980).
176 Ramsteiner, K., Hoermann, W.D. and Eberle, D. *Journal of the Association of Official Analytical Chemists*, **57**, 192 (1974).
177 Karlhuber, B., Hormann, W. and Rainsteiner, K. *Analytical Chemistry*, **47**, 2450 (1975).
178 Laseter, J.L., Dehion, I.R. and Remele, P.C. *Analytical Chemistry*, **50**, 1169 (1978).
179 Kaiser, K.L.E. *Science*, **185**, 523 (1974).
180 Michael, L.C., Pellizzari, E.D. and Wiseman, R.W. *Environmental Science and Technology*, **22**, 565 (1988).

181 Water Research Centre (UK). Notes on Water Pollution No. 70. Analytical methods for organic compounds in sewage effluents (1980).
182 Warner, J.S., Jungclaus, G.A., Engel, T.M., Riggin, R.M. and Chuang, C.C. US National Technical Information Service, Springfield, Virginia. Report PB 80–00793. US Environmental Protection Agency Report EPA–600/2–80030L (1980).
183 Supplement to the 15th edn. of *Standard Methods for the Examination of Water and Wastewater*. Selected Analytical Methods. Approved and cited by the US Environmental Protection Agency. American Public Health Association Method S51 (1978).
184 Burlingham, A.I. *Ecotoxicology Environmental Safety*, **1**, 111 (1979).
185 Langenfeld, J.J., Hawthorne, S.B. and Miller, D.J. *Analytical Chemistry*, **68**, 1444 (1996).
186 Shackleford, W.M., Cline, D.M., Faas, L. and Kurth, G. *Analytica Chimica Acta*, **146**, 15 (1983).
187 Matsumoto, K. and Tsuruho, K. *Mizu Shori Gijutsu*, **28**, 161 (1987).
188 Lee, H.B. and Peart, T.E. *Analytical Chemistry*, **67**, 1976 (1995).
189 Burgasser, A.J. and Calaruotolo, J.F. *Analytical Chemistry*, **49**, 1508 (1977).
190 Ericksen, M.D., Stanley, J.S., Turman, J.K. *et al. Environmental Science and Technology*, **22**, 71 (1988).
191 Betowski, L.D., Yinon, J. and Voyksner, R.D. *Environmental Chemistry of Dyes and Pigments*, 255 (1996).
192 Burleson, J.L., Peyton, G.R. and Glaze, W.H. *Environmental Science and Technology*, **14**, 1354 (1980).
193 Hartmetz, G. and Shemrova, J. *Bulletin of Environmental Contamination and Toxicology*, **25**, 106 (1980).
194 Trehy, M.L., Gledhill, W.E., Mieure, J.P. *et al. Environmental Toxicology Chemistry*, **15**, 233 (1996).
195 Onuska, F.I. and Boos, W.R. *Analytical Chemistry*, **45**, 967 (1973).
196 Bishop, D.F. US Environmental Protection Agency, Cincinnati, Ohio. Report No. EPA 600/52–80-196. *GC–MS methodology for measuring priority organics in municipal waste water treatment* (1980).
197 Thomas, A.V., Stork, J.R. and Lammert, S.L. *Journal of Chromatographic Science*, **18**, 583 (1980).
198 Hornbroak, W.R. and Ode, R.H. *Journal of Chromatographic Science*, **25**, 206 (1987).
199 Pauluis, C.D.A., Anderson, S., Bonmetti, G. *et al.* CONCAWE, The Hague. Report No. 6/82. *Analysis of trace substances in aqueous effluents from petroleum refineries* (1982).
200 *Determination of Vinyl Monomer in Aqueous Effluents*. Analytical Chemistry Branch, Southeast Environmental Research Laboratory, Environmental Protection Agency, Athens, GA, USA (1974).
201 Bjorseth, A., Carlberg, G.E. and Moller, M. *Science of the Total Environment*, **11**, 197 (1979).
202 Lindstrom, K. and Nordin, J. *Journal of Chromatography*, **128**, 13 (1976).
203 Carlberg, G.E., Gjos, N., Maller, M. *et al. Science of the Total Environment*, **15**, 3 (1980).
204 Tamilarason, R., Morabito, P.L., Lamparski, L., Hazelwood, P. and Butt, A. *Journal of High Resolution Chromatography*, **17**, 689 (1994).
205 Spanggord, R.J., Gibson, B., Keek, R.G. and Thomas, D.W. *Environmental Science and Technology*, **16**, 229 (1982).
206 Lacorte, S., Malino, C. and Barcelo, D. *Analytica Chimica Acta*, **281**, 71 (1993).

207 Jungelaus, G.A., Game, L.H. and Hites, R.A. *Analytical Chemistry*, **48**, 1894 (1976).
208 Kumata, H., Takada, H. and Ogura, N. *Analytical Chemistry*, **68**, 1976 (1996).
209 Perry, D.L., Jungelaus, G.A. and Warner, J.S. US National Technical Information Service, Springfield, Virginia. Report No. PB 294794, *Identification of organic compounds in industrial effluent discharges* (1979).
210 Sugar, J.W. and Conway, R.A. *Journal of Water Pollution Control Federation*, **40**, 1622 (1968).
211 Lussi-Schlatter, B. and Brandenberger, H. In: *Advances in Mass Spectrometry in Biochemistry and Medicine*, pp. 231–243. Spectrum Publications, New York (1976).
212 Odanaka, Y., Tsuchlya, W., Matono, O. and Goto, S. *Analytical Chemistry*, **55**, 929 (1983).
213 Meinema, H.A., Burger Wersina, T., Verslius-Dehaan, G. and Geners, E.S. *Environmental Science and Technology*, **12**, 288 (1978).
214 Maguire, R.J. and Hunealt, H. *Journal of Chromatography*, **209**, 458 (1981).
215 Mueller, M.D. *Fresenius Zeitschrift für Analytische Chemie*, **317**, 32 (1984).
216 Matthias, C.L., Bellama, J.M., Olsen, G.J. and Brinkman, F.E. *Environmental Science and Technology*, **20**, 609 (1986).
217 Unger, M.A., MacIntyre, W.G., Greaves, J. and Huggett, R.J. *Chemosphere*, **15**, 461 (1986).
218 Greaves, J. and Unger, M.A. *Biomedical Environmental Mass Spectrometry*, **15**, 565 (1988).
219 Colby, S.M., Stewart, M. and Reilly, J.P. *Analytical Chemistry*, **62**, 2400 (1990).
220 Plzak, Z., Polanska, M. and Suchanek, M. *Journal of Chromatography*, **699**, 241 (1995).
221 Talmi, Y. and Bostick, D.T. *Analytical Chemistry*, **47**, 2145 (1975).
222 Gorecki, T. and Pawliszyn, J. *Analytical Chemistry*, **68**, 3008 (1996).
223 Ishibashi, M. and Shigematsu, T. *Bulletin of the Institute of Chemical Research, Kyoto University*, **23**, 59 (1950).
224 Chuecas, L. and Riley, J.P. *Analytica Chimica Acta*, **35**, 240 (1966).
225 Fukai, R. *Nature (London)*, **213**, 901 (1967).
226 Mykytiuk, A.P., Russell, D.S. and Sturgeon, R.E. *Analytical Chemistry*, **52**, 1281 (1980).
227 Sturgeon, R.E., Berman, S.S., Willie, S.N. and Desauliniers, J.A.H. *Analytical Chemistry*, **53**, 2337 (1981).
228 Ahnoff, M. and Josefsson, B. *Analytical Chemistry*, **6**, 1083 (1973).

Index

Milton Keynes UK
Ingram Content Group UK Ltd.
UKHW021918071024
449327UK00022B/1678